Handbooks in Radiology

Interventional Radiology and Angiography

HANDBOOKS IN RADIOLOGY SERIES

Series Editors

ANNE G. OSBORN, M.D.
Director of Neuroradiology and Professor of Radiology, University of Utah School of Medicine, Salt Lake City, Utah

DAVID G. BRAGG, M.D.
Professor and Chairman, Department of Radiology, University of Utah School of Medicine, Salt Lake City, Utah

---

NEURORADIOLOGY: SKULL AND BRAIN
Anne G. Osborn, M.D., H. Ric Harnsberger, M.D., Wendy R. K. Smoker, M.D.

SKELETAL RADIOLOGY
B. J. Manaster, M.D., Ph.D.

CHEST RADIOLOGY
Howard Mann, M.D., David G. Bragg, M.D.

NEURORADIOLOGY: HEAD AND NECK
H. Ric Harnsberger, M.D., Wendy R. K. Smoker, M.D., Anne G. Osborn, M.D.

ULTRASONOGRAPHY
William J. Zwiebel, M.D.

NUCLEAR MEDICINE
Frederick L. Datz, M.D.

NEURORADIOLOGY: SPINE AND SPINAL CORD
Wendy R. K. Smoker, M.D., B. J. Manaster, M.D., Ph.D. Anne G. Osborn, M.D., H. Ric Harnsberger, M.D.

Handbooks in Radiology

Interventional Radiology and Angiography

Myron Wojtowycz, M.D.
Associate Professor
Department of Radiology
Angiography and Interventional Radiology
  Section
University of Wisconsin at Madison Medical
  School
Madison, Wisconsin

YEAR BOOK MEDICAL PUBLISHERS, INC.
Chicago • London • Boca Raton • Littleton, Mass.

Copyright© 1990 by Year Book Medical Publishers, Inc. All rights reserved. No part of this publication may be reproduced, stored in a retrieval system, or transmitted, in any form or by any means—electronic, mechanical, photocopying, recording, or otherwise—without prior written permission from the publisher. Printed in the United States of America.

Permission to photocopy or reproduce solely for internal or personal use is permitted for libraries or other users registered with the Copyright Clearance Center, provided that the base fee of $4.00 per chapter plus $.10 per page is paid directly to the Copyright Clearance Center, 21 Congress Street, Salem, MA 01970. This consent does not extend to other kinds of copying, such as copying for general distribution, for advertising or promotional purposes, for creating new collected works, or for resale.

2   3   4   5   6   7   8   9   0   M   R   94   93   92   91

Library of Congress Cataloging-in-Publication Data
Wojtowycz, Myron.
    Interventional radiology and angiography / Myron Wojtowycz.
        p.   cm.—(Handbooks in radiology)
    Includes bibliographical references.
    ISBN 0-8151-5868-8
    1. Radiology, Interventional—Handbooks, manuals, etc.
    2. Angiography—Handbooks, manuals, etc. I. Title. II. Series:
Handbooks in radiology series.
    [DNLM: 1. Angiography—handbooks. 2. Radiography,
Interventional–handbooks.   WG 39 W847i]
    RD33.55.W64   1990   90-12202
    617′.05—dc20   CIP
    DNLM/DLC
    for Library of Congress

Sponsoring Editor: James D. Ryan
Associate Managing Editor, Manuscript Services: Deborah Thorp
Production Project Coordinator: Yvette L. Sellers
Proofroom Supervisor: Barbara M. Kelly

*To Olena, Lyuba, and all others
who have suffered in concentration camps.*

Preface

Over the past decade, cross-sectional imaging has revolutionized the practice of radiology. One consequence of cross-sectional imaging has been the contraction of the indications for diagnostic angiography. At the same time, a broad range of percutaneous interventions has been made possible. Catheter and wire technologies have advanced, and the refined instruments have been applied to novel purposes. Baskets, balloons, needles, thrombolytic drugs, lasers, fiberoptics, and lithotriptors of various shapes and sizes have become available, creating a bewildering array of options for treating conditions once requiring open surgery. Even more bewildering can be the enormous amount of information pertaining to radiologic diagnosis and percutaneous interventions.

The aim of this book is to provide residents with a quick and coherent introduction to the field of vascular radiology and percutaneous interventions as it stands on the threshold of the 1990s. It may also serve as a convenient review for those about to take their radiology board examinations. This text is not meant to be used alone; rather, it should supplement the many distinguished and more encyclopedic books that have been published recently. In addition, I hope general radiologists will find this handbook a useful reference.

Aside from relying on my personal experience over the past seven years in vascular and interventional radiology, I have attempted to integrate a great deal of information recently presented in medical, surgical, and radiologic publications. It is inevitable that some of the information will soon be outdated in the more rapidly progressing areas, such as caval filters, biliary stone removal, vascular stenting, laser angioplasty, and percutaneous atherectomy. One can only ask for the reader's understanding.

It should be acknowledged what this book is *not*. Although basic techniques are presented, the gamut of "tricks of the catheterization trade" is not addressed. Neither is the institution and management of an admitting interventional radiology service (although it must be emphasized that with the proper collegial environment, percutaneous angioplasty and other interventions can be performed safely on patients referred directly from primary care physicians). Pediatric conditions and procedures are not specifically addressed. Perhaps the most important omission is a directed discussion of cardiopulmonary resuscitation, and the treatment of serious contrast medium reactions. Rather than superficially treat a vital topic, I direct the reader to the recommendations of the American Heart Association and the American College of Radiology.[1-3] *Anyone engaging in invasive and angiographic procedures must be versed in resuscitation and life support!*

With a recognition of these limitations, I hope this book will meet the reader's approval. I owe a great debt to Holly Jackson and Joan Kozel for assistance in preparation of the manuscript and illustrations. I also wish to thank Drs. Andrew Crummy, John McDermott, and Phil Carlson for their review of the manuscript and helpful suggestions. I alone am responsible for any defects that remain.

Myron Wojtowycz, M.D.

1. American Heart Association: Standards and guidelines for cardiopulmonary resuscitation and emergency cardiac care. *JAMA* 1986; 255:2841—3044.

2. Albaran-Sotelo R, Atkins JM, Bloom RS, et al: *Textbook of Advanced Cardiac Life Support.* Dallas, American Heart Association, 1987.

3. Commission on Public Health and Radiation Protection, Committee on Drugs, with the cooperation of the Committee on Professional Liability (Commission on Standards in Radiologic Practice), American College of Radiology: *Prevention and management of adverse reactions to intravascular contrast media.* Chicago, American College of Radiology, 1977.

Editor's Introduction

Interventional Radiology and Angiography, by Myron Wojtowycz, can still be claimed as one of the *Handbooks in Radiology* series despite Myron's current location at the University of Wisconsin. Myron was a member of our faculty, and left to rejoin the University of Wisconsin faculty in November of 1987. He has constructed a beautifully written and well-organized handbook of the evolving field of interventional radiology and angiography. I was always impressed with Myron's well-organized, computer-based acquisition of current articles, classics, and teaching cases, which made each of his lectures a treat for our residents.

The chapters in this handbook cover the spectrum of this rapidly advancing field in a most elegant and readable style. This handbook will serve the purposes of not only the beginning resident, but also of the practicing radiologist. As with all of our handbook series, the small size and portability of this format should keep this book off the bookshelf and in pockets and near viewboxes, where the action occurs. I hope you enjoy this well-prepared handbook.

David G. Bragg, M.D.

Contents

1

Basic Principles of Angiography

KEY CONCEPTS

1. Know before starting what questions are to be answered or aims are to be achieved by angiography.
2. Essential preangiographic information includes history of hypersensitivity to contrast media, presence of coagulopathy or use of anticoagulant medications, and signs of renal insufficiency.
3. Catheters and wires should not be advanced against resistance.
4. A test injection by hand should precede any angiographic run to confirm that the proper vessel is selected and the catheter tip is free in the lumen.
5. Low-osmolality contrast media are better tolerated by patients and are associated with fewer adverse reactions, but cost considerably more than conventional ionic media.
6. Good hydration is the most important factor for preventing contrast material-induced renal dysfunction.

Angiography has been a valuable diagnostic tool for decades. Now, as a purely diagnostic measure, its scope has become limited, but with percutaneous angioplasty and therapeutic embolization, its applications have expanded in a therapeutic role. This chapter is meant to acquaint the reader with some techniques and instruments commonly used in arteriographic procedures as well as hazards of the procedures. Many of the principles described apply equally to selective venous catheterizations.

ARTERIAL ACCESS

The Seldinger Technique

Selective catheter angiography became possible on a wide scale only after Sven Ivar Seldinger described a method of puncturing an artery and introducing a catheter safely without the need for surgical exposure of the vessel.[1] His technique remains the basis for arteriography today.

After the arterial pulse has been identified and the proper site for puncture chosen, the skin over the vessel is shaved, scrubbed with povidone-iodine, and draped; lidocaine 1% is injected for local anesthesia (8 to 10 mL usually suffices). A small incision is made in the skin to allow easy entry of the needle and catheter. In thin individuals care must be taken to prevent the scalpel from injuring the vessel.

An 18-gauge hollow needle with a sharp stylet (Seldinger needle, or one of its many variations) is advanced into the vessel, and the stylet is removed. If the needle tip is in arterial lumen, a stream of pulsating blood will be seen. If there is no blood return, the needle is slowly withdrawn until it is removed completely or until pulsatile blood is encountered. When blood returns, a guidewire is passed through the needle and into the arterial lumen. With a good length of wire in the vessel, the needle is removed and the artery compressed, while the diagnostic catheter is threaded onto the wire. As the wire is held taut ("clotheslined"), the catheter is advanced into the vessel under fluoroscopic control.

If any resistance to wire introduction is noted, it is best to remove the wire from the needle to see if pulsatile blood continues to return. Should backflow be absent, the needle must be repositioned or removed and a new needlestick performed. When a needle is removed after successful arterial puncture but a guidewire is not introduced, compression of the artery for 5 to 10 minutes is required. The presence of a hematoma will not only make repeated arterial puncture more difficult but also afford less control for manipulations of the catheter and make effective arterial compression after catheter removal more difficult.

If blood continues to flow through the needle after the wire is withdrawn, the angle of the needle may be adjusted and another attempt at wire introduction made. When there is good backflow of blood but continued difficulties with wire insertion, injection of a small amount of contrast material through the needle (most easily performed through a plastic connecting tube) during fluoroscopy is often revealing. The needle tip may be in a small branch artery, it may be angled against the wall of the vessel or only partially within lumen, or it may be against an atherosclerotic plaque. Appropriate adjustments can then be made.

Needles Without Stylets

Alternatives to Seldinger-type needles are sharp, hollow needles lacking stylets. When these are advanced into an artery, tip entry can be recognized immediately. One variation has a short Teflon sheath that can be slipped directly into the vessel. A disadvantage of hollow needles is that small pieces of tissue or clot may occlude the needle, preventing blood return. For this reason, such needles must be flushed before repeated puncture attempts.

Angle of Entry

For entering an artery, the direction of the needle should match the palpated or expected course of the vessel as much as possible. The actual angle of incidence is best about 45°. Too steep or too shallow an angle can cause problems in catheter exchanges and other manipulations. However, if it is anticipated that the direction of entry might be switched during the procedure (as in balloon angioplasty of the superficial femoral artery after bilateral runoff arteriography from the ipsilateral femoral approach), some angiographers prefer to come straight down with the needle, making a 90° angle of entry. Subsequently, catheters can be used to reverse the direction of wire placement.

Dilators

If a wire is securely in the vessel but catheter introduction is difficult, use of one or more Teflon dilators is recommended. The tapering and stiffness of dilators allow their introduction when other catheters cannot be passed, such as in groins scarred from previous surgery or angiographic procedures. It may be necessary to "overdilate," that is, to pass a 6-F* dilator before inserting a 5-F catheter. This is often the case when entry is obtained through a synthetic vascular graft. If a dilator can be inserted but there continue to be problems with catheter passage, a heavier, stiffer wire can be placed through the dilator.

Sheaths

A vascular sheath is recommended when multiple changes of catheters are anticipated, when manipulation of the catheter is difficult because of obesity or other local factors, or when bleeding occurs about

---

*French (F) is a measure of circumference, translating to 3 F/mm diameter for round tubes.

a catheter in place. Sheaths are supplied fitted over short dilators. Most sheaths have hemostasis valves and flushing sideports. If a sheath does not enter the artery easily, it must be removed and inspected to ensure that crimping or tip flaring has not occurred.

Puncturing Grafts

Catheters can be safely introduced through synthetic graft materials, such as Dacron or Gore-Tex, as long as the grafts have been in place for at least 2 months.[2] Introduction tends to be more difficult because of local scarring from the surgical procedure, as well as the toughness of the graft material. Overdilation of the tract by 1 to 2 F may be needed, and use of a low-friction Teflon catheter or sheath is recommended. If no sheath is used, manipulations and removal are best performed with a guidewire through the catheter. Catheter separation occurs with extreme rarity but is possible if proper care is not taken.[3]

Passing Obstructions

If an obstruction is encountered during wire placement, a catheter should be advanced to the vicinity of the obstruction. The catheter is then used to inject contrast material to delineate the nature and severity of the obstruction. At times, use of digital subtraction "roadmap" angiography can be extremely helpful.[4] A hooked-tip catheter, such as a Berenstein or H1H shape, can direct the guidewire through tortuous vessels and past eccentric plaques. When a hooked-tip catheter is used, a soft-tip straight or steerable guidewire is often the most appropriate choice for successful negotiation of the vessel. The patient should be given 5,000 units of heparin intravenously or intra-arterially to prevent acute occlusion if a very tight stenosis is catheterized. If an intimal flap is raised during passage attempts, access from that site should be abandoned for all but the most exceptional of circumstances.

Catheter Removal and Hemostasis

At the close of a procedure, pigtail catheters, and those catheters with back-curves, such as Simmons or sidewinder catheters, should be straightened before removal. This usually requires passage of a guidewire beyond the tip of the catheter. As the catheter is removed, firm pressure is placed over the site of arterial puncture (not necessarily at the site of skin incision itself) with compression of the proximal vessel as well. Pressure should not be so firm as to obliterate distal pulses, but there should be no bleeding or oozing from the incision or hematoma formation in the soft tissues. Compression is maintained continuously for at least 10 minutes before pressure is gingerly released to check for effective hemostasis. A longer period of compression is prudent if manipulations are difficult, if large catheters are used, or if a hematoma is present. If bleeding recurs, compression must be reinstituted for a similar time period. Any hypertension should be controlled by nifedipine or other medications before an arterial catheter is removed.

Common Femoral Artery Puncture

The vessel most commonly used for arterial access in angiography is the common femoral artery. The pulse is palpated below the inguinal ligament (which courses between the pubis and the anterior superior iliac spine). The inguinal crease is *not* a reliable marker for the location of the ligament! In obese individuals the inguinal crease may be as much as 11 cm caudal to the inguinal ligament.[5] If the patient is heavy or if the pulse is not easily palpable, it is useful to check the site fluoroscopically before administering local anesthesia. For retrograde catheterization the skin incision can be made over the inferior margin of the femoral head, with the intention of entering the artery at the level of the midfemoral head. If an antegrade puncture is required, skin entry can be made at the level to the acetabulum, again aiming to enter the mid common femoral artery.

It is important to puncture the artery over the femoral head, for the common femoral artery is larger than its branches, the superficial femoral and profunda femoris (deep femoral) arteries, and is thus less prone to injury. Also, manual compression of the puncture site is more effective over the femoral head. Low puncture with larger catheters has been implicated in pseudoaneurysm formation.[6] On the other hand, puncture above the inguinal ligament prevents effective compression and exposes the patient to the risk of major, life-threatening hemorrhage.

Puncture of the femoral limb of an aortobifemoral graft can sometimes be perplexing, for if a Seldinger double-wall technique is employed one may enter the native femoral artery deep to the end-to-side graft insertion. In such a case, the guidewire will meet resistance in the vessels of the pelvis or near the aortic bifurcation. Injection of contrast medium through the needle, or through an inserted dilator, will confirm that the diseased native vessel has been entered. The needle or dilator is then slowly withdrawn until pulsatile blood no longer returns. If the limb of the graft has been traversed, a second squirt of pulsatile blood will be noted with further withdrawal. The guidewire is then inserted into the graft and advanced into the aorta.

When a femoral pulse is poorly palpable and there is no feasible alternative access site, fluoroscopy may detect mural calcification. If present, calcium can guide needle passage. Otherwise, real-time sonography can be used to mark the course of the vessel and to exclude the possibility of occlusion. If the artery is not entered but venous blood returns through the needle, passage of a guidewire into the vein (coursing medial to the artery) can serve as a marker for further needle placements. In exceptional cases, a venous catheter can be used to inject contrast material for a digital subtraction "roadmap" of the artery.

Axillary or Brachial Artery Puncture

The only other arteries used routinely for selective catheterization studies are the axillary and brachial arteries. Upper extremity arterial access is necessary in the face of bilateral femoral or iliac artery or aortic occlusion, as well as for certain selective arterial studies most easily approached from above. Among the latter are celiac, mesenteric, or renal artery catheterizations in severely angled vessels. Entry through the left upper extremity is preferred, to minimize the possibility of stroke. A catheter placed from the right exposes both carotid and vertebral arteries to potential emboli; a left-sided catheter crosses only the left vertebral artery origin. Care should be taken that a catheter placed down the descending aorta does not buckle into the ascending aorta and form a loop in the left ventricle during manipulations!

Besides the risk of cerebral embolization, upper extremity arterial puncture poses problems with needle entry (owing to the greater mobility and smaller size of the vessels in comparison with the common femoral artery), effective compression of the artery (especially for "high" axillary punctures), and patient discomfort from the prolonged immobilization in abduction needed after angiography by way of axillary artery puncture. Neural injury can arise from needle puncture or from compression by hematoma. The complication rate of axillary puncture is twice as high as that of common femoral artery catheterization.[7] Placement of even 4-F and 5-F catheters through the brachial artery can result in major complications in 11% of patients, mostly from spasm or arterial thrombosis.[8, 9] For this reason, upper extremity access should be used only if femoral artery catheterization is not possible.

Translumbar Aortography

If no femoral pulse is palpable, but abdominal aortography and runoff angiography are needed, translumbar puncture remains a valuable option. A special 18-gauge needle bearing a 6-F sheath is placed directly into the abdominal aorta from a left flank approach. It is *critical* that the patient have normal coagulation status and no problems with hypertension to be a candidate for this approach. A certain degree of retroperitoneal hemorrhage is expected after translumbar aortography, and a large amount of blood may extravasate before such hemorrhage becomes clinically evident.[10]

For translumbar study the patient is placed prone on the angiography table, and a skin entry site is chosen approximately one hand's breadth to the left of the spinous process, below the level of the 12th rib. Local anesthesia is administered superficially, and lidocaine is also injected into the deeper soft tissues with a 20-gauge spinal needle. The sheath needle is advanced under fluoroscopic guidance to the appropriate level (see "Low Translumbar Approach" and "High Translumbar Approach," later in this chapter). If the needle is too horizontal, vertebral body will be encountered. When this happens, the needle must be withdrawn almost completely from the patient and redirected in a slightly more vertical angle until its tip passes anterior to the vertebral column. Often the pulsatile wall of the aorta can be felt as the needle tip comes in contact with it. The needle is then advanced 1 to 2 cm in a quick jab to pierce the wall, with the operator taking care not to cross the midline.

If the sheath has only an endhole, both the trocar and metal cannula of the needle are removed; if there are sideholes in the sheath, only the trocar is removed. The sheath (with or without cannula) is slowly withdrawn until pulsatile blood returns. A short J-tip guidewire is then inserted, monitored fluoroscopically to ensure that it passes easily into aortic lumen, and the sheath is introduced over the wire. After the sheath is in place, it is flushed, and a small amount of contrast material is injected to confirm proper positioning. The sheath is secured by sterile tape or towel clips to prevent inadvertent withdrawal. No more than 15 mL/sec must be injected through an endhole-only sheath, because the high-velocity jet formed at the tip can injure endothelium.

The potential of major unrecognized hemorrhage dictates that catheter exchanges not be performed through an aortic puncture. If arterial blood returns through the needle but wire placement is unsuccessful, a separate puncture may be attempted. However, multiple repeated aortic entries are dangerous.

Low Translumbar Approach

For low puncture, the sheath needle is directed in the axial plane, 35° to 45° from the horizontal, toward the level of L-3. Low entry has the advantage of providing better opacification of the lower extremity runoff vessels, particularly if the sheath tip can be directed caudally during placement. However, the infrarenal abdominal aorta can be variable in location due to tortuosity, sometimes actually lying to the right of midline. Also, with the L-3 approach there is a higher likelihood that the needle might enter an aneurysm or an occluded aorta. Preangiographic examination with ultrasound can alert one to the necessity for high translumbar puncture.

High Translumbar Approach

If aortic occlusion or aneurysm is suspected, a high translumbar route is the appropriate choice. The needle is directed cranially toward the pedicle of T-12, again making a 35° to 45° angle with the horizontal (coronal) plane. The L-1 level or T12-L1 interspace must be avoided, to prevent injury to the renal artery. Because of the steep angle of entry with respect to the course of the aorta, direction of the sheath tip caudally is rarely possible, and injection must be made into the descending thoracic aorta. Because much of the injected contrast material enters the mesenteric circulation and the renal arteries, opacification of distal vessels tends to be worse than that provided by the low lumbar approach. One particular risk of high translumbar puncture is pneumothorax, especially in patients with obstructive lung disease.

CENTRAL VENOUS ACCESS

Selective central venous studies can be performed by way of the upper extremity or right jugular veins, but most are approached from the common femoral vein. The femoral vein is posteromedial to the artery, so the skin medial to the femoral pulse is anesthetized. The puncture technique is identical to that employed for arterial studies, except that a syringe containing saline should be used for gentle aspiration. When central catheters are placed (particularly if the tip is positioned in the chest or right atrium), the lumen must not be exposed to atmospheric pressure. Carelessness may result in air embolism.

GUIDEWIRES

Guidewires are designed to allow the safe introduction of the catheter to the vessel and selective positioning of catheters within the vascular system. Most conventional wires are composed of a stainless steel coil of wire supported by a stiff inner mandril running the length of the coil and tapering over 5 to 15 cm to a soft tip. There is also another fine supporting wire in the core, joining both ends of the coil; this wire is meant to prevent complete transection of the guidewire should the coil break. Standard wires have a soft leading end and a stiff trailing end that is not ordinarily inserted into a catheter.

In the United States, wire sizes are usually expressed as diameter in inches. The wires most commonly used in adult angiography have 0.035- or 0.038-inch diameters. Much finer wires, such as 0.018-inch guidewires, are used in special applications, including percutaneous angioplasty of small vessels. However, initial insertion of the catheter normally requires a heavier wire (or vascular sheath), because fine wires tend to coil or kink in the extravascular soft tissues.

Standard wires are 145 cm long, allowing insertion of 100-cm catheters without difficulty. Exchange wires up to 300 cm long are used when a selectively placed catheter must be exchanged for another without losing access to the vessel of interest. The wire must be long enough for the catheter to be completely removed without withdrawal of the wire tip from its selective position.

Wires are further characterized by the length of taper to the inner mandril, tip configuration (straight, 1.5 mm-, 3.0 mm-, 15 mm-diameter "J"), stiffness (standard, heavy-duty, "coat-hanger"), and coating. J-tip wires are valuable in atherosclerotic vessels, because straight wires tend to catch on the edge of plaques with the attendant danger of producing arterial dissection. J-tip wires are more likely to deflect from obstructions. Heavy-duty or extra-stiff wires may be needed to traverse scarred puncture sites or very tortuous vessels. Steerable wires are designed for efficient transmission of torque to a flexible curved or angled tip. These may be employed to traverse tight eccentric stenoses. Teflon and heparin coatings decrease the friction and thrombogenicity of guidewires. Coating with hydrophilic polymers has produced wires with extremely low coefficients of friction. Special deflecting systems are available for directing catheter tips. The ends of deflecting wires are quite stiff and they should not be passed beyond the tip of the catheter being deflected (see Chapter 14, Pulmonary Angiography). One should become familiar with a variety of wires and develop preferences on the basis of experience.

CATHETERS

As with guidewires, angiographic catheters come in a range of shapes and sizes. As previously stated, catheter size is conventionally designated in "French," where 1 F = 3 mm diameter. Packaging labels will list the size, shape, material, and length of the catheter; the presence of sideholes; and the maximum diameter of guidewire that can be used with the catheter. Among the materials used are polyethylene, polyurethane, nylon, and Teflon, and the catheter handling characteristics vary accordingly. For example, Teflon is smooth, stiff, and has a relatively low coefficient of friction. It is useful when insertion of other catheters through a graft or scarred puncture site is difficult. Other polymers are designed to accommodate high injection rates through a thin-walled tube. Some catheters are constructed with a steel mesh incorporated into the wall, providing ready translation of torque to the catheter tip. Review of manufacturers' catalogues, discussion with other angiographers and manufacturers' representatives, and experience will provide an idea of the catheters most useful for a given indication.

For flush aortography a pigtail tip is recommended (Fig 1–1). Pigtail catheters contain multiple sideholes, permitting dispersal of contrast medium in a relatively compact bolus while preventing a strong "jet" effect from producing injury or perforation of a vessel. Pigtail catheters must be removed over a guidewire.

Selective catheterization requires a shaped tip to engage the orifice of the artery being selected. A Cobra (C_2) catheter has an angled tip joined to a gentle curve, making it suitable for selection of many renal and mesenteric arteries (see Fig 1–1). Brachiocephalic vessels approached from the femoral artery are often easily engaged by a Headhunter (H_1) tip. Highly curved catheters, such as Simmons or sidewinder catheters, must have their primary curves reconstituted by manipulation in the aortic arch or brachiocephalic arteries. They are most useful in "hooking" sharply angled vessels (celiac axis, renal artery to a ptotic kidney, and so forth). For secure positioning of a selective catheter, it is often necessary to pass a guidewire well into the vessel and advance the catheter over the wire.

Most catheters now in use are available in 5- to 7-F sizes. The 5-F catheters are more flexible and tend to advance better over a wire coursing through tortuous vessels. Larger catheters allow better control of torque and movement in patients whose size or anatomy makes selective angiography difficult.

Once a catheter is introduced into a vessel and the guidewire removed, the catheter must be immediately flushed with a solution of heparinized saline (2,000 units of heparin in 500 mL 0.9N saline). Allowing blood to remain stagnant within a catheter for any length of time invites thrombosis. Flushing should be repeated every 2 to 3 minutes throughout the procedure. Before each flush a small amount of blood should be freely aspirated into the syringe.

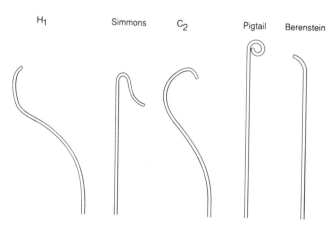

H₁   Simmons   C₂   Pigtail   Berenstein

FIG 1–1.
Various catheter shapes useful in arteriography.

If a catheter has no sideholes, the tip of the catheter may sometimes rest against the vessel wall and no blood can be aspirated. In such a situation the catheter must be slowly withdrawn until blood does freely return through the lumen. Placement of one or two sideholes near the catheter tip may prevent this problem, as well as decrease the intensity of the tip jet created with injection. However, sideholes should not be placed in catheters used in the brachiocephalic vessels. There is a small but finite risk that thrombus may form in a sidehole (due to the relative ineffectiveness of flushing in keeping sideholes clear). Embolization of such a thrombus is commonly inconsequential, but can be disastrous in the coronary or cerebral circulation. For the same reason "double-flushing" (forceful aspiration with one syringe followed by forward flushing with a different syringe) is advised in the brachiocephalic vessels.

An occluded catheter *must not* be opened by forward injection of fluid or by passage of a guidewire. This turns the occluding material into emboli. Instead, the catheter hub may be cut and a vascular sheath of appropriate size slipped over the catheter into the site of entry. In this manner the catheter can be removed without loss of vascular access.

Whenever a wire is inserted into a catheter, the first few millimeters of wire exiting catheter tip will behave as a stiff instrument, no matter how soft the guidewire tip actually is. Therefore, if the catheter endhole is wedged against an atherosclerotic plaque or other obstruction, the catheter should be slightly withdrawn. In small vessels it is prudent to form the tip of a tight J wire by advancing the wire just to the end of the catheter and then withdrawing catheter to unsheath the wire.

FILMING

Conventional Film Angiography

Basic Equipment

The type of equipment available in a given department dictates the form of the angiographic images obtained. Conventional films may be individual "cut" sheets or continuous rolls placed in special magazines and rapid film changers. These devices permit up to six films per second to be obtained. For lower extremity runoff angiography, special long-leg films and ceiling-mounted tubes have been used in some institutions. Alternatively, moving step-tables can be used to shift patient position over several stations during a single injection, in essence "following" the opacification of vessels after contrast medium injection.

Spot films using 70-mm or 100-mm cameras are sometimes used in angiography, but the small size of the frames makes the appreciation of fine detail difficult. Cineradiography is customarily performed in cardiac angiography, but the high radiation doses and cumbersome film viewing make this an unpopular technique for noncardiac applications. Most peripheral arteriograms do not call for extremely rapid frame rates.

Setting Up

After the catheter is placed, a test injection under fluoroscopy must be performed prior to filming to ensure that the tip is free (not "wedged") and in the correct vessel. Selective catheter placement assures the best opacification of vessels with the least amount of contrast medium and it minimizes confusion from the undesired filling of other overlapping vessels. The rapidity of blood flow in the artery can be gauged by the test injection, and injection parameters are set to nearly match that flow. Injection rates of 20 mL/sec and 30 mL/sec are typical for abdominal and thoracic aortography, respectively. Selective renal and mesenteric studies usually call for rates of 6 to 10 mL/sec. Inferior mesenteric artery injection is commonly performed at 2 to 3 mL/sec.

Duration of injection is at least 2 seconds and may require up to 10 or 15 seconds, depending on whether step-table angiography is being employed or dense organ opacification is desired.

With conventional film angiography a scout view is obtained to check positioning of the patient and exposure factors. The catheter is attached to a power injector, and the injection parameters are set. If the need for subtraction (analog) views is anticipated, the first exposure occurs before the injection of iodinated contrast material. Exposures must be made at a frame rate and duration appropriate to the problem at hand. For rapid flow of blood through an arteriovenous fistula, rates of three or more frames per second are needed for evaluation. For flow through aneurysmal vessels in patients with poor cardiac output, one exposure every 2 to 3 seconds for a protracted period may be needed. Delayed films (through 20 to 40 seconds) are necessary for assessing venous opacification or extravasation in cases of trauma or gastrointestinal bleeding.

In some settings, flow can be manipulated by medications or other means. Distal extremity vessel opacification is enhanced by warming of the extremity, reactive hyperemia after release of pressure cuffs, or direct injection of nitroglycerin or tolazoline.[11] Tolazoline can be injected into the superior mesenteric artery immediately prior to angiography to improve visualization of the portal venous system. Dilute epinephrine has been administered into renal arteries to decrease flow through normal vessels and make abnormal vessels more conspicuous.

A single projection is often insufficient for diagnostic angiography. The position of atherosclerotic plaques cannot be anticipated, and use of two orthogonal projections can improve detection of stenosis in 30% to 40% of patients.[12, 13] Two or more projections are mandatory for the proper evaluation of aortic aneurysms or thoracic aortic trauma. If the angiography suite is equipped with only one radiographic tube, a separate injection of contrast material must be performed for each projection. However, if two tubes are available, simultaneous biplane filming can be performed, and the total dose of contrast medium injected is minimized.

All studies obtained must be developed and reviewed before the termination of angiography. Conventional films give one the advantages of a large field of view, high spatial resolution, and the ability to use a step-table and simultaneous biplane filming. Disadvantages include relatively low contrast resolution, inability to view images immediately, and the large number of superfluous films usually exposed.

Digital Subtraction Angiography

Digital subtraction angiography (DSA) stores images electronically and uses a dedicated computer to subtract images in real time, ideally showing nothing but those vessels filled with contrast medium.[14] The resultant angiogram can be reviewed without delay. The high contrast resolution of DSA allows dilute contrast medium (diluted 1:1 with saline) to be employed. Another advantage is that the images can be electronically manipulated to change the contrast and window level, integrate frames, and compensate for a certain amount of motion. Other special features are available on various imager models.

High contrast resolution is valuable in the study of the vessels of the ankle and foot.[15] In upper extremity angiography, use of dilute contrast material decreases the considerable discomfort that has been associated with conventional angiography in the past. In many applications DSA can be used interchangeably with conventional film angiography according to the discretion of the angiographer, and many of the procedures outlined earlier (see "Setting Up") also apply for DSA studies. However, if a patient is uncooperative or is unable to remain motionless for the duration of exposures, conventional filming is preferred. Also, DSA is inappropriate in viewing the abdomen if a large amount of bowel gas is present, especially if delayed exposures are needed, and should not be employed in the search for a gastrointestinal bleeding source.

Originally, DSA used intravenous injection of contrast medium, but it soon became apparent that intra-arterial injection overcame many of the problems encountered.[16] Intra-arterial studies are much less subject to motion artifact, smaller doses of contrast medium can be used, and the study is much less dependent on cardiac output. Arterial injections are routinely performed through 4-F or 5-F catheters, minimizing the risk of postangiographic bleeding. In our department, intravenous DSA has virtually disappeared, but it does remain an option if no arterial access site is available.

CONTRAST MEDIA

Conventional Ionic and Low-Osmolality Iodinated Media

Vessels subjected to angiography must be rendered radiopaque by injection of iodinated contrast medium. For many years the prime agents in use were ionic derivatives of fully-substituted tri-iodobenzoic acid, namely diatrizoate and iothalamate. The ions released in solution by these compounds are meglumine and/or sodium cations and the radiopaque anions. Recently, new agents have become available in the United States which are either dimeric ionic compounds (ioxaglate) or nonionic derivates of tri-iodobenzoic acid (iohexol, iopamidol, ioversol). The newer agents, by virtue of their lower osmolality at diagnostic concentrations, cause fewer physiologic alterations during injection, produce less pain or discomfort, and are associated with fewer adverse reactions than the conventional ionic contrast media.

Individual agents are often referred to in terms of concentration (percentage weight per volume), for example, Renografin 76 (diatrizoate sodium and diatrizoate meglumine). Perhaps a better term for comparison is iodine content (in milligrams per milliliter). "Full-strength" agents for angiography typically range from 300 to 370 mg/mL iodine content. Contrast media of lower concentration may be employed for DSA.

Conventional ionic contrast media are usually five to seven times the osmolality of blood in the concentrations injected.[17] A transient sensation of heat is reported by most patients following contrast media injection, which can be quite painful with selective arteriography of the extremities. Lidocaine has been added to ionic contrast material for lower extremity arteriography in the past. Use of lower osmolar ionic or nonionic media (one-third to one-half the osmolality of conventional ionic contrast media) makes patient discomfort much less of a problem, and lidocaine is not combined with the newer media. Lidocaine should never be injected into the brachiocephalic vessels!

Iothalamate and diatrizoate have anticoagulant properties that are not shared by the new nonionic agents.[18] It has been common practice to leave ionic contrast material in a catheter during manipulations. However, because the margin of safety with respect to thrombus formation is decreased, catheters containing nonionic agents should be flushed with heparinized saline immediately after contrast material injection.

There is little question that low-osmolality agents decrease adverse reactions and mortality rates are also lower (see "Risks Related to Contrast Media," later in this chapter). However, their use has elicited great controversy at present because of their cost, 12 to 15 times higher than that of conventional ionic media.[19] If the new agents are adopted universally for all vascular contrast studies in the United States, it has been estimated that fatalities might be prevented at the cost of about $3,400,000 per life saved![19] Scoring systems or other methods of high-risk patient selection have been proposed to limit use of the new expensive agents to those individuals most likely to benefit.[17, 19] However, malpractice issues make it unlikely that such proposals will be widely accepted. No doubt these issues will take years to be settled. At the University of Wisconsin-Madison we have decided to retain conventional ionic media for intravenous pyelography and computed tomographic applications, but nonionic media are now used for all venographic and arteriographic studies.

Carbon Dioxide

Not all angiographic examinations have utilized iodinated contrast material. Carbon dioxide has seen limited application as a vascular contrast agent, and the advent of DSA has enhanced its use.[20] A satisfactory delivery system remains to be perfected, but carbon dioxide injection can be considered for selected patients with a history of life-threatening reaction to iodinated media or very marginal renal function.

PATIENT PREPARATION AND MONITORING

Routine Measures

At the time of consultation by the patient's clinical service, the precise indications and aims of angiography should be elicited. If the same diagnostic information can be gained by less invasive, less expensive means, or if angiographic study is unlikely to yield the information sought, this should be thoroughly discussed with the referring clinician. Before angiography is scheduled, pertinent history—including use of anticoagulants or presence of bleeding tendency, status of renal function, and prior reactions to contrast media administration—must be obtained. This is especially important if the patient will be examined on an outpatient basis or if admission is to occur on the same day as the arteriogram.

When possible, the patient is seen the evening before angiography. The radiologist explains the purpose, nature, and risks of the study in order to obtain informed consent and establish a rapport with the patient. History of symptoms, previous pertinent surgical procedures (such as bypass graft placement), and allergies is obtained. Physical examination includes assessment of peripheral pulses to determine if a common femoral artery approach is possible, or if alternative arterial access must be pursued.

Orders are written for intravenous hydration and restriction of oral intake to clear liquids or nothing by mouth for at least 6 to 8 hours prior to the procedure. In most cases no routine premedications are given on the ward. When the patient is brought to the angiography suite on a cart, the peripheral pulses are checked and marked, and premedication is given at that time. Benzodiazepines are highly effective at relieving anxiety and are quite safe. Diazepam can be given slowly intravenously at a total dose of 5 to 10 mg. Midazolam is a shorter-acting drug, which can be administered either intravenously or intramuscularly; the usual dosage is 2 to 4 mg. Both drugs should be administered in small increments to avoid oversedation, and particular care must be exercised in elderly patients. Fentanyl is a short-acting (30 to 60 minutes) synthetic opiate that can be given in 50-μg increments for pain. However, respiratory depression can occur at doses as low as 50 to 100 μg.[21]

Physiologic monitoring is mandatory for those undergoing arteriography. Electrocardiography, automatic blood pressure measurement, and pulse oximetry are highly recommended. Oxygen, intubation equipment, and medications for treating possible life-threatening reactions (such as atropine, diphenhydramine, epinephrine, phentolamine, naloxone) must be at hand. Examining physicians must be familiar with cardiopulmonary resuscitation, as well as with treatment of more common untoward reactions.

After arteriography the catheter is removed, and hemostasis is achieved by manual compression. It is sometimes useful to move the patient onto the transporting cart or bed prior to catheter removal, for any extra patient motion can cause puncture site bleeding. Any hematoma present should be noted in the chart and outlined in ink on the skin. The patient is instructed to remain at bedrest with the relevant extremity immobilized for a minimum of 4 hours. During this time the patient's vital signs are monitored frequently and the peripheral pulse and puncture site checked (e.g., every 15 minutes during the 1st hour following arteriography, and decreasing in frequency over the course of several hours). Intravenous hydration is continued, and oral intake is encouraged. The patient is later visited on the ward by the angiographer at the end of the working day.

Selected Problems

If a patient has a history of hypersensitivity to contrast media, one must ascertain the nature of the previous episode. Although repeated reactions do not always occur, such patients are at increased risk from contrast medium exposure.[22] Nonionic or reduced osmolality agents are less likely to produce adverse reactions, but fatal events have also occurred with these agents.[19] For this reason, 1 to 3 days of premedication with oral corticosteroids and use of low-osmolar contrast media are advised for those patients with previous serious reactions[23] (see "Risks Related to Contrast Media").

Patients on heparin should have its administration stopped prior to arterial puncture. The effective half-life of heparin is about 90 minutes, so that active reversal of anticoagulation may not be needed in many cases.[24] Otherwise, protamine sulfate can be given by slow intravenous injection at a dose of 10 mg protamine per 1,000 units of heparin being neutralized.[24] The effects of heparin are reflected by prolongation of the activated partial thromboplastin time (PTT). Heparin anticoagulation should not be reinstituted for at least 1 hour after the completion of angiography.

Reversal of Coumadin (warfarin) anticoagulation is more of a problem, for the effective half-life is measured in days. If immediate arteriography is needed, fresh frozen plasma can be used to replenish the hepatic coagulation factors affected by Coumadin. The prothrombin time (PT) reflects the efficacy of Coumadin anticoagulation. As a rule of thumb, arteriography should be avoided if PT or PTT are more than 150% of control, or if a patient has fewer than 50,000 platelets/mm³.

Outpatient Arteriography

Arteriography can be performed on an outpatient basis as long as the patient has no attendant major medical problems, is cooperative and reliable, and has a companion for transportation and overnight assistance should problems arise after discharge. Standard evaluation and monitoring procedures must be arranged in a dedicated area, for example, through an ambulatory surgery service. Patients are kept supine and immobile for at least 4 hours, then engage in limited ambulation before being discharged by the angiographer. Instructions are provided for controlling recurrent bleeding and obtaining assistance for emergency problems. The patient's phone number is recorded for a follow-up call the following day. Necessity for hospital admission has been reported in 3% to 6% of those undergoing outpatient angiography.[25, 26]

RISKS OF ARTERIOGRAPHY

The safety of arteriography has been well-documented over its many years of use, but because of the invasive nature of the study and exposure of the patient to iodinated contrast media, there are real risks of major complication. The incidence of problems requiring directed therapy has been reported as 1.7%, 2.9%, 3.3%, and 7% after femoral, translumbar, axillary, and brachial artery punctures, respectively.[7, 8]

The 0.025% mortality rate described in the survey by Hessel and Adams resulted from arterial dissection, aneurysm rupture, vasovagal reactions, cardiac and neurologic complications, and renal failure.[7] If more than 800 arteriograms are obtained annually in an institution, the rate of nonfatal complications is significantly lower than if fewer studies are performed.[7]

Direct Injury

The most common problem related to catheterization is puncture site bleeding, which may require surgical repair or transfusion. Major hematomas are more likely to occur in hypertensive patients. Subintimal passage of the catheter or wire can produce vascular occlusion. Local endothelial trauma or catheter-related thrombi can do the same. One potentially fatal but fortunately rare complication is cholesterol embolization, which results in widespread small vessel occlusion from a shower of cholesterol crystals in patients with severe atherosclerotic disease.[27]

Risks Related to Contrast Media

Idiosyncratic Reactions

Idiosyncratic or hypersensitivity reactions to iodinated contrast media are unpredictable and vary from sneezing, nausea and vomiting, or urticaria, to laryngeal edema, cardiovascular collapse, and death. The great majority of reactions are mild and require no treatment, but severe reactions complicate about 1 study per 1,000, and the risk of death ranges from 1 per 12,000 to 1 per 75,000.[23] Mild reactions can be expected in about 2% to 4% of patients exposed to conventional ionic media, with urticaria accounting for more than two thirds of the reactions.[28, 29]

A history of any allergy approximately doubles the risk for adverse reaction to the contrast medium and increases the chance of severe reaction fourfold.[22] No definite dose-response effect to contrast media has been demonstrated, but there may be a positive correlation between the dose administered and severe and fatal reactions.[23] Repeated exposure in those with documented reactions to contrast produces another reaction 15% to 60% of the time, but premedication with corticosteroids and diphenhydramine can decrease this rate to less than 10%.[22]

A controlled study by Lasser et al. confirmed that premedication with oral methylprednisolone 32 mg given the evening before and repeated the morning of contrast injection decreased by 33% to 62% the incidence of reaction to ionic contrast agents administered intravenously.[23] Limiting premedication to a morning dose of methylprednisolone alone did not show a protective effect. A comparable reduction in risk can be produced by use of nonionic contrast media instead of the conventional ionic agents.

If a patient gives a history of severe hypersensitivity to contrast media, a 3-day course of premedication with oral corticosteroids is prudent. If prolonged premedication is not possible, at least one dose given 6 to 12 hours before angiography is recommended, together with use of low-osmolality contrast agents. However, one must be prepared to treat any and all reactions that might occur.

Nephrotoxicity

Risk factors for renal failure after contrast administration include preexisting renal insufficiency, dehydration, diabetes mellitus, hyperuricemia, and multiple myeloma. Those patients at particular risk are diabetics with renal disease (especially if serum creatinine exceeds 3.5 mg/dL), and those with combined hepatic and renal dysfunction.[22, 30, 31] The reported incidence of renal dysfunction after arteriography has varied greatly, from 0.5% to 38%, reflecting differences in patient selection and criteria for renal dysfunction.[30–33] Few of those with disturbances of function will have overt oliguria, and fewer still need support by hemodialysis. Any creatinine rise will become apparent within 24 hours, peak in several days, and usually reverse completely by 2 weeks. Nevertheless, permanent loss of renal function is a small but real risk.

Injection of contrast medium produces an initial vasodilatation which is followed by a reactive vasoconstriction in the renal circulation, an effect implicated in renal toxicity. These physiologic changes are less marked with nonionic media.[34] However, early clinical studies have not shown any clear decrease in nephrotoxicity with the use of low-osmolality media.[30, 32] What is clear is that patient hydration is an important protective measure. Liberal administration of intravenous and oral fluids is highly recommended unless specifically contraindicated by cardiac or other disease.[35] The amount of contrast material administered during a single study is not related to renal complications in those without underlying disease, but a dose limit of 4 mL/kg body weight is commonly observed by angiographers.[36] Furosemide and mannitol infusions have been proposed for maintaining renal blood flow and preventing toxicity, but benefit from either medication remains to be established in practice.[37]

REFERENCES

1. Seldinger SI: Catheter replacement of the needle in percutaneous arteriography. *Acta Radiol [Diagn]* 1953; 39:368–376.

2. Wade GL, Smith DC, Mohr LL: Follow-up of 50 consecutive angiograms obtained utilizing puncture of prosthetic vascular grafts. *Radiology* 1983; 146:663–664.

3. Weinshelbaum A, Carson SN: Separation of angiographic catheter during arteriography through vascular graft. *AJR* 1980; 134:583–584.

4. McDermott JC, Babel SG, Crummy AB, et al: Review of the uses of digital "road map" techniques in interventional radiology. *Ann Radiol* 1989; 32:11–13.

5. Lechner G, Jantsch H, Waneck R, et al: The relationship between the common femoral artery, the inguinal crease, and the inguinal ligament: A guide to accurate angiographic puncture. *Cardiovasc Intervent Radiol* 1988; 11:165–169.

6. Altin RS, Flicker S, Naidech HJ: Pseudoaneurysm and arteriovenous fistula after femoral artery catheterization: Association with low femoral punctures. *AJR* 1989; 152:629–631.

7. Hessel SJ, Adams DF: Complications of angiography. *Radiology* 1981; 138:273–281.

8. Grollman JH Jr, Marcus R: Transbrachial arteriography: Techniques and complications. *Cardiovasc Intervent Radiol* 1988; 11:32–35.

9. Moran KT, Halpin DM, Zide RS, et al: Long-term brachial artery catheterization: Ischemic complications. *J Vasc Surg* 1988; 8:76–78.

10. Yandow D, Wojtowycz M, Alter A, et al: Detection of retroperitoneal hemorrhage after translumbar aortography by computerized tomography. *Angiology* 1980; 31:655–659.

11. Cohen MI, Vogelzang RL: A comparison of techniques for improved visualization of the arteries of the distal lower extremity. *AJR* 1986; 147:1021–1024.

12. Sethi GK, Scott SM, Takaro T: Multiple plane angiography for more precise evaluation of aortoiliac disease. *Surgery* 1975; 78:154–159.

13. Crummy AB, Rankin RS, Turnipseed WD, et al: Biplane arteriography in ischemia of the lower extremity. *Radiology* 1978; 126:111–115.

14. Crummy AB, Strother CM, Lieberman RP, et al: Digital video subtraction angiography for evaluation of peripheral vascular disease. *Radiology* 1981; 141:33–37.

15. Gavant ML: Digital subtraction angiography of the foot in atherosclerotic occlusive disease. *South Med J* 1989; 82:328–334.

16. Crummy AB, Stieghorst MF, Turski PA, et al: Digital subtraction angiography: Current status and use of intra-arterial injection. *Radiology* 1982; 145:303–307.

17. Swanson DP, Thrall JH, Shetty PC: Evaluation of intravascular low-osmolality contrast agents. *Clin Pharm* 1986; 5:877–891.

18. Corot C, Perrin JM, Belleville J, et al: Effect of iodinated contrast media on blood clotting. *Invest Radiol* 1989; 24:390–393.

19. Jacobson PD, Rosenquist J: The introduction of low-osmolar contrast agents in radiology: Medical, economic, legal, and public policy issues. *JAMA* 1988; 260:1586–1592.

20. Hawkins IF: Carbon dioxide digital subtraction angiography. *AJR* 1982; 139:19–24.

21. Lind LJ, Mushlin PS: Sedation, analgesia, and anesthesia for radiologic procedures. *Cardiovasc Intervent Radiol* 1987; 10:247–253.

22. Cohan RH, Dunnick NR: Intravascular contrast media: Adverse reactions. *AJR* 1987; 149:665–670.

23. Lasser EC, Berry CC, Talner LB, et al: Pretreatment with corticosteroids to alleviate reactions to intravenous contrast material. *N Engl J Med* 1987; 317:845–849.

24. Bookstein JJ, Moser KM, Hougie C: Coagulative interventions during angiography. *Cardiovasc Intervent Radiol* 1982; 5:46–56.

25. Rogers WF, Moothart RW: Outpatient arteriography and cardiac catheterization: Effective alternatives to inpatient procedures. *AJR* 1985; 144:233–234.

26. Saint-Georges G, Aube M: Safety of outpatient angiography: A prospective study. *AJR* 1985; 144:235–236.

27. Fine MJ, Kapoor W, Falanga V: Cholesterol crystal embolization: A review of 221 cases in the English literature. *Angiology* 1987; 38:769–784.

28. Shehadi WH, Toniolo G: Adverse reactions to contrast media: A report from the Committee on Safety of Contrast Media of the International Society of Radiology. *Radiology* 1980; 137:299–302.

29. Sigstedt B, Lunderquist A: Complications of angiographic examinations. *AJR* 1978; 130:455–460.

30. Parfrey PS, Griffiths SM, Barrett BJ, et al: Contrast material–induced renal failure in patients with diabetes mellitus, renal insufficiency, or both: A prospective controlled study. *N Engl J Med* 1989; 320:143–149.

31. Swartz RD, Rubin JE, Leeming BW, et al: Renal failure following major angiography. *Am J Med* 1978; 65:31–37.

32. Gomes AS, Lois JF, Baker JD, et al: Acute renal dysfunction in high-risk patients after angiography: Comparison of ionic and nonionic contrast media. *Radiology* 1989; 170:65–68.

33. Mason RA, Arbeit LA, Giron F: Renal dysfunction after arteriography. *JAMA* 1985; 253:1001–1004.

34. Golman K, Almén T: Contrast media-induced nephrotoxicity: Survey and present state. *Invest Radiol* 1985; 21:S92–S97.

35. Eisenberg RL, Bank WO, Hedgock MW: Renal failure after major angiography can be avoided with hydration. *AJR* 1981; 136:859–861.

36. Miller DL, Chang R, Wells WT, et al: Intravascular contrast media: Effect of dose on renal function. *Radiology* 1988; 167:607–611.

37. Cruz C, Hricak H, Samhouri F, et al: Contrast media for angiography: Effect on renal function. *Radiology* 1986; 158:109–112.

2 | Angiography of the Abdominal Aorta, Pelvis, and Lower Extremities

---

KEY CONCEPTS

1. Arteriography remains the simplest, most reliable examination for precisely defining the extent of aortic and lower extremity vascular disease.
2. Digital subtraction angiography, intra-arterial injection of vasodilators, reactive hyperemia, and distal catheter placement are used to optimize distal vessel filling.
3. Abdominal aortic aneurysms are studied preoperatively to establish the relationship of renal arteries to the aneurysm, the status of the mesenteric circulation, and involvement of the iliac or common femoral arteries.
4. Aneurysms are prone to rupture, occlusion, and distal embolization.
5. Vascular grafts should be studied at the first sign of any problem.
6. Arteriography is highly accurate for study of traumatic vascular injury, but is rarely needed in the absence of specific signs.

INDICATIONS

Abdominal aortography and lower extremity runoff anteriography are the "bread and butter" of the peripheral angiographer. Despite recent improvements and innovations in the noninvasive evaluation of lower extremity vascular disease, angiography with contrast material remains the definitive diagnostic measure before surgery or transluminal intervention. Symptomatic peripheral vascular disease, whether occlusive or aneurysmal, continues to afflict 2% to 3% of men and 1% of women above the age of 60.[1] Aside from this large population of patients, arteriography of the lower extremities is often needed in evaluation of severe pelvic or extremity injuries, both penetrating and nonpenetrating, and a variety of other less common disorders such as tumors, arteriovenous malformations, thromboangiitis obliterans, frostbite, popliteal entrapment syndrome, and cystic adventitial disease.

ARTERIAL APPROACH

Common Femoral Artery

The standard approach for abdominal aortography and lower extremity angiography is common femoral artery catheterization. Assuming femoral pulses are present, the femoral approach has the advantage of providing optimal opacification of vessels at lowest risk. After abdominal aortography, the flush catheter is withdrawn to the aortic bifurcation so that injection of contrast material will fill the runoff vessels bilaterally, with little of the injected volume diverted to the kidneys or viscera outside the field of interest. In general, the side with the strongest femoral pulse is chosen for arterial catheterization.

Alternatives in the Face of Occlusive Disease

When femoral pulses are lacking, catheterization may be attempted through an axillary artery, a brachial artery, or by a translumbar approach. Upper extremity arterial access allows catheters to be placed selectively when needed. However, in older patients with tortuous vessels it may be very difficult to direct a catheter into the descending thoracic aorta. The left upper extremity should always be preferred, in order to minimize the possibility of embolic stroke. Translumbar aortography does not allow selective placement of the catheter, but is a reliable method that has a somewhat lower complication rate than upper extremity arterial puncture.[2] Both approaches are fraught with hazard in the presence of poorly controlled hypertension or coagulation disorder. Because major hemorrhage can occur without early warning signs, particular care must be taken in selecting patients for translumbar study (for more details about arterial access, see Chapter 1, Basic Principles of Angiography).

When a vascular bypass graft is present, direct entry should be avoided if there is a pulse in the opposite groin. However, grafts may be punctured at no great risk as long as dilators are used and the catheter is manipulated and removed over a guidewire.[3] Among those needing direct graft entry are patients with aortobifemoral or axillo-femoral bypasses.

Obtaining Optimal Distal Vessel Opacification

With the increased use of saphenous vein bypass grafting for disease below the knee (which marks the practical limit for the use of synthetic materials), there is need to obtain good preoperative definition of the runoff vessels in the distal calf, ankle, and foot.[4] Patency of the distal vessels is predictive of subsequent graft patency. Conventional filming and contrast medium injections at the aortic bifurcation yield inadequate images in approximately one quarter of distal arterial examinations.[5] For this reason various techniques have been used to improve filling of distal arteries in the face of occlusive disease.

Digital Subtraction Angiography

One of the simplest measures to apply is digital subtraction angiography (DSA). When conventional films are insufficient, the foot (or the fluoroscopy C-arm) can be rotated into a lateral projection and a 6-inch or 10-inch fluoroscopic field centered over the talus. A total injection of 20 to 25 mL of contrast material (300 mg iodine/mL) will usually provide the arterial detail needed. Digital subtraction angiography has been shown superior to conventional film angiography in this clinical setting.[5, 6]

Reactive Hyperemia

If DSA alone fails to demonstrate distal anatomy, vasodilatation may be produced by various means. Placement of an occlusive cuff on the thigh with inflation above arterial pressure for 7 minutes will produce a reactive hyperemia immediately after cuff deflation. Flow becomes more rapid, and vascular opacification improves in nearly 90% of patients.[7] Although this technique is quite safe, it lengthens the procedure, can be quite uncomfortable for the patient, and does carry a finite risk of producing vascular occlusion. Compression of a bypass graft must be avoided.[8]

Vasodilators and Other Measures

Tolazoline has long been used as an adjunct to arteriography. It is a potent vasodilator with a duration of action of several minutes when injected intra-arterially. It is customarily given at a dose of 25 to 50 mg diluted in 10 mL of saline, injected slowly through the catheter immediately prior to filming. The presence of heart disease is a contraindication to the use of tolazoline, for the drug has the potential of inducing hypotension, tachycardia, arrhythmias, and myocardial infarction.[9]

A safe alternative medication is dilute nitroglycerin. Injected into the arterial catheter as a bolus of 200 µg in 10 mL of saline just prior to contrast material injection, nitroglycerin is more effective than tolazoline in producing vasodilatation, and it is approximately equivalent in effect to reactive hyperemia.[10] Extrinsic warming of the feet and legs is also valuable in obtaining distal vascular opacification.[9]

When all other measures fail and the decision must be made to amputate or place a graft, antegrade catheterization of the superficial femoral artery with injection of contrast material directly into the popliteal artery may be appropriate.[9] If long occlusions are present in the iliac arteries with poor collateral flow through the pelvis, puncture of the common femoral artery on the symptomatic side allows a greater concentration of contrast medium to reach the distal arteries. Failure to show arterial fill despite all these measures generally means that the distal vessels are occluded. If there is reason to doubt this conclusion, intraoperative arteriography can be performed as a last resort.

Proper length of filming time is important when evaluating distal disease, up to 50 or 60 seconds in extreme cases. When reactive hyperemia or vasodilatation is used, contrast material injection rates should be increased, and any filming delay must be shortened; the time needed for contrast medium to appear in the distal vessels may be substantially decreased with these techniques.[7]

FILMING ALTERNATIVES

Step-Table Angiography

Stepping tables have been widely employed in lower extremity arteriography. When conventional films are used, contrast material injections and table steps can be combined in a variety of ways. One may choose to have the table step during the abdominal aortic run to also obtain views of the pelvis. The runoff vessels may then be filmed over three or four stations during a subsequent injection of contrast medium. During translumbar angiography a single injection may be made with filming at four different stations, from the abdominal aorta to the level of the knees. Filming routines may be easily adapted to suit the particular patient or the angiographer's preference.

A technique that has led to satisfactory results—with repeated injections (for failure in filling or film timing) needed in fewer than 10% of patients examined—starts with a single-stage set of exposures of the abdominal aorta. The pigtail catheter is placed at the level of the renal arteries in order to prevent major reflux into celiac or superior mesenteric artery branches. Contrast medium is injected at 20 mL/sec for 2 seconds, while films are obtained at a rate of two exposures per second for 3 seconds, followed by one exposure per second for 6 more seconds. This program allows the kidneys to be evaluated through their nephrographic phase.

After aortic filming is complete the catheter is withdrawn to the distal abdominal aorta, leaving the sideholes superior to the aortic bifurcation. A long injection of a large volume of contrast medium is programmed, with the flow gauged by a hand test injection under fluoroscopic guidance. A typical injection in a 60-year-old man might be 7 mL/sec for a total of 12 seconds. After a filming delay of 3 seconds, the first three exposures are obtained at the level of the pelvis at a rate of one exposure per second. The table moves and two films are exposed at the level of the thighs, followed by one step and four film exposures at the knees. After the third step, filming is slowed to one exposure every 2 seconds for six films at the level of the calf. If the very distal vessels of the ankle and foot must be imaged (as in the case of diabetics, patients with rest pain, or those with ulceration or gangrene), a fifth station can be added to the run.

With a step-table angiogram, timing is critical. A long, high-volume injection provides a good margin for error. With the program described it is impossible for the injected contrast medium to "get ahead" of filming, because the injection will finish only at the time exposures are being made at the popliteal level. A filming delay is set to allow adequate pelvic arterial opacification. A longer delay may be needed for patients with severe occlusive disease, poor cardiac output, or large aneurysms.

While the films are being developed, DSA may be used expeditiously to "fill in" regions of interest. Areas that often merit imaging in more than one projection include the iliac arteries, the common femoral bifurcations, renal artery origins, and graft anastomoses.

Digital Subtraction Runoff Technique

Digital subtraction angiography may be used for complete lower extremity arteriography if a large image intensifier (14-inch diameter) is available. As a rule, only a single field may be imaged during one injection. A DSA study has the advantage that one may observe in real time the arrival of contrast medium and the adequacy of filling. Because table position does not change during a run, individual injection volumes may be kept to a minimum (for example, 14 mL injected over 2 seconds for popliteal filming).

Because of edge cutoff with the round field of view of DSA, it is sometimes advisable to examine the lower extremities separately in larger patients. A greater segment of the leg can be imaged, limiting the amount of overlap necessary between fields. Also, it is easier to set proper exposure factors when examining a single leg than if both legs (including a radiolucent gap) are viewed together. Exposure factors can also constrain DSA study of the abdomen or pelvis in large patients, limiting the degree of obliquity or the frame rate that can be achieved.

Although individual injection volumes are smaller with DSA, the total amount of contrast medium used does not differ greatly from that given during step-table film runoff angiography. Because of the multiple table positionings and repeated injections of contrast material needed, the duration of a DSA study is also not substantially different. In lower extremity angiography the technique chosen depends on the equipment installed, preference of the angiographer, and acceptability of images to referring clinicians.

Use of Multiple Projections

Atherosclerosis does not produce uniform concentric stenoses. If a plaque is located on the posterior wall of the aorta or common iliac artery, it may be poorly appreciated on the anteroposterior (AP) projection. Biplane aortography is absolutely necessary in cases of lower extremity embolization or abdominal aortic aneurysm (AAA). In suspected embolization, an irregular or ulcerated plaque can be the source of embolized clot, and such a plaque must be diligently sought. The lateral projection is invaluable in evaluating aneurysms, for it most easily demonstrates the neck of the aneurysm as well as its relationship to the renal arteries.

Use of both oblique views for pelvic angiography improves detection of plaque and the grading of stenosis severity in over a third of patients with significant atherosclerosis.[11] The right posterior oblique projection best profiles the origin of the right internal iliac artery and left deep femoral artery. The left posterior oblique does the same for the contralateral vessels. Multiple views must absolutely be obtained whenever symptoms or noninvasive pressure measurements indicate more severe disease than that apparent on AP angiography.

POINTS OF ANATOMY

The abdominal aorta bifurcates at the level of L-3, and its branches, the common iliac arteries, are of variable length (Fig 2–1). The internal iliac (hypogastric) artery divides into a posterior trunk, which gives off the superior gluteal, iliolumbar, and lateral sacral arteries, and an anterior trunk, which supplies the pelvic viscera, including urinary bladder, rectosigmoid colon, and uterus. The external iliac arteries

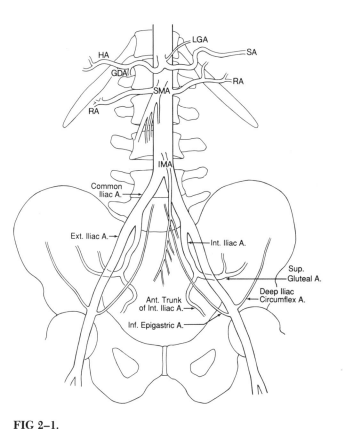

FIG 2–1.
Anatomy of the abdominal aorta and iliac arteries. *HA* = hepatic artery; *LGA* = left gastric artery; *SA* = splenic artery; *GDA* = gastroduodenal artery; *RA* = renal artery; *SMA* = superior mesenteric artery; *IMA* = inferior mesenteric artery.

have no major branches except in the region of the inguinal ligament, where the inferior epigastric and deep circumflex iliac arteries arise.

The common femoral artery (Fig 2–2) normally bifurcates near the inferior margin of the femoral head, but occasionally the point of division is several centimeters more cephalad. The profunda femoris (deep femoral artery) provides major branches about the femoral neck (medial and lateral femoral circumflex arteries) as well as large muscular branches to the upper and mid-thigh. The superficial femoral artery (SFA) extends to the adductor canal, at which point it becomes the popliteal artery. The SFA has no major branches in the absence of obstructive disease.

The popliteal artery gives off the anterior tibial artery near the head of the fibula, as well as the tibioperoneal trunk. The latter vessel divides into the posterior tibial and peroneal (fibular) arteries. The anterior tibial artery angles laterally and then curves inferiorly to cross the interosseous membrane, forming a characteristic "knee." The vessel gradually swings medially near the ankle and continues into the foot as the dorsalis pedis. The tibioperoneal trunk continues the axis of the popliteal artery, and the peroneal artery is the more lateral of its branches. The posterior tibial artery courses into the foot where it becomes the plantar arch. The peroneal is not directly continuous with any major pedal vessel, but it supplies lateral and medial malleolar arteries, which may become important collateral pathways.

Major Collateral Pathways

In the face of abdominal aortic occlusion, the lower intercostal and lumbar arteries dilate and supply blood to the pelvis and lower extremities by way of internal iliac branches (iliolumbar, superior, and middle gluteal arteries) or over the abdominal wall by way of the inferior epigastric and deep iliac circumflex arteries. The inferior mesenteric artery, by virtue of the anastomoses between its superior hemorrhoidal (rectal) branches with the inferior hemorrhoidal arteries, can be a major source of collateral perfusion. In some cases, distal flow is maintained by way of the anterior abdominal wall through internal thoracic (mammary)-inferior epigastric artery communications.

In iliac or common femoral artery occlusive disease, the internal iliac branches often play a crucial role. The gluteal and obturator arteries serve as bridges, connecting with the lateral and medial femoral circumflex arteries. In unilateral obstruction there are multiple pathways through pelvic branches of the internal iliac.

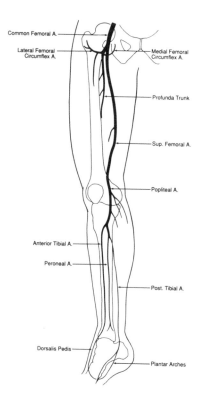

FIG 2–2.
Anatomy of arterial supply to the lower extremity.

Perhaps the single most critical vessel for the preservation of distal perfusion in the leg is the profunda femoris. Muscular branches provide fill of the distal SFA or popliteal artery in the presence of SFA obstruction. Because the origin of the profunda femoris is prone to stenosis and this area is poorly seen on an AP projection, the femoral bifurcation should be studied in more than one view.

Geniculate branches form a frame about the knee joint, and are important pathways when popliteal flow is obstructed. At the ankle, the three runoff arteries communicate through malleolar vessels. Additional distal anastomoses connect the dorsalis pedis to the plantar arch.

Variants

The sciatic artery, a rare and sometimes perplexing variant, represents the failure of regression of a primitive fetal vessel.[12] The proximal portion of the sciatic artery normally is incorporated into the inferior gluteal artery. More distal segments become parts of the popliteal and peroneal arteries. The external iliac, common femoral artery, and SFA may be occluded, small, or normal in size when a persistent sciatic artery is present. The internal iliac artery is large, does not taper in a normal fashion, and a large branch takes a lateral course at the femoral head. The persistent sciatic artery tends to be ectatic, and it joins the distal SFA by taking a medial bend near the adductor canal. Patients with a sciatic artery often present with claudication, but nearly half of those with symptoms will have aneurysmal dilatation in the gluteal segment of the vessel.[12] Because flow is typically slow, care must be taken to extend filming for a sufficient duration if the diagnosis is suspected.

Both the branching and size of the trifurcation vessels of the calf are subject to variation. Occasionally, a true trifurcation is present and all three vessels arise from the terminal popliteal artery. In other individuals the anterior tibial artery may originate above the knee from the proximal popliteal. Rarely, the posterior tibial may be the first terminal branch of the popliteal, with the peroneal and anterior tibial arteries sharing a common trunk. Although the posterior tibial artery is the primary runoff vessel in most adults, either of the other calf arteries may be dominant on a congenital basis.

MANIFESTATIONS OF ATHEROSCLEROSIS

Atherocclusive Disease

Patients with signs of peripheral atherosclerosis have a 24% risk of developing critical lower limb ischemia within 5 years.[13] However, because atherosclerosis is a systemic disease, death from myocardial infarction or cerebrovascular accident is more likely than limb loss. Many patients with symptomatic claudication will benefit from a structured exercise program. It is those with persistent limiting symptoms, rest pain, or nonhealing ulcers who require runoff arteriography to define the precise morphology of disease prior to surgical or percutaneous transluminal revascularization.

Occlusive atherosclerotic disease tends to follow characteristic patterns. It may primarily affect the infrarenal abdominal aorta and common iliac arteries, sparing more distal vessels. In men this is manifested by the classic Leriche syndrome of thigh and buttock claudication accompanied by impotence. A similar distribution of lesions may be seen in relatively young women who smoke and have small aortoiliac vessels.[14] The greatest danger of aortic occlusion is from thrombus extension above the level of the renal arteries, leading to renal failure and even mesenteric infarction.

In other patients the vessels affected most severely are the SFA or the more distal arteries in the calf. Diabetics tend to have disease involving medium and small arteries, but they are subject to severe diffuse atherosclerosis of larger vessels as well.

Atherosclerotic plaques are characteristically irregular and eccentric; however, short smooth concentric lesions and membranes are also seen. Ulceration may be suspected by an excavated appearance to a plaque, but this is more properly a histologic diagnosis. Uncommonly, lesions may grossly project into the vessel lumen, simulating clot or other filling defect. On the other hand, thrombus may form either upstream and downstream of a significant stenosis. Atherosclerosis is likely to produce diffuse plaque, and arteries in the vicinity of occlusions or stenoses will rarely have entirely smooth lumens of normal caliber.

A positive diagnosis of arterial embolus is possible only when an intraluminal thrombus is uncovered in an otherwise unremarkable vessel. A convex margin to a vascular occlusion is not alone sufficient to warrant the diagnosis of embolism, because in situ thrombosis can have an identical appearance. Most peripheral emboli arise from the heart, but aortic, iliac, and femoral plaques may also be to blame. One should also not forget the (unlikely) possibility of paradoxic venous embolization through a patent foramen ovale. Emboli tend to lodge at arterial bifurcations, where the vessel lumen tapers. "Blue toe" syndrome is one manifestation of peripheral embolization.

Aneurysms

The second major manifestation of atherosclerosis is aneurysmal disease, and the most common site of involvement is the infrarenal abdominal aorta. Such AAAs are typically fusiform, have a well-defined neck arising distal to the origin of the renal artery, and often extend directly into the common iliac arteries. Aortoiliac aneurysms sometimes are continuous with internal iliac aneurysms. Extension into the external iliacs is distinctly unusual.

Risks of Abdominal Aortic Aneurysm

The overriding concern with aneurysms of any sort is rupture. Distal embolization is a serious complication which occurs less frequently. An unusual form of rupture is development of an aortocaval fistula. Such patients may present with lower extremity edema and high-output cardiac failure.

The risk of AAA rupture is estimated to be 10% for aneurysms up to 4 cm in diameter, 25% for those between 4 and 7 cm in diameter, and even higher for those larger than 7 cm.[15] Elective surgery is generally recommended for all symptomatic aneurysms and for asymptomatic aneurysms larger than 6 cm in diameter, because the mortality of elective repair is under 5% but emergency surgery for rupture carries a 50% risk of death.[15] Standard treatment involves resection and graft replacement. Nonresective treatment, ligation with extra-anatomic bypass, is an option for high-risk patients, but it can result in severe late morbidity.[16]

Role of Aortography

The role of angiography has been curtailed in establishing the diagnosis of AAA. Because many aortic aneurysms are lined by mural clot, aortography is not especially useful for showing the true size of a lesion. Ultrasound, computed tomography (CT), and magnetic resonance imaging (MRI) are superior in this regard, and ultrasound is the procedure of choice for following aneurysm growth. However, study with contrast material continues to have certain advantages over cross-sectional imaging. Aortography more precisely defines the relationship of the renal arteries to the aneurysm, delineates mesenteric perfusion, and is considered essential for preoperative planning by many surgeons.

The surgical approach may be modified by angiographic findings. Rösch et al. determined that important surgical decisions were made on the basis of aortography in 75 of 100 patients studied.[17] Particular problems are posed by aberrant renal arteries, suprarenal aneurysm extension, renal or mesenteric artery stenosis, and continued patency of the inferior mesenteric artery (IMA). The IMA is occluded in the great majority of AAAs, but failure to recognize that the vessel is patent may result in failure to reimplant it surgically, with consequent mesenteric ischemia. Postoperative visceral infarction complicates fewer than 3% of aneurysm repairs, but when it does occur, mortality is high.[15, 17]

Technique of Examination

When examining an AAA, biplane aortography must be performed. The catheter should be placed above the level of the celiac axis in order to fill all abdominal visceral arteries. The lateral projection is essential for demonstrating the celiac axis, as well as the proximal superior mesenteric artery and IMA (Fig 2–3). It is also useful for showing the number and location of renal arteries. Filming is extended through at least 12 seconds to allow renal perfusion and possible mesenteric collateral circulation to be assessed. The pelvic arteries and common femorals should also be studied in two planes to exclude aneurysms or occlusive disease in these vessels.

If the renal arteries are not delineated or the IMA is not filled, the catheter may be withdrawn into the neck or body of the aneurysm itself and the study repeated. Injection proximal to a large aneurysm often results in dependent layering of contrast medium, and the catheter may appear to be extravascular on the lateral projection. Power injection into an aneurysm does not pose undue risk to the patient, while it does provide the best opportunity to show small aberrant renal vessels and a patent IMA.

Inflammatory Abdominal Aortic Aneurysms

Inflammatory aneurysms are associated with marked mural thickening of the aorta and retroperitoneal adhesions. Up to 10% of AAAs may show inflammatory components, but the diagnosis cannot be made angiographically.[18] Computed tomographic scanning is capable of showing retroperitoneal fibrotic plaque in this condition, and is also useful in cases of suspected leaking aneurysm, because it can demonstrate periaortic or retroperitoneal hematoma. It is extraordinarily rare to show leakage by means of an aortogram. If brisk bleeding is suspected, a patient is best taken immediately to the operative suite without expending time for diagnostic imaging.

Iliac and Peripheral Aneurysms

Aneurysms of the iliac arteries constitute about 2% of all intra-abdominal aneurysms and are usually associated with aortic aneu-

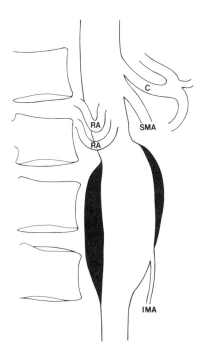

FIG 2–3.
Lateral view of an abdominal aortic aneurysm. Shaded portion represents mural thrombus. *C* = celiac axis; *SMA* = superior mesenteric artery; *RA* = renal arteries; *IMA* = inferior mesenteric artery.

rysms.[19] One third of iliac lesions are isolated, and because they tend to be asymptomatic or produce nonspecific complaints, many are not found until rupture. For this reason, some surgeons recommend repair of any lesion larger than 3 cm in diameter.[19]

Atherosclerotic aneurysms also frequently affect the common femoral and popliteal arteries. Less common sites include the SFA and adductor canal. Diffuse vascular ectasia (arteriomegaly) is another unusual expression of aneurysmal disease. When a large aneurysm or diffuse ectasia is present, contrast medium injection and filming must be prolonged because of the extremely slow flow which may be encountered.

A particular risk of popliteal artery aneurysm is repeated embolization to the small vessels of the calf. Such emboli often lead to amputation if their source is not recognized and treated. An aneurysm at any one location increases the likelihood that a contralateral lesion is present or that other vessels are affected. If a patient is found to have popliteal aneurysms on physical examination, an AAA should be excluded by ultrasonography.

Bypass Grafts

Autologous vein, umbilical vein, and synthetic bypass grafts have become a mainstay of vascular surgery. Common types of grafts seen in the abdomen, pelvis, and lower extremities (described by proximal and distal anastomoses) include aortic tube, aortobiiliac, aortobifemoral, axillofemoral, femorofemoral (crossover), femoropopliteal, and femorodistal bypasses. Each graft has an expected long-term patency, which depends on location, graft material, and certain host factors (e.g., patency of distal vessels).

Graft Stenosis and Thrombosis

Acute graft thrombosis can be approached by infusion of thrombolytic enzymes, aspiration thrombectomy, or surgical thrombectomy (see Chapter 11, Local Thrombolytic Infusion). In most cases there is an underlying abnormality also needing correction by surgical revision or percutaneous angioplasty. However, detecting and repairing an abnormality before occlusion occurs can substantially improve subsequent patency of a graft.[20] For this reason, angiography is indicated as soon as noninvasive studies or symptoms indicate a problem. Because many potential problems can be uncovered soon after surgery, some advocate routine postoperative intravenous DSA for patients undergoing graft placement.[21] A less invasive study that would likely be more acceptable to patients is duplex ultrasonography. It can detect the majority of graft stenoses even before they become evident by segmental blood pressure measurements.[20] Stenoses are typically the result of fibrointimal hyperplasia, which tends to occur at anastomoses.

Other Complications

Other assorted complications of bypass graft include infection (of prosthetic grafts), pseudoaneurysms, and in the case of in situ saphenous grafts, arteriovenous fistulas due to unligated perforating veins.

In situ saphenous vein grafts, unlike reversed saphenous grafts, are not completely mobilized. Rather, only the ends of the vein are freed and anastomosed to donor and recipient arteries, their valves are disrupted by a valvulotome, and communicating perforators are ligated. An in situ graft is more technically demanding for the surgeon, but there have been reports of improved patency and limb-salvage rates for femorodistal grafts.[22] However, an ongoing cooperative study being conducted by Veterans Administration Hospitals has failed to confirm any advantage to in situ grafts thus far.[4]

Leaking Anastomoses and Arterioenteric Fistulas

A serious problem with infected grafts is the development of an arterioenteric fistula. This is most likely to affect the third portion of the duodenum and an adjacent infected aortic graft. The most direct diagnostic study is endoscopy, but if endoscopy is impossible or inconclusive, aortography may sometimes demonstrate focal irregularity or ulceration at the anastomotic site. Unfortunately, angiography is usually not helpful in detecting fistula or anastomotic disruption. Computed tomography is a more productive examination, for it may uncover retroperitoneal hemorrhage or an abscess adjacent to or surrounding the graft.

Special Examination Techniques

If a failing femoropopliteal or femorodistal graft cannot be visualized by conventional arteriography, the patient may be given heparin and a catheter positioned at the proximal anastomosis. Sprayregen et al. found a number of grafts obstructed, but not actually thrombosed, when they were injected directly.[23] For grafts that cross the knee joint, examination of the patient in a lateral projection with the knee extended, partially flexed, and fully flexed has been recommended.[24] Kinks or loops found in flexion have been correlated with subsequent graft occlusion. Axillofemoral grafts are most easily examined by direct puncture. The proximal anastomosis can be demonstrated either by retrograde catheter placement to the anastomosis or brief manual compression of the graft during contrast material injection.

OTHER CONDITIONS

Thromboangiitis Obliterans and Various Lesions

Thromboangiitis obliterans or Buerger's disease is an uncommon inflammatory vasculitis of unknown cause, usually limited to the small- and medium-sized vessels of the extremities.[25] It is an obliterating process that produces intimal hyperplasia and thrombosis. There is a strong association with smoking, and those affected are predominantly men 30 to 40 years of age. Patients may complain of claudication, thrombophlebitis, acral ulceration, and intense pain. The condition is characterized by exacerbations and remissions, and the upper extremities are frequently involved. Raynaud's phenomenon is commonly observed. Many will eventually need amputation to some extent, but mortality is not increased. Thromboangiitis obliterans is distinctive in sparing the coronary and cerebral circulations.[25] Acute episodes are treated by anti-inflammatory medications and anticoagulation. Bypass surgery has disappointing long-term results in these patients.

The majority of cases will show bilateral lower extremity involvement with segmental occlusions, sharp cutoffs, tortuous (corkscrew) collateral vessels, and at times, vasospasm and arteriovenous shunting. Recanalization may produce "direct collaterals." Unlike atherosclerotic disease, thromboangiitis obliterans shows relatively normal vascular segments between areas of disease ("skip" lesions). Occasionally the iliac arteries and aorta may be affected, and the angiographic findings may be equivocal. The diagnosis is best made histologically by punch biopsy, but biopsy itself may initiate ulceration.

The use of ergot alkaloids for migraine headache, or poisoning with the toxin of the fungus *Claviceps purpurea* can produce diffuse or segmental arterial spasm and lower extremity ischemia.[26] The narrowing seen on arteriography is smooth and may reverse with time. Takayasu arteritis may compromise the abdominal aorta, and is usually confined to its proximal portion.[27] Fibromuscular disease has been found in the iliac arteries, the third most common site of involvement.[28] The angiographic appearance is similar to that of lesions in the renal or carotid arteries. Frostbite or electrical injury can cause vascular occlusion, and angiography may be necessary to select a level for amputation.

Popliteal Entrapment Syndrome

The popliteal artery normally courses between the medial and lateral heads of the gastrocnemius muscle. If the vessel takes an anomalous route medial to the tibial head of the muscle or if abnormal fascial slips are present, the popliteal artery may be subject to compression.[29] When compression causes claudication or thrombosis, the syndrome is termed popliteal entrapment. Those typically affected are active men of heavy build between 20 and 40 years of age.

Classical angiographic findings are medial deviation of the popliteal artery and stenosis. Unfortunately, no narrowing or abnormal course may be evident in many with the clinical syndrome.[30] A combination of passive dorsiflexion of the foot with active plantar flexion against resistance will elicit marked narrowing or occlusion. Post-stenotic dilatation, mural irregularity, or aneurysm may also be observed, and the condition is often bilateral. False positive diagnosis is possible if based on noninvasive studies, so angiography is recommended whenever popliteal entrapment is suspected.[30]

Cystic Adventitial Disease

Another cause of claudication, predominantly in young men, is cystic adventitial disease. Arterial compromise of the popliteal artery results from compression by a unicameral or multilocular cyst of the vessel wall. The cause of the cyst is unclear, but microtrauma or congenital mucous cell rests may be responsible. Symptoms may be sudden in onset, and confusion with popliteal entrapment syndrome is not unusual.[31] Ultrasonography and CT scanning have been of value in making the diagnosis.[32, 33] Angiography will demonstrate a smooth, scalloped, or hourglass-shaped narrowing of the popliteal artery without post-stenotic dilatation.

Tumors and Arteriovenous Malformations

Osseous and soft tissue tumors are diagnosed by plain radiography, scintigraphy, CT, ultrasound, or MRI. Angiography is reserved for preoperative evaluation, defining tumor vascularity and feeding vessels. In selected cases treatment may call for local arterial infusion of chemotherapeutic agents or embolization (see Chapter 12, Embolotherapy).

Congenital arteriovenous malformations of the pelvis and extremities can cause bleeding, ulceration, and disfigurement. They may behave in a quite malignant fashion, requiring extensive surgery or amputation for control. Unfortunately, many recur after treatment. As in malignant tumors, arteriography serves to define the vascular anatomy prior to surgery, or it may be used to palliate by injection of particulate emboli or sclerosing agents. Chronic arteriovenous fistulas may develop multiple feeding arteries and draining veins, simulating arteriovenous malformations.[34] Evaluation must include rapid serial-film angiography with superselective injection. The distinction of arteriovenous fistula from a vascular malformation is important, because embolic occlusion of the arteriovenous communication in a fistula is definitive therapy and the lesion should not recur.

TRAUMA

When a patient has sustained major trauma, and an enlarging extremity, hematoma, hypotension, absent or diminished pulse, or neurologic deficit is present, emergency angiography is indicated to define arterial disruption or occlusion. Positive findings include compression, thrombosis, spasm, extravasation, pseudoaneurysm, intimal injury, and arteriovenous fistula. Digital subtraction angiography, despite its lower spatial resolution, can be as accurate for a diagnosis as conventional film angiography.[35] Scout views are indispensable, and each site of suspected injury must be examined in at least two projections! Entry and exit sites of bullets or other projectiles should be indicated by radiopaque markers. When properly performed, angiography has 92% to 98% accuracy in extremity trauma.[36, 37]

Except for a few specific forms of trauma (such as dislocation of the knee or high-power gunshot wound), lack of suggestive physical findings is highly predictive of normal angiography. If a patient does not have absent or diminished pulse, neurologic deficit, bruit, murmur, or signs of active bleeding, and angiography is requested merely because of the proximity of penetrating trauma to a major vessel, the chance of uncovering a significant abnormality is less than 5%.[38–40] What is more, if an asymptomatic intimal flap *is* found, it can be treated conservatively with little risk to the patient.[41]

Vessel narrowing may represent spasm or external compression from a compartment syndrome. Radiolucent angiographic strips in a longitudinal orientation within a vessel often represent flow or layering phenomena and must not be mistaken for intimal flaps. Filling defects must be constant and visible on more than one projection for the diagnosis of intimal injury to be made. Flaps typically appear as thin transverse lines or globules attached to the vessel wall in at least one view.[42] When arterial occlusion is found, its precise cause cannot reliably be determined by angiography.

NONINVASIVE STUDIES

Nonangiographic examinations are valuable in screening patients for vascular disease and in selecting who might benefit by arteriography. Segmental Doppler blood pressure measurements, which are easy to perform and generate an ankle-brachial pressure index, are extremely useful in the detection of significant arterial obstruction. However, noninvasive imaging examinations are being actively developed in the hope of increasing sensitivity and perhaps someday replacing catheter angiography. At present, velocity ratios determined by duplex ultrasonography are detecting graft stenoses before they are suggested by symptoms or segmental pressure changes.[20]

Various strategies are being employed to develop magnetic resonance angiography.[43-45] However, the technical obstacles to be overcome are formidable. Contrast aortography and lower extremity arteriography are unlikely to be displaced as primary diagnostic measures in the near future.

REFERENCES

1. Genton E, Clagett GP, Salzman EW: Antithrombotic therapy in peripheral vascular disease. *Chest* 1986; 89(suppl):75–81.
2. Hessel SJ, Adams DF: Complications of angiography. *Radiology* 1981; 138:273–281.
3. Wade GL, Smith DC, Mohr LL: Follow-up of 50 consecutive angiograms obtained utilizing puncture of prosthetic vascular grafts. *Radiology* 1983; 146:663–664.
4. Veterans Administration Cooperative Study Group 141: Comparative evaluation of prosthetic, reversed, and in situ vein bypass grafts in distal popliteal and tibial-peroneal revascularization. *Arch Surg* 1988; 123:434–438.
5. Gavant ML: Digital subtraction angiography of the foot in atherosclerotic occlusive disease. *South Med J* 1989; 82:328–334.
6. Hol PK, Heldaas J, Skjennald A: Demonstration of pedal arterial arcades in occlusive arteriosclerotic disease. *Acta Radiol* 1989; (fasc 1)30:61–63.
7. Kahn PC, Boyer DN, Moran JM, et al: Reactive hyperemia in lower extremity arteriography: An evaluation. *Radiology* 1968; 90:975–980.
8. Zagoria RJ, D'Souza VJ, Sharling ES: Prosthetic arterial graft occlusion: A complication of tourniquet use during arteriography. *Radiology* 1988; 167:121–122.
9. Kozak BE, Bedell JE, Rösch J: Small vessel leg angiography for distal vessel bypass grafts. *J Vasc Surg* 1988; 8:711–715.
10. Cohen MI, Vogelzang RL: A comparison of techniques for improved visualization of the arteries of the distal lower extremity. *AJR* 1986; 147:1021–1024.

11. Sethi GK, Scott SM, Takaro T: Multiple plane angiography for more precise evaluation of aortoiliac disease. *Surgery* 1975; 78:154–159.

12. Mandell VS, Jaques PF, Delany DJ, et al: Persistent sciatic artery: Clinical, embryologic, and angiographic features. *AJR* 1985; 144:245–249.

13. Rosenbloom MS, Flanigan DP, Schuler JJ, et al: Risk factors affecting the natural history of intermittent claudication. *Arch Surg* 1988; 123:867–870.

14. Holmes DR Jr, Burbank MK, Fulton RE, et al: Arteriosclerosis obliterans in young women. *Am J Med* 1979; 66:997–1000.

15. Trede M, Storz LW, Petermann C, et al: Pitfalls and progress in the management of abdominal aortic aneurysms. *World J Surg* 1988; 12:810–817.

16. Cho SI, Johnson WC, Bush HL Jr, et al: Lethal complications associated with nonrestrictive treatment of abdominal aortic aneurysms. *Arch Surg* 1982; 117:1214–1217.

17. Rösch J, Keller FS, Porter JM, et al: Value of angiography in the management of abdominal aortic aneurysm. *Cardiovasc Radiol* 1978; 1:83–94.

18. Gmelin E, Burmester E, Valesky A, et al: Das sogenannte "inflammatorische Aneurysma" der Bauchaorta. *ROFO* 1984; 141:56–60.

19. Richardson JW, Greenfield LJ: Natural history and management of iliac aneurysms. *J Vasc Surg* 1988; 8:165–171.

20. Grigg MJ, Nicolaides AN, Wolfe JHN: Detection and grading of femorodistal vein graft stenoses: Duplex velocity measurements compared with angiography. *J Vasc Surg* 1988; 8:661–666.

21. Teeuwen C, Eikelboom BC, Ludwig JW: Clinically unsuspected complications of arterial surgery shown by post-operative digital subtraction angiography. *Br J Radiol* 1989; 62:13–19.

22. Carney WI Jr, Balko A, Barrett MS: In situ femoropopliteal and infrapopliteal bypass: Two-year experience. *Arch Surg* 1985; 120:812–816.

23. Sprayregen S, Veith FJ, Bakal CW: Catheterization and angioplasty of the nonopacified peripheral autogenous vein bypass graft. *Arch Surg* 1988; 123:1009–1012.

24. Waneck R, Jantsch H, Lechner G, et al: Dynamic angiography in below-knee bypass grafts. *Vasc Surg* 1988; 23:95–101.

25. Hagen B, Lohse S: Clinical and radiologic aspects of Buerger's disease. *Cardiovasc Intervent Radiol* 1984; 7:283–293.

26. Bagby RJ, Cooper RD: Angiography in ergotism: Report of two cases and a review of the literature. *AJR* 1972; 116:179–186.

27. Sano K, Aiba T, Saito I: Angiography in pulseless disease. *Radiology* 1970; 94:69–74.

28. Lüscher TF, Lie JT, Stanson AW, et al: Arterial fibromuscular dysplasia. *Mayo Clin Proc* 1987; 62:931–952.

29. Gaines VD, Ramchandani P, Soulen RL: Popliteal entrapment syndrome. *Cardiovasc Intervent Radiol* 1985; 8:156–159.

30. Greenwood LH, Yrizarry JM, Hallet JW Jr: Popliteal artery entrapment: Importance of the stress runoff for diagnosis. *Cardiovasc Intervent Radiol* 1986; 9:93–99.

31. Jasinski RW, Masselink BA, Partridge RW, et al: Adventitial cystic disease of the popliteal artery. *Radiology* 1987; 163:153–155.

32. Jantsch J, Lechner G, Kretschmer G: Sonographische Diagnose einer zystischen Adventitia-degeneration (ZAD). *ROFO* 1985; 143:600–601.

33. Wilbur AC, Woelfel GF, Meyer JP, et al: Adventitial cystic disease of the popliteal artery. *Radiology* 1985; 155:63–64.

34. Lawdahl RB, Routh WD, Vitek JJ, et al: Chronic arteriovenous fistulas masquerading as arteriovenous malformations: Diagnostic considerations and therapeutic implications. *Radiology* 1989; 170:1011–1015.

35. Sibbitt RR, Palmaz JC, Garcia F, et al: Trauma of the extremities: Prospective comparison of digital and conventional angiography. *Radiology* 1986; 160:179–182.

36. Rose SC, Moore EE: Emergency trauma angiography: Accuracy, safety, and pitfalls. *AJR* 1987; 148:1243–1246.

37. Snyder WH, Thal ER, Bridges RA, et al: The validity of normal arteriography in penetrating trauma. *Arch Surg* 1978; 113:424–428.

38. Lipchik EO, Kaebnick HW, Beres JJ, et al: The role of arteriography in acute penetrating trauma to the extremities. *Cardiovasc Intervent Radiol* 1987; 10:202–204.

39. McDonald EJ Jr, Goodman PC, Winestock DP: The clinical indications for arteriography in trauma to the extremity. *Radiology* 1975; 116:45–47.

40. Hartling RP, McGahan JP, Blaisdell EW, et al: Stab wounds to the extremities: Indications for angiography. *Radiology* 1987; 162:465–467.

41. Frykberg ER, Vines FS, Alexander RH: The natural history of clinically occult arterial injuries: A prospective evaluation. *J Trauma* 1989; 29:577–583.

42. Rose SC, Moore EE: Angiography in patients with arterial trauma: Correlation between angiographic abnormalities, operative findings, and clinical outcome. *AJR* 1987; 149:613–619.

43. Wedeen VJ, Meuli RA, Edelman RR et al: Projective imaging of pulsatile flow with magnetic resonance. *Science* 1985; 230:946–948.

44. Shimizu K, Matsuda T, Sakurai T, et al: Visualization of moving fluid: Quantitative analysis of blood flow velocity using MR imaging. *Radiology* 1986; 159:195–199.

45. AMA Council on Scientific Affairs, Report of the Magnetic Resonance Imaging Panel. Magnetic resonance imaging of the cardiovascular system: Present state of the art and future potential. *JAMA* 1988; 259:253–259.

3 | Renal Angiography

KEY CONCEPTS

1. Arteriography is performed in prospective renal donors to define arterial supply, which is subject to variation, and to detect unsuspected disease.
2. Angiography in renal tumor is used primarily to demonstrate tumor vascularity, arterial supply, and the extent of any venous invasion.
3. Renal adenoma or oncocytoma can be suggested by the absence of malignant features on the angiogram, but a positive diagnosis requires surgical resection.
4. Post-stenotic dilation, collateral vessels, and a pressure gradient of >20 mm Hg imply a significant renal artery stenosis.
5. Lateralized selective renal vein renin elevation helps predict response of hypertension to revascularization, but lack of lateralization does not mean a patient will not benefit from surgery or percutaneous angioplasty.

INDICATIONS

Angiography was a valuable diagnostic tool in the study of renal abnormalities in the 1960s, but even at that time its limitations were being recognized.[1] Today, arteriography remains an essential part of the evaluation of those with suspected renovascular hypertension, renal artery stenosis, and acute arterial occlusion. This holds true for renal transplants as well. Angiography also still plays a useful role for many patients with renal neoplasms, not so much to make a diagnosis as to define arterial anatomy and to assess the vascularity of the tumor prior to surgery or palliative embolization. Tumor extension into renal vein or inferior vena cava (IVC) usually requires invasive studies for confirmation and precise delineation of involvement.

Depiction of vascular anatomy is also the major reason for study of prospective renal donors. Other indications for angiography include hematuria of unresolved cause, trauma, renal arteriovenous malformation, arteriovenous fistula, or aneurysm. Rarely, a request may be made for examination of a patient with suspected polyarteritis nodosa or other systemic vasculitis. In general, arteriography has no place in the evaluation of diffuse renal parenchymal disease.

ANATOMIC CONSIDERATIONS AND RENAL DONOR ANGIOGRAPHY

Each kidney is usually supplied by a single vessel arising from the abdominal aorta caudal to the superior mesenteric artery origin, near the level of the L1–2 interspace. However, in a large number of people multiple renal arteries supply one or both kidneys, sometimes arising as caudally as the iliac arteries. The incidence of multiple vessels varies between 33% and 44%.[2, 3] The renal artery normally provides branches to the adrenal gland and upper ureter. There is often a major anteroposterior bifurcation before further ramifications form the segmental, interlobar, arcuate, and interlobular arteries.

Although the renal arterial circulation is often considered in isolation from surrounding organs, communications with the aorta and with intercostal, lumbar, hepatic, and inferior mesenteric arteries do occur normally.[4] They are important for providing collateral flow in cases of renal artery obstruction. These communications should also be kept in mind in the arteriographic staging of renal cell carcinoma and when therapeutic embolization is pursued.

Because of the proximity of the right kidney to the IVC, the right renal vein is short. Two to four right renal veins may be present in up to 15% of the population.[5] A single preaortic vein usually drains the left kidney, but a circumaortic venous ring is seen in 7%, and an isolated retroaortic vein, characteristically lower in position than L-1 or L-2 (the normal level of renal veins), can be found in 2% of patients.[5] Rich retroperitoneal communications exist on the venous side of the renal circulation, explaining why some patients with chronic renal vein thrombosis are asymptomatic.[6]

In renal transplantation the left kidney is usually selected from the donor because of the greater length of its vein. Any arterial ureteral branches or bifurcations within 1 cm of the aorta pose technical problems, and such kidneys, as well as those with multiple arteries, are less desirable for transplantation. After intravenous urography and ultrasound are used to screen out those donors with an obvious renal abnormality, arteriography is performed to define the blood supply to each kidney, as well as to uncover the presence of atherosclerosis, fibromuscular disease, infarct, or unsuspected tumor prior to surgery. In the series studied by Walker et al., arteriography led to the choice of the right kidney in 24% of prospective donors.[2]

Renal donors are best examined by aortography in the anteroposterior projection. Intravenous digital subtraction angiography (DSA) has little advantage over outpatient arteriography with small-bore catheters, and is subject to greater diagnostic error. Intra-arterial DSA can allow one to accurately identify multiple renal vessels[7]; nevertheless, the superior spatial resolution provided by conventional film studies is preferred in order to exclude lesions of fibromuscular dysplasia and other subtle abnormalities. However, DSA can be used as a supplement to film arteriography by allowing oblique views to be obtained and assessed rapidly, if there are any questionable abnormalities.

When assessing the renal arteries by aortic injection, the sideholes of the pigtail catheter must be placed below the origin of the superior mesenteric artery, and the injection rate should not exceed 20 mL/ sec. Otherwise, opacification of celiac and mesenteric branches may confuse interpretation. When doubts about renal vascular anatomy cannot be resolved by aortography, selective injections are indicated.

TUMORS

Renal Cell Carcinoma

Any solid renal mass in an adult without another known primary tumor must be considered renal cell carcinoma until proved otherwise. Percutaneous needle biopsy is not particularly useful in this situation, and it is not obtained in potentially resectable tumors.[8] Computed tomography (CT) is helpful in demonstrating fat in benign angiomyolipomas, but other neoplasms have nonspecific findings on cross-sectional imaging. Cortical nodules (enlarged columns of Bertin) will show normal excretory function on radionuclide scans. As noted earlier, angiography is no longer used primarily for diagnosis, but for operative planning.

Vascularity

Most renal cell carcinomas are hypervascular, with bizarre enlarged tumor vessels and arteriovenous shunting typically present. However, about 10% of these tumors are hypovascular or avascular, and the diagnosis of malignancy is not made by angiographic criteria alone. The common dilemma of the past, renal carcinoma vs. cyst, has been essentially resolved by sonography. Years ago Emmett et al. found only one case of malignancy intimately associated with a simple renal cyst among hundreds that had been subjected to surgical exploration, and in that case cyst aspiration would have been diagnostic[9] (see Chapter 21, Genitourinary Interventions). Tumors with markedly increased vascularity may be treated by preoperative embolization to decrease blood loss, but routine embolization of carcinomas undergoing resection has generally fallen from favor (see Chapter 12, Embolotherapy).

Venous Invasion

Most preoperative angiograms include study of the IVC, because renal cell carcinoma has a definite tendency to invade the renal veins, and caval extension may complicate up to 9% of those submitting to surgery.[10] Because the right renal vein is shorter, right-sided tumors are more likely to involve the IVC. Tumor may grow up the vena cava into the right side of the heart, and angiography performed through a catheter placed from above is needed to define intrathoracic extent in exceptional cases. At times, arterial injection shows neovascularity within the renal vein or IVC. Caval tumor thrombus does not adversely influence prognosis if nodal metastases are absent, but it does change the surgical approach.[11, 12] It is noteworthy that most patients with tumor invasion of the cava do not have symptoms of IVC obstruction.

Tumor Staging

Clinical staging of renal cell carcinoma as described by Robson is as follows: stage 1, neoplasm confined within renal capsule; stage II, extracapsular extension without penetration of the perinephric fascia; stage IIIA, venous invasion; stage IIIB, involvement of regional lymph nodes; stage IIIC, venous and lymph node involvement; stage IVA, invasion of an adjacent organ; and stage IVB, distant metastases.[8] Other than for confirming the presence and defining the extent of venous invasion, angiography is not as good as CT scanning in staging. Parasitic blood supply from retroperitoneal and other arteries does not imply extracapsular extension, for it is seen in many tumors confirmed histologically to be entirely intracapsular.[4]

Angiographic Approach

Intravenous DSA has been proposed for
Such a study can be carried out by placem
IVC from the femoral approach, with early
and later imaging during the same injection
parenchymal, and renal venous phases. If t
adequate, an arterial catheter can then be
travenous injection is unsatisfactory for DSA i
and even IVC injection does not allow tumor
reliably.[14] Even so, IVC-DSA can provide
needed for surgery, but direct arteriography
tial nephrectomy is under consideration.

If an arterial approach is chosen, aortograp
renal artery catheterization. A relatively lar
dium can be injected into the kidney to be
30 mL often allows the renal vein to be vis
DSA is employed. If there is any hint of abnor
kidney, a selective arteriogram should be obt
cell carcinomas are not rare, particularly in p
Lindau syndrome.

Injection of epinephrine (10 to 25 µg) dire
has been used to accentuate neovascularity wi
Such enhancement is felt to represent a dim
response of abnormal arteries in comparison t
sels. Even at the time when arteriography ha
portance, the value of this maneuver was que
epinephrine has also been used to temporaril
flow for selective renal venography. Venous en
are commonly present in tumors with no dem
ity.[17] Even so, renal venography is now rarely

Other Malignancies

Urothelial tumors and most neoplasms metast
hypovascular. Although vascular encasement fro
cinoma is occasionally seen, angiography does
role in the evaluation of these tumors. Lymphor
kidney, but usually when it is widespread elsewh
likely to be multinodular, and can diffusely infilt
a single renal mass is the exception, and angiograp
features, only stretching of vessels and little, if ar.y, neovascularity.[18]

Benign Masses

Renal Adenomas/Oncocytomas

Adenomas, which are usually small and asymptomatic, are fairly common benign tumors of the kidney. Renal oncocytomas are adenomas arising from the proximal tubular epithelium.[19] They can grow quite large, but necrosis and hemorrhage are unusual. Flank pain may be the presenting symptom, but most lesions are found incidentally. On arteriography there may be peripheral hypervascularity, but the majority of adenomas are hypovascular.[19–21] Parenchymal enhancement tends to be homogeneous and approximates that of adjacent normal kidney. A wheel-spoke arterial pattern has been described as characteristic of oncocytoma, but it is seen in fewer than one third of lesions and does not exclude the diagnosis of renal cell carcinoma when it is present.[19] Features that favor oncocytoma include a sharply circumscribed homogeneous lesion and the absence of findings typical for renal cell carcinoma: bizarre tumor vessels, arteriovenous shunting, venous invasion, and vascular encasement. The diagnosis of benign lesion can be suggested, but malignancy must be assumed until tissue is examined histologically. The use of needle biopsy is controversial, and complete resection is standard therapy.[21] When adenoma/oncocytoma is suspected, surgery can be limited to partial nephrectomy.

Other Nonmalignant Masses

Angiomyolipomas can grow large and bleed. Nowadays, smaller asymptomatic lesions are being detected by ultrasound and CT. They occur much more frequently in women, and although small berry-like aneurysms are sometimes present, most lesions have nonspecific arteriographic findings.[22] Fortunately, the characteristic fat content in almost all angiomyolipomas can be demonstrated by CT.

By their clinical presentation, abscesses and inflammatory masses are not a differential diagnostic problem. Needle aspiration and catheter drainage can confirm the presence of abscess. In the past, the peripheral neovascularity common to renal carbuncles could be confused with that of carcinoma.

Renal cortical nodules represent ectopic cortical hyperplasia or persistent renal lobulation. Cortical nodules tend to occur at the junction of the upper and middle thirds of the kidney, and may be associated with partial or complete duplication of the renal collecting system.[23] If adjacent scarring is present from a previous infarct, a cortical nodule can appear quite exophytic.[24] Angiography shows a normal or dense homogeneous blush with no abnormal feeding vessels. Almost all cortical nodules can now be diagnosed by noninvasive means.

ANGIOGRAPHY IN HYPERTENSION

Nearly one quarter of U.S. population is estimated to be hypertensive, and perhaps 250,000 have renovascular problems causing hypertension.[25] Finding this latter set of patients is important, for renovascular hypertension is potentially curable by percutaneous angioplasty or surgery. The major problem is to determine which patients should be subjected to arteriography. Physical examination, history, and screening tests such as excretory urography and radionuclide scanning lack sensitivity and accuracy.[26] Renovascular hypertension must be suspected in hypertensive patients under 30 years of age, those with onset of hypertension after 50 years of age, or in anyone with rapid deterioration and poor control of blood pressure.

Atherosclerosis vs. Fibromuscular Disease

Most renovascular hypertension results from atherosclerosis, but fibromuscular dysplasia (FMD) is a common cause among younger patients. Atherosclerotic stenoses are likely to become worse and carry the risk of producing arterial occlusion and renal insufficiency.[25] Risk of renal functional impairment is low in FMD. Other conditions that can produce hypertension by renal artery narrowing include Takayasu arteritis and neurofibromatosis.

Atherosclerotic lesions are more likely to affect patients over 60 years of age. They may be concentric or eccentric, smooth or irregular, and they are often bilateral. Stenosis is most common in the proximal 2 cm of a renal artery, and a lesion within 1 cm of the aorta (particularly if aortic plaques are present) can be considered an ostial lesion for purposes of balloon angioplasty. Percutaneous transluminal angioplasty is less effective in treating ostial lesions. In performing arteriography in patients with atherosclerosis, great care should be taken to be sure that the origin of each renal artery is profiled. Oblique views must be routinely obtained to avoid missing an ostial stenosis.

Fibromuscular dysplasia is responsible for 20% to 50% of renovascular hypertension, and it is mainly a disease of young white women.[27] In comparison with atherosclerotic stenoses, the lesions of FMD tend to be more peripheral, and they may affect branch arteries. Balloon angioplasty is particularly well suited for the treatment of these patients.

There are a number of histologic varieties of FMD, and some have characteristic angiographic findings. The majority of patients with FMD have medial fibroplasia with mural aneurysms, a lesion producing the classic "string of beads" appearance. It is to be noted that aneurysms of FMD rarely rupture. Perimedial fibroplasia (15% to 25% of all FMD) has a similar appearance, but it usually lacks true aneurysms.[28] The rarer varieties of FMD, intimal fibroplasia and periarterial fibroplasia can produce smooth stenoses or webs. Medial dissection, a complication of FMD, affects 5% to 10% of patients.[28]

Prediction of Response to Correction of Stenosis

When a stenotic lesion ($\geq$70% luminal narrowing) is discovered, its relationship to hypertension is not always clear, particularly in those with atherosclerosis. Many stenoses are caused by exacerbation of atherosclerosis in patients with essential hypertension. If renal function is marginal, treatment of significant occlusive disease can be argued, even when improvement in blood pressure is unlikely. In other patients the risks of nonmedical management must be justified by expected correction of hypertension. It is for this reason that selective renal vein renin sampling has been developed.

High blood pressure from ischemia results from excessive renin secretion by the kidney affected. Renin stimulates the production of angiotensin II and aldosterone, promoting sodium and fluid retention. Normally, each kidney adds about 24% to the renal artery renin, but excessive production by one kidney induces a feedback suppression of renin production by the opposite kidney.[5] A ratio of $\geq$1.5:1.0 of renin levels obtained from selective renal venous blood might be expected in this situation. Problems arise if both kidneys are affected by vascular disease. Lateralization of renins is predictive of response of hypertension to revascularization in 70% to 95% of patients, but 20% to 50% of patients in whom lateralization does *not* occur will also improve.[5, 25]

The presence of enlarged retroperitoneal, ureteral, and capsular arteries correlates well with ischemia.[29, 30] Such collateral vessels are a good sign that intervention will be beneficial. Unfortunately, collateral arteries are demonstrated in only a portion of patients with renovascular hypertension. Post-stenotic dilatation or measurement of a gradient of at least 20 mm Hg across the lesion can also be taken as indicative of significant narrowing.

Selective Venous Sampling for Renin

Selective samples are obtained from a catheter introduced from the femoral vein. It is best to have one or two sideholes placed near the tip of the catheter, otherwise problems are encountered with "sidewalling" during aspiration. Cobra or Simmons catheters can be used, as can a multipurpose angled catheter. A deflecting guidewire is necessary to reform the curve of the Simmons catheter at the junction of the iliac veins. A deflecting wire is also usually needed for selecting the renal veins with a multipurpose catheter.

Contrast material is injected before each set of samples is obtained in order to check the position of the catheter tip. After 1 to 2 mL is withdrawn to clear the catheter of flush solution or contrast medium, two samples of 7 mL each are drawn at each sampling position. Appropriate collection tubes are filled and kept on ice until the samples reach the laboratory. Each renal vein is sampled, in addition to mixed venous blood from the IVC and aortic blood (if an arterial catheter is in place for renal arteriography). If multiple veins drain one kidney, each should be sampled separately. If a branch renal artery stenosis is found, superselective sampling of the renal vein is in order.

Various maneuvers have been recommended to accentuate differences in renin secretion; among these are premedication with captopril, salt depletion, controlled hypotension, or upright posture (maintained at least 20 minutes before sampling).[5] Selective renal arteriography has not been shown to affect renin levels substantially, so samples can be taken immediately after arterial study. Beckman and Abrams recommend stimulatory maneuvers only if baseline supine sampling gives normal or borderline values.[5]

Angiography and Pheochromocytoma

Another potentially correctable cause of hypertension is pheochromocytoma, found in fewer than 1% of hypertensives.[31] About 10% are part of the constellation of tumors called multiple endocrine neoplasia syndrome-type II, a condition inherited as a dominant trait. The majority of tumors are benign, but the incidence of malignancy varies from 5% to 33%.[31] All but 10% arise in the adrenal medulla. Most extra-adrenal pheochromocytomas (paragangliomas) are found in the abdomen.

Angiography is not normally indicated when CT scanning shows an adrenal mass and when urinary catecholamine values are consistent with the diagnosis. A new radionuclide study utilizing metaiodobenzylguanidine may even be better than CT scanning for tumor localization.[31] Angiography has an accuracy of 80% to 90%, usually showing a hypervascular mass, but it should be reserved for cases in which noninvasive examinations are equivocal. Patients must be premedicated with alpha- and beta-adrenergic blockers (phenoxybenzamine, 10 to 40 mg twice daily, and propranolol, 10 mg three times daily) to prevent a possible hypertensive crisis. Blood pressure must be carefully monitored during the procedure, and an acute hypertensive episode can be treated with intravenous phentolamine.

TRANSPLANTS

Problems with renal transplantation persist despite improvements in patient care. Suspected rejection can accurately be assessed by radionuclide perfusion and excretion studies, but arteriography is recommended if a primary vascular problem is under consideration.[32] Anastomotic or nonanastomotic stenosis, occlusion, infarct, and arteriovenous fistula are best demonstrated by catheter studies, and intraarterial DSA can minimize the amount of contrast medium needed. Biplane film arteriography is an alternative, with the lateral view often providing critical information. One should be aware of the type of anastomosis present (end-to-end to internal iliac artery or end-to-side to external iliac) in order to tailor the study and provide for any necessary catheter intervention (see Chapter 10, Percutaneous Angioplasty, Recanalization, and Vascular Stents).

Angiographic results can be normal in cases of acute tubular necrosis, but study using contrast material is unnecessary and detrimental for such patients. Hyperacute rejection may be difficult to distinguish from acute arterial thrombosis by radionuclide studies, and transplant arteriography is diagnostic.[33] Arterial transit time of less than 2 seconds is normal, but delayed contrast clearance, lack of small artery opacification ("pruned-tree" appearance of the intrarenal vessels), and absence of nephrogram are diagnostic of rejection. Findings of chronic rejection include small vessel irregularity, cortical infarcts, and small artery aneurysms.

RENAL ARTERY ANEURYSM

The incidence of renal artery aneurysm found at autopsy is 0.3% to 0.7%.[34] Despite this, renal artery aneurysms rarely cause clinical problems, and many are found incidentally because of calcification seen on plain abdominal films or on angiograms performed for other reasons. Aneurysms typically involve the main renal artery or a first-order branch. They occur with equal frequency in men and women, and mean age of presentation is 60 years. Tham et al. described 69 patients with aneurysms not treated surgically, and found no case of rupture in a mean 4-year follow-up.[34] They believe the risk of rupture has been exaggerated, and asymptomatic lesions can be treated conservatively, even if they are larger than 2 cm in diameter. Aneurysms should be resected if they are associated with hematuria (all other potential causes being excluded), flank pain, or hypertension.

MISCELLANEOUS CONDITIONS

Vasculitis

Polyarteritis nodosa is a systemic vasculitis that has been associated with small artery aneurysms in multiple viscera, but particularly in the kidney. The disease is characterized by fibrinoid medial necrosis of medium and small arteries. Although aneurysms are virtually pathognomonic, only a minority of patients with suspected polyarteritis nodosa will have aneurysms, and angiography is not justified as a screening measure.[35] It should be noted that aneurysms may regress with clinical improvement of the patient.

When present, microaneurysms must be distinguished from mycotic lesions. Identical aneurysms can be found in hypersensitivity or drug-induced angiitis.[36] Other findings of vasculitis include irregularity, stenosis, and occlusion of intrarenal branches, and nephrograms may have a spotted or striated pattern.[37, 38]

Renal Trauma

The effects of renal trauma can usually be determined by excretory urography, radionuclide scanning, or CT scanning. Renal infarcts, tears, and contusions can all be accurately detected with CT, but arterial injury is easily missed.[39] If a patient deteriorates clinically and arterial injury is suspected, renal arteriography should be performed to detect arterial laceration, intimal flap, arteriovenous fistula, or occlusion.

Hematuria of Unresolved Origin

Angiography is not usually helpful in cases of unilateral hematuria that cannot be diagnosed by other means. Mitty and Goldman reviewed 48 patients with documented unilateral gross hematuria and found only six angiographic abnormalities, including emboli and vascular malformations.[40] Forniceal and suburethelial veins are thin-walled and may be a source of bleeding, but they cannot be demonstrated reliably by angiography. Compression of the left renal vein between aorta and superior mesenteric artery (the "nutcracker" phenomenon) has been implicated in some cases of hematuria. Nishimura et al. found a significant elevation in left renal vein pressure in patients with left renal bleeding vs. controls, and they define a left renal vein-IVC pressure gradient of 3.0 mm Hg or more as abnormal.[41] The filling of ureteral or capsular veins on arterial or venous injection may alert one to the diagnosis.

Renal Venography and Renal Vein Thrombosis

Renal venography has lost much of its past importance, especially in the diagnosis of malignant neoplasms. Today it is most often performed prior to placement of splenorenal shunts or to make the diagnosis of renal vein thrombosis. The main renal vein is usually well demonstrated by selective catheterization with a multisidehole catheter and injection of contrast medium at 15 mL/sec for 2 seconds. Rapid filming or high DSA frame rate (2 to 3 frames per second) gives best results. Demonstration of intrarenal branches requires a similar injection of contrast material immediately after the administration of epinephrine (10 µg) directly into the renal artery.[5]

Acute renal vein thrombosis results in a tense, swollen kidney, but occlusion of more gradual onset, or chronic occlusion, may produce few symptoms. Most cases of renal vein thrombosis in adults are the result of trauma or an underlying disease such as pancreatitis, retroperitoneal fibrosis, tumor, or diffuse parenchymal disease (e.g., glomerulonephritis).[6] It appears that nephrotic syndrome predisposes one to renal vein thrombosis, not vice versa. Renal vein thrombosis is often overlooked in the absence of nephrotic syndrome or anuria, but it should be included in the differential diagnosis of sudden deterioration in renal function in renal transplant patients or those with known renal disease.[42] Standard treatment is systemic anticoagulation, but local thrombolytic infusion has been reported.[43]

REFERENCES

1. Chait A: Current status of renal angiography. *Urol Clin North Am* 1985; 12:687–698.
2. Walker TG, Geller SC, Delmonico FL, et al: Donor renal angiography: Its influence on the decision to use the right or left kidney. *AJR* 1988; 151:1149–1151.
3. Spring DB, Salvatierra O Jr, Palubinskas AJ, et al: Results and significance of angiography in potential kidney donors. *Radiology* 1979; 133:45–47.
4. Wilkins RA, Sandin B, Price A, et al: Extrarenal arterial connections of the normal renal artery. *Cardiovasc Intervent Radiol* 1986; 9:119–122.
5. Beckman CF, Abrams HL: Renal venography: Anatomy, technique, applications, analysis of 132 venograms, and a review of the literature. *Cardiovasc Intervent Radiol* 1980; 3:45–70.
6. Keating MA, Althausen AF: The clinical spectrum of renal vein thrombosis. *J Urol* 1985; 133:938–945.

7. Petty W, Spigos DG, Abejo R, et al: Arterial digital angiography in the evaluation of potential renal donors. *Invest Radiol* 1986; 21:122–124.

8. Levine E: Renal cell carcinoma: Radiological diagnosis and staging. *Semin Roentgenol* 1987; 22:248–259.

9. Emmett JL, Levine SR, Woolner LB: Coexistence of renal cyst and tumour: Incidence in 1007 cases. *Br J Urol* 1963; 35:403–410.

10. Kadir S, Coulam CM: Intracaval extension of renal cell carcinoma. *Cardiovasc Intervent Radiol* 1980; 3:180–183.

11. Selli C, Barbanti G, Barbagli G, et al: Caval extension of renal cell carcinoma: Results of surgical treatment. *Urology* 1987; 30:448–452.

12. Henriksson C, Aldenborg F, Haljamäe H, et al: Renal cell carcinoma with vena cava extension: Diagnosis and surgical features of 41 cases. *Scand J Urol Nephrol* 1987; 21:291–296.

13. Ford KK, Braun SD, Miller GA Jr, et al: Intravenous digital subtraction angiography in the preoperative evaluation of renal masses. *AJR* 1985; 145:323–326.

14. Zabbo A, Novick AC, Risius B, et al: Digital subtraction angiography for evaluating patients with renal carcinoma. *J Urol* 1985; 134:252–255.

15. Abrams HL: The response of neoplastic renal vessels to epinephrine in man. *Radiology* 1964; 82:217–223.

16. Meaney TF: Errors in angiographic diagnosis of renal masses. *Radiology* 1969; 93:361–366.

17. Smith JC Jr, Rösch J, Athanasoulis CA, et al: Renal venography in the evaluation of poorly vascularized neoplasms of the kidney. *AJR* 1975; 123:552–556.

18. Seltzer RA, Wendlund DE: Renal lymphoma: Arteriographic studies. *AJR* 1967; 101:692–695.

19. Ambos MA, Bosniak MA, Valensi QJ, et al: Angiographic patterns in renal oncocytomas. *Radiology* 1978; 129:615–622.

20. Holt RG, Neiman HL, Korsower JM, et al: Angiographic features of benign renal adenoma. *Urology* 1975; 6:764–767.

21. Drüber C, Schweden F, Klose K-J, et al: Das Onkozytom der Niere. *ROFO* 1988; 148:227–233.

22. Bret PM, Bretagnolle M, Gaillard D, et al: Small, asymptomatic angiomyolipomas of the kidneys. *Radiology* 1985; 154:7–10.

23. Popky GL, Bogash M, Pollack H, et al: Focal cortical hyperplasia. *J Urol* 1969; 102:657–660.

24. King MC, Friedenberg RM, Tena LB: Normal renal parenchyma simulating tumor. *Radiology* 1968; 91:217–222.
25. Working Group on Renovascular Hypertension: Detection, evaluation, and treatment of renovascular hypertension. *Arch Intern Med* 1987; 147:820–829.
26. Greminger P, Schneider E, Siegenthaler W, et al: Renovaskuläre Hypertonie. *Internist* 1988; 29:246–251.
27. Lüscher TF, Lie JT, Stanson AW, et al: Arterial fibromuscular dysplasia. *Mayo Clin Proc* 1987; 62:931–952.
28. Harrison EG Jr, McCormack LJ: Pathologic classification of renal arterial disease in renovascular hypertension. *Mayo Clin Proc* 1971; 46:161–167.
29. Abrams HL, Cornell SH: Patterns of collateral flow in renal ischemia. *Radiology* 1965; 84:1001–1012.
30. Meyers MA, Friedenberg RM, King MC: The significance of the renal capsular arteries. *Br J Radiol* 1967; 40:949–956.
31. Samaan NA, Hickey RC, Shutts PE: Diagnosis, localization, and management of pheochromocytoma: Pitfalls and follow-up in 41 patients. *Cancer* 1988; 62:2451–2460.
32. Prager P, Clorius JH, Dreikorn K: Beitrag der digitalen Subtraktionsangiographie zur Diagnostik von Abstoßungsreaktionen nach Nierentransplantation. *ROFO* 1985; 143:426–431.
33. Hamway S, Novick A, Braun WE, et al: Impaired renal allograft function: A comparative study with angiography and histopathology. *J Urol* 1979; 122:292–297.
34. Tham G, Ekelund L, Herrlin K, et al: Renal artery aneurysms: Natural history and prognosis. *Ann Surg* 1983; 197:348–352.
35. Sellar RJ, Mackay IG, Buist TAS: The incidence of micro-aneurysms in polyarteritis nodosa. *Cardiovasc Intervent Radiol* 1986; 9:123–126.
36. Fisher RG: Renal artery aneurysms in polyarteritis nodosa: A multiepisodic phenomenon. *AJR* 1981; 136:983–985.
37. Vázquez JJ, San Martin P, Barbado FJ, et al: Angiographic findings in systemic necrotizing vasculitis. *Angiology* 1981; 32:773–779.
38. Warren BH, Rösch J: Angiography in the diagnosis of renal scleroderma. *Radiologia Clin* 1977; 46:194–202.
39. Lang EK, Sullivan J, Frentz G: Renal trauma: Radiological studies. *Radiology* 1985; 154:1–6.

40. Mitty HA, Goldman H: Angiography in unilateral renal bleeding with a negative urogram. *AJR* 1974; 121:508–517.

41. Nishimura Y, Fushiki M, Yoshida M, et al: Left renal vein hypertension in patients with left renal bleeding of unknown origin. *Radiology* 1986; 160:663–667.

42. Clark RA, Wyatt GM, Colley DP: Renal vein thrombosis: An underdiagnosed complication of multiple renal abnormalities. *Radiology* 1979; 132:43–50.

43. Robinson JM, Cockrell CH, Tisnado J, et al: Selective low-dose streptokinase infusion in the treatment of acute transplant renal vein thrombosis. *Cardiovasc Intervent Radiol* 1986; 9:86–89.

4

Hepatic Angiography

KEY CONCEPTS

1. Hepatic arterial anatomy is variable; partial supply from superior mesenteric and left gastric arteries is common.
2. Normal liver is perfused by the hepatic artery and portal vein; malignant tumors are supplied by the hepatic artery only.
3. Hepatocellular carcinoma is associated with hepatitis and cirrhosis. It has a propensity for venous invasion.
4. Infusion angiography, computed tomographic (CT)-arteriography, CT-portography, and arterial injection of lymphographic oily contrast agents can improve tumor detection.
5. Hepatic venous obstruction causes Budd-Chiari syndrome.

INDICATIONS

With the development of noninvasive diagnostic imaging techniques, the role of hepatic angiography has narrowed and become more strictly defined. One of the chief uses of arteriography remains in the evaluation of hepatic neoplasms when radionuclide scanning, computed tomography (CT), ultrasound (US), magnetic resonance imaging, or needle biopsy fail to produce a satisfactory diagnosis. Such may be the case if a large regenerating nodule needs to be differentiated from possible hepatocellular carcinoma. Similarly, a tumor not clearly malignant on histologic sampling may have unequivocally malignant features on angiography (vascular encasement, portal occlusion). Occasionally cavernous hemangioma may have an atypical appearance on CT scanning, but can be confidently diagnosed arteriographically.

Often the nature of a hepatic malignancy is known, but the precise location, number of satellite nodules or metastases, and presence of vascular invasion must be defined before surgical resection for cure can be pursued. Arteriography remains a prime diagnostic tool in this situation. If a tumor is unresectable, one may still choose to treat by transcatheter embolization or selective arterial infusion of chemotherapeutic agents.

Other current indications include preoperative evaluation for portal hypertension (see Chapter 5, Portal Venography in Portal Hypertension), liver transplantation, vascular malformations, trauma, or Budd-Chiari syndrome.

ANATOMY

The anatomy of the liver differs from that of the other viscera by its dual blood supply: arterial and portal venous. Within the liver the branches parallel one another and terminate in the acinus, the organ's basic functional unit. Blood flows through the hepatic sinusoids into the venules at the periphery of each acinus. These small vessels drain into the three major hepatic veins (right, middle, and left), which enter the inferior vena cava (IVC) and define surgical lines of cleavage.

The morphologic division between right and left lobes corresponds to a line drawn between the gallbladder and IVC, not to any surface features. The right lobe is comprised of anterior and posterior segments, each with superior and inferior subsegments. The left lobe contains medial and lateral segments (medial segment representing the quadrate lobe of older terminology). The caudate lobe receives arterial and portal branches from both the major lobar divisions, and drains by small veins directly into the IVC.

The arterial supply of the liver is quite variable (Fig 4–1), and ten anatomic categories have been described by Michels.[1] He found that only 55% of the cadavers studied had complete arterial supply through a common hepatic artery arising from the celiac axis. A branch of the superior mesenteric artery (SMA) provided either a complete blood supply (replaced right hepatic artery) or incomplete supply (accessory right hepatic artery) to the right lobe in 17%. Replaced or accessory left hepatic arteries arose from the left gastric artery in 25%. Rarely one may find the hepatic artery originating entirely from the SMA or left gastric artery. The middle hepatic (artery to the medial segment of the left lobe) is as likely to come from the right hepatic artery as

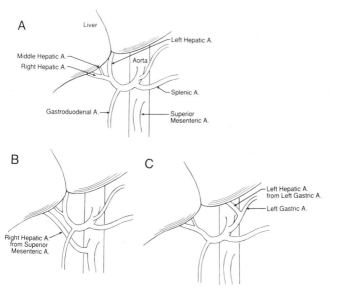

FIG 4–1.
Variations in hepatic arterial supply. **A**, most common celiac and hepatic arterial anatomy. **B**, right hepatic artery "replaced" to superior mesenteric artery. **C**, left hepatic artery arising from the left gastric artery.

from the left. These variations obviously have relevance for planning surgical resection or transplantation and for performing therapeutic embolization or chemoinfusion. One must be prepared to perform SMA and left gastric injections for complete arteriographic study of the liver.

TECHNICAL ASPECTS

Aortography is not ordinarily necessary prior to selective catheterization, but if celiac occlusion is suspected and the celiac artery cannot be entered, a lateral aortogram (by digital subtraction or conventional film studies) may be helpful.

Catheters most often used in hepatic arterial studies from the femoral approach are those with downward or reverse curves such as Cobra, sidewinder, or MK-2 shapes. If the celiac artery takes a right-angle or cephalad course, an H1H catheter can be employed successfully at times. Depending on the anatomy, preformed catheters may not enter the celiac or hepatic artery, and it may be necessary to custom form a tip configuration by steam. Rarely, a narrow or acutely angled celiac can be engaged only from an axillary or brachial artery approach. An upper extremity approach may also allow much more selective placement of the catheter if embolization is indicated.

The smaller, high-torque catheters now available tend to pass farther peripherally for superselective injections. If distal placement is desired, torque-control guidewires and coaxial catheter systems may be tried. One should be aware of the potential for arterial spasm during manipulations. A sidehole near the catheter tip prevents problems with "sidewalling" during aspiration, and decreases the jet effect from the tip during power injections. For embolization, however, only endhole catheters are used in order to ensure that no injected particles remain lodged within the catheter.

Conventional Hepatic Arteriography

In conventional hepatic arteriography, flow within the vessel selected should be approximated by the injection rate (generally between 6 and 10 mL/sec). A test injection with fluoroscopic observation can give an idea of the flow that is present, as well as stability of catheter position. Sensitivity of arteriography is much improved if the common or proper hepatic artery is selected, rather than just the celiac axis.[2] Between 30 and 40 mL of contrast medium is injected. Filming should be rapid (one or two films per second) during the first few seconds, and then may be slowed to cover a total of 25 to 30 seconds filming in order to include both parenchymal and venous phases.

Infusion Angiography

Infusion angiography of the liver is designed to provide better opacification of tumor nodules during the parenchymal phase and is accomplished by a slower but prolonged injection of the common hepatic or proper hepatic artery.[3] Rate of injection is 3 to 5 mL/sec of 76% contrast medium, for a total of 36 to 60 mL. Filming may be slow (one exposure every 2 seconds) and extended through at least 24 seconds. Infusion angiography is complementary to conventional arteriography and is usually performed only after the latter technique because it does not provide as much arterial detail.

Computed Tomography–Angiography

Combined CT-angiography shows increased sensitivity for detection of neoplasm. It immediately follows standard film angiography and may be performed in one of two ways. In CT-arteriography, the hepatic artery is injected with 8 to 12 mL of 30% contrast medium per CT slice.[4] Most lesions appear hypervascular by this technique, unlike CT-portography in which they appear hypovascular in comparison with surrounding normal liver. In CT-portography, scanning begins 30 seconds after contrast injection of the SMA (70 mL total) is initiated at a rate of 0.5 mL/sec.[5]

Oil-based iodinated contrast agents such as Lipiodol (iodized oil) or Ethiodol (ethiodized oil) have a propensity to accumulate in primary hepatic tumors. They can be injected into the hepatic artery (2 to 20 mL, with or without chemotherapeutic agents in emulsion) and followed by plain radiography or CT scanning.[6]

Portal Venography

The portal venous system may be opacified by various techniques (see Chapter 5), the most widely used being arterial portography. The splenic artery or SMA is injected with a large volume of contrast medium (50 to 60 mL) and filming carried out to 30 seconds. If opacification of the portal vein is inadequate, SMA injection may be repeated after tolazoline (40 mg) is administered into the artery.

Hepatic Venography

Hepatic venography is performed with an endhole or occlusion balloon catheter. If introduced from the femoral vein, the catheter is guided into a major hepatic vein by means of a tip-deflecting guidewire. Small volumes of contrast medium are injected by hand to gauge size of the vein as well as flow rate. A free hepatic venogram is obtained by injection with unobstructed flow around the catheter. Wedge hepatic venography requires either occlusion balloon inflation or such peripheral tip placement that the catheter itself effects occlusion. In a wedge injection, no more than several milliliters of contrast medium should be injected by hand. Digital imaging allows very dilute contrast material to be used, lessening the chance of hepatic infarction resulting in the area of parenchymal stain. Pressure measurements are routinely obtained during hepatic vein catheterization (see Chapter 5).

MASS LESIONS

The diagnostic accuracy of angiography in the liver has been variously reported as between 74% and 96%.[7] The presence of cirrhosis or obstructive jaundice can make diagnostic interpretation considerably more difficult, particularly for small focal lesions. Many tumors can have a nonspecific angiographic appearance, and needle or open biopsy may be necessary to establish a diagnosis. Nevertheless, certain lesions may have a characteristic enough appearance to make angiography worthwhile.

Benign Tumors

Cavernous Hemangiomas

Cavernous hemangiomas are encountered very frequently in CT, US, or radionuclide examinations. An incidence of up to 7% has been noted in autopsy studies; metastases are the only mass lesions seen more commonly in the liver.[8] Cavernous hemangiomas are almost always incidental findings and not the reason for the imaging procedure. Very rarely a large lesion may cause symptoms by mass, bleeding, or arteriovenous (AV) shunting. Cavernous hemangioma occurs more commonly in women, and multiple lesions may be detected in 10% of cases. Appearance on dynamic contrast material–enhanced CT is quite characteristic: initially hypodense, then showing peripheral areas of focal enhancement followed by diffuse hyperdensity at 2 minutes, and finally isodense on delayed scans. Red blood cell radionuclide scanning can also be used to make a reliable diagnosis. If a cavernous hemangioma does not exhibit conventional features, aspiration needle biopsy may be safely performed as long as thin needles (no larger than 20 gauge) are employed. One should alert the examining pathologist to the suspected diagnosis, so that endothelial elements can be specifically sought in the biopsy sample.[9]

If a cavernous hemangioma is studied by arteriography, the following are pathognomonic when seen as a constellation of findings:

1. Normal size feeding artery.
2. Absence of neovascularity or AV shunting.
3. Early peripheral ring or C-shaped accumulation of contrast material.
4. Persistent "cotton-wool" pooling of contrast material to at least 20 to 30 seconds.

Focal Nodular Hyperplasia

Focal nodular hyperplasia is not a neoplasm, but rather a hamartomatous lesion containing hepatocytes, Kupffer cells, and biliary elements. It is generally found in young women, is asymptomatic, and may be associated with the use of oral contraceptives, although the association is not very strong.[10] There have been no reliable reports of malignant degeneration, and severe complications arise only exceptionally (bleeding in less than 3%), so that routine surgical intervention is not indicated.[10] Multiple foci are found in 13% to 45%, and because of the presence of reticuloendothelial cells, there is significant uptake of radionuclide within the tumor in roughly half of cases examined.[10–12] If the lesion is "cold" by scintigraphy, angiography may be the next step in evaluation. Characteristic angiographic findings of focal nodular hyperplasia include:

1. Hypervascular lesion with very tortuous vessels, but without frankly malignant neovascularity.
2. Dense capillary stain of a sharply marginated mass.
3. Septations visible within the capillary stain (in about one third), radiating from a small fibrous core (wheel-spoke pattern).

Previous reports that a central feeding vessel is typical of focal nodular hyperplasia have not proved useful in the differentiation of this entity from hepatic adenoma in most cases.[10–12] Although the features listed are quite specific of the category as a whole, two thirds of "cold" hypervascular lesions will still require biopsy to differentiate focal nodular hyperplasia from hepatic adenoma.[11]

Hepatic Adenoma

Hepatic adenoma is a true neoplasm of clinical importance in that it is often complicated by necrosis and may present with catastrophic bleeding. In addition, there appears to be a small potential for malignant degeneration.[11] For these reasons, adenomas are not ordinarily treated conservatively. The great majority arise in younger women, and there is a strong association with the use of oral contraceptives or anabolic steroids. Because of hemorrhage and necrosis, most adenomas are symptomatic when discovered, and the CT and US appearances are quite heterogeneous and nonspecific. Hepatic adenomas do not contain Kupffer cells and do not accumulate tagged sulfur colloid. On angiography, they manifest the following:

1. Well-circumscribed mass somewhat larger (8 to 15 cm) than typical focal nodular hyperplasia.
2. Hypovascularity in about 50%.
3. Blush rather homogeneous (somewhat less intense than in focal nodular hyperplasia) except for hypovascular areas corresponding to hemorrhage or necrosis.

In many cases the diagnosis is positively made only with open biopsy and resection.

Malignant Tumors

An important point to keep in mind about both primary and secondary malignant tumors in the liver is that they almost uniformly obtain blood supply from the arterial circulation only, not from portal vessels. This has practical implications for both diagnostic studies and therapeutic catheter interventions. The reason that selective hepatic arterial injection of contrast medium is better than celiac injection is that in the former, contrast between a neoplastic focus and normal liver is enhanced because of unopacified portal blood "washing out" normal parenchyma in the face of persistent capillary stain in the tumor.[2] With infusion arteriography, only a very few malignant lesions will not show an appreciable stain. At times, a treated lesion will persist on US or CT despite the fact that no viable tissue is present. Lack of capillary stain on a post-treatment angiogram appears reliable in confirming a positive response.[2]

Hepatocellular Carcinoma

Hepatocellular carcinoma is the most frequent primary tumor of the liver. In the majority of patients it arises within a previously diseased and cirrhotic liver, and there is a strong propensity for those with a history of hepatitis B to develop this tumor. Hepatocellular carcinoma is one of the most common neoplasms in the Orient. It is a highly malignant lesion, and 5-year survival of those not undergoing tumor resection is 0%.[13] At present surgery provides the only hope of cure, with 21% to 30% 5-year survival after resection.[13, 14] At the time of tumor discovery, many patients will already have satellite nodules or metastases, and these findings (or vascular invasion) may preclude surgical therapy.

Hepatic angiography is quite sensitive in the detection of hepatocellular carcinomas larger than 2 cm diameter, with 93% of such tumors identified by conventional selective arteriography.[15] False negative studies occur in smaller lesions or in diffusely infiltrating tumors. The latter tend to be hypovascular, lack mass effect, and are very difficult to recognize angiographically, but they comprise only 5% to 7% of all hepatocellular carcinomas.[16]

Angiographic features of hepatocellular carcinoma include:

1. Mass effect and hypervascularity.
2. Irregular tumor vessels.
3. Nodular or irregular tumor stain.
4. Arteriovenous shunting.
5. "Threads and streaks" sign of tumor thrombus in portal or hepatic veins.
6. Portal vein occlusion.

Portal hypertension, new or abruptly worsening, may be the first clinical sign of malignancy. Massive AV shunting can produce portal hypertension, as can tumor obstructing portal flow. The portal vein does not normally opacify with selective hepatic artery injection. The "threads and streaks" sign described in portal vein invasion represents the vascularity of the tumor thrombus itself.[17] Segmental staining after an arterial injection is an indirect sign of tumor with focal portal obstruction.[18]

Hepatocellular carcinoma may be confused angiographically with hypervascular metastases from another primary tumor. False positive studies can also result in those with macronodular cirrhosis.[19] The lack of sensitivity of angiography in detection of lesions smaller than 2 cm, as well as the invasive nature of the study, does not permit it to be used as an aggressive screening measure for those at high risk for hepatocellular carcinoma. The best available screens at present appear to be serum alpha-fetoprotein determination and US scanning.

When angiography is performed, it can be enhanced by selective intra-arterial injection of an lymphographic oily contrast agent (Lipiodol or Ethiodol). Injection of 2 to 5 mL of the agent results in its accumulation and persistence in hepatic tumors, particularly in hepatocellular carcinoma, while normal liver parenchyma is clear of contrast material within 1 month.[20, 21] With performance of CT at intervals after oily contrast injection, more tumor nodules can be recognized than with angiography alone.[21, 22]

Fibrolamellar Hepatocellular Carcinoma

Fibrolamellar hepatocellular carcinoma is a distinct clinical variant of primary liver cancer that occurs in a much younger group of patients (mean age at presentation is 20 years), and has a distinctly better prognosis.[14] These tumors tend to be quite large, and they often contain calcification visible on plain radiographs. On angiograms they usually appear hypervascular, but the presence of fibrous septae or areas of necrosis can make their differentiation from focal nodular hyperplasia or hepatic adenoma difficult. Detection of satellite nodules may be helpful, but in most cases surgery is needed to establish the diagnosis.[14] Pleomorphism of tissue within the tumor makes confident diagnosis by needle biopsy unlikely.

Cholangiocarcinoma

Cholangiocarcinoma is the only liver tumor that commonly causes vascular encasement and arterial occlusion. The tumor tends to be hypovascular and infiltrating, and obstructive jaundice is the typical clinical presentation.[23] With availability of US, CT, percutaneous biliary drainage, and percutaneous biopsy, angiography is rarely needed for diagnosis or determining extent of disease.

Hepatic Angiosarcoma

This is an aggressive malignancy associated with occupational exposure to vinyl chloride. True tumor vessels may be difficult to identify.[24] Hepatic angiosarcomas tend to have hypervascular rims and hypovascular centers, features that are far from specific. These lesions are associated with cirrhosis and periportal fibrosis.

Metastases

Vascularity of hepatic metastases is related to that of the tumor of origin. Most are hypervascular and manifest a capillary stain. Metastases do not characteristically produce AV shunting or portal vein occlusion, but these features are sometimes encountered.[25] Angiography today is generally reserved for those in whom surgical resection is contemplated. More than half of cases in which the patient is thought to have a single metastasis will actually demonstrate multiple lesions.[4]

Infusion arteriography can improve visualization of metastases in the capillary phase, and Rösch et al. found it essential for making the diagnosis in 7% of cases.[3] Sensitivity of conventional or infusion angiography is further improved by CT-angiography and CT-portography, which may uncover up to 84% of additional metastases detectable at surgery.[4,5] Such CT examinations are very helpful in assessing the left lobe, an area of diagnostic difficulty in film angiography.[26] Arterial injection of oily contrast media does not seem as useful for metastases detection as it is in the evaluation of hepatocellular carcinoma.

One particular situation in which angiography remains useful is in the patient with focal fatty liver.[27] Fat can mask hypodense metastases on CT scanning, and even when tumor is recognized, its full extent may be hidden.

Embolotherapy of Hepatic Malignancies

Hepatic arterial embolization is possible for palliation because of the unique dual blood supply to the liver. As mentioned previously, primary and metastatic liver tumors obtain their blood supply almost exclusively from the hepatic arterial circulation. As long as the portal vein is shown to be patent, there is little mortality or morbidity from therapeutic occlusion of arterial branches supplying tumor. Embolization for hepatocellular carcinoma, with or without subsequent surgery, has been shown to improve survival.[28,29] By preceding gelatin sponge (Gelfoam) embolization with an injection of an oily contrast agent-chemotherapeutic agent mixture, Ohishi et al. claim further improvement in patient survival, with 69% alive at 1 year[6] compared with 7% 1-year survival cited for chemotherapy alone in another study.[29]

Therapeutic embolization has produced similar beneficial effects in patients with tumors metastatic to liver.[30] Patients with carcinoid tumor refractory to other therapy can experience considerable palliation by such treatment.[31] (See Chapter 12, Embolotherapy.)

Non-neoplastic Masses

A major problem encountered in patients with cirrhosis is the nature of a focal mass that might be found on physical examination, US, CT, or other imaging procedure. Alcoholism induces micronodular cirrhosis, characterized by small uniform nodules, 2 to 3 mm in diameter. Occasionally, regenerating nodules may reach a large size and be confused for hepatocellular carcinoma.[32] Although radionuclide uptake tends to be normal in a regenerating nodule, it may be diminished. Needle biopsy results are difficult to interpret because of the varying histologic features of hepatic primary tumors and the fear that a study that is negative for disease may represent sampling error. Therefore, in rare cases angiography may be called for prior to surgical exploration.

Regenerating hepatic nodules may be either hypervascular or hypovascular. A major artery often courses to the center of the nodule, and although its branches may be stretched, they are not increased in number.[32] No neovascularity is present, and the parenchymal phase may show a homogeneous opacification comparable to that in the rest of the liver. Demonstration of a patent portal branch within a large nodule helps distinguish a regenerating nodule from hepatocellular carcinoma.

Subacute inflammatory disease in the form of pyogenic hepatitis or abscess was more of a diagnostic problem in the past. Lesions investigated by angiography show hypervascular margins, hypovascular centers, and occasional AV shunting.[33, 34] Diagnosis today is simplified by cross-sectional imaging and early needle aspiration with drainage.

DIFFUSE DISEASE, TRAUMA, VASCULAR MALFORMATIONS, AND TRANSPLANTATION

Arteriography has limited value in cases of diffuse liver disease. Acute hepatitis may produce hepatomegaly, hypervascularity, and an inhomogeneous parenchymal phase.[35] Arterial flow is increased in cirrhosis, as antegrade portal flow is diminished. The liver may be small, and arterial branches may be crowded and have a "corkscrew" appearance.

Computed tomography has displaced angiography as a primary diagnostic study for those with trauma and suspected liver injury. Angiography may be used for those who continue bleeding postoperatively.[36] Embolization can be valuable in such a situation.

Arteriovenous malformations can affect liver, and discrete areas of hypervascularity with early venous filling have been described in hereditary hemorrhagic telangiectasia (Rendu-Osler-Weber syndrome).[37,38] However, most AV fistulas are iatrogenic or traumatic in origin. They tend to be asymptomatic and close spontaneously, but embolization can be employed to close symptomatic fistulas.

Peliosis hepatis describes a condition, seen in chronic wasting diseases (cancer, acquired immunodeficiency syndrome, tuberculosis) or in patients taking anabolic steroids, in which there are multiple blood-filled spaces in the liver. These spaces have diameters of 2 to 6 mm, can be distributed focally or diffusely, and lack an endothelial lining or communication with venules.[39,40] The spaces fill with contrast material on the late arterial phase and persist through the venous phase of an ateriogram. When focal, a sharply-defined collection can appear similar to a hypervascular nodule. There is some risk for intrahepatic hemorrhage to occur.

Peliosis hepatis must be distinguished from periportal sinusoidal dilatation, a condition described only in conjunction with pregnancy or use of oral contraceptives.[41] Angiography can demonstrate punctate areas of contrast accumulation in periportal sinusoidal dilatation, but these are not as pronounced or as well circumscribed as the collections in peliosis hepatis. The lesions are dilated sinusoids, spaces possessing an endothelial lining. There is no risk of hemorrhage, and the condition resolves after pregnancy or discontinuance of oral contraceptives.

Angiography is used in prospective liver transplant recipients for determination of arterial anatomy, IVC patency, and portal vein patency.[42] Transplant recipients may demonstrate arterial stenosis or thrombosis postoperatively, or diffuse intrahepatic arterial narrowing may be present in cases of graft rejection.[43] Diffuse arterioportal shunting indicates severe rejection. Arteriovenous fistulas can result from biopsies in transplant patients. Another potential cause for hepatic failure in such a setting is IVC and hepatic vein thrombosis.

BUDD-CHIARI SYNDROME

The Budd-Chiari syndrome is a clinical condition caused by obstruction of hepatic venous outflow, resulting in congestion, portal hypertension, and progressive liver failure. It has been associated with malignancy, use of oral contraceptives, paroxysmal nocturnal hemoglobinuria, polycythemia rubra vera, or other myeloproliferative disorders, but one third of patients will have no evident predisposing condition.[44] Onset is usually insidious; the affected individual develops hepatomegaly, pain, ascites, and edema. Obstruction can be due to thrombosis of hepatic veins or the IVC, invasion of the vessels by tumor, or vascular webs (this latter condition seen more often in the Orient). Venous obstruction may be partial or complete, and the most effective treatment for complete obstruction is surgical placement of a portocaval shunt, when possible.[44]

If the entire liver is affected, arteriography demonstrates diffuse straightening and narrowing of intrahepatic arterial branches, slowed flow, and diminished filling of peripheral vessels.[35] The hepatogram may be prolonged and diffusely mottled. Often, despite involvement of the rest of the liver, the caudate lobe will have unimpeded outflow into the IVC and will show compensatory hypertrophy.

When only portions of the liver are involved, the unaffected segments enlarge and portal blood is shunted from the affected to the unaffected segments.[45] The affected segments may manifest crowding of arterial branches and an intense hepatogram. Retrograde portal filling is sometimes observed.

Although arteriography or noninvasive studies may suggest the diagnosis, the most direct and conclusive evidence for Budd-Chiari syndrome is provided by IVC cavography and hepatic venography with pressure measurements in all vessels studied. Even when all hepatic veins are thrombosed, one can usually cannulate at least one. Contrast material injection then shows the typical hepatic vein-to-hepatic vein "spider-web" collaterals. If a hepatic vein cannot be entered, power injection of contrast material into the IVC with the patients in the Valsalva maneuver may be performed. Failure to opacify any veins by reflux then gives the presumptive diagnosis of hepatic vein occlusion.[46] Multiple veins should be catheterized when possible to exclude the possibility of partial Budd-Chiari syndrome.

There have been isolated reports of percutaneous treatment of webs or thrombus by balloon angioplasty catheters or thrombolytic agents, with at least temporary success.[47, 48] Further experience is necessary to determine the effectiveness of such interventions.

REFERENCES

1. Michels NA: Newer anatomy of the liver and its variant blood supply and collateral circulation. *Am J Surg* 1966; 112:337–347.

2. Chuang VP: Hepatic tumor angiography: A subject review. *Radiology* 1983; 148:633–639.

3. Rösch J, Freeny P, Antonovic R, et al: Infusion hepatic angiography in diagnosis of liver metastases. *Cancer* 1976; 38:2278–2286.

4. Freeny PC, Marks WM: Computed tomographic arteriography of the liver. *Radiology* 1983; 148:193–197.

5. Matsui O, Takashima T, Kadoya M, et al: Liver metastases from colorectal cancers: Detection with CT during arterial portography. *Radiology* 1987; 165:65–69.

6. Ohishi H, Uchida H, Yoshimure H, et al: Hepatocellular carcinoma detected by iodized oil. *Radiology* 1985; 154:25–29.

7. Gutierrez OH, Rösch J: Limitations of angiographic differential diagnosis in major hepatic processes. *ROFO* 1977; 127:1–8.

8. Brant WE, Floyd JL, Jackson DE, et al: The radiological evaluation of hepatic cavernous hemangioma. *JAMA* 1987; 257:2471–2474.

9. Cronan JJ, Esparza AR, Dorfman GS, et al: Cavernous hemangioma of the liver: Role of percutaneous biopsy. *Radiology* 1988; 166:135–138.

10. Rogers JV, Mack LA, Freeny PC, et al: Hepatic focal nodular hyperplasia: Angiography, CT, sonography, and scintigraphy. *AJR* 1981; 137:983–990.

11. Casarella WJ, Knowles DM, Wolff M, et al: Focal nodular hyperplasia and liver cell adenoma: Radiologic and pathologic differentiation. *AJR* 1987; 131:393–402.

12. Welch TJ, Sheedy PF, Johnson CM, et al: Radiographic characteristics of benign liver tumors. Focal nodular hyperplasia and hepatic adenoma. *RadioGraphics* 1985; 5:673–682.

13. Marks WM, Jacobs RP, Goodman PC, et al: Hepatocellular carcinoma: Clinical and angiographic findings and predictability for surgical resection. *AJR* 1979; 132:7–11.

14. Friedman AC, Lichtenstein JE, Goodman Z, et al: Fibrolamellar hepatocellular carcinoma. *Radiology* 1985; 157:583–587.

15. Takayasu K, Shima Y, Muramatsu H, et al: Angiography of small hepatocellular carcinomas: Analysis of 105 resected tumors. *AJR* 1986; 147:525–529.

16. Freeny PC: Angiography of hepatic neoplasms. *Semin Roentgenol* 1983; 18:114–122.

17. Okuda K, Musha H, Yoshida T, et al: Demonstration of growing casts of hepatocellular carcinoma in the portal vein by celiac angiography: The thread and streaks sign. *Radiology* 1975; 117:303–309.

18. Matsui O, Takashima T, Kadoya M, et al: Segmental staining on hepatic arteriography as a sign of intrahepatic portal vein obstruction. *Radiology* 1984; 152:601–606.

19. Shumida M, Ohto M, Ebara M, et al: Accuracy of angiography in the diagnosis of small hepatocellular carcinoma. *AJR* 1986; 147:531–536.

20. Nakakuma K, Tashiro S, Hiraoka T, et al: Hepatocellular carcinoma and metastatic cancer detected by iodized oil. *Radiology* 1985; 154:15–17.

21. Yumoto Y, Jinno K, Tokuyama K, et al: Hepatocellular carcinoma detected by iodized oil. *Radiology* 1985; 154:19–24.

22. Hayashi N, Yamamoto K, Tamaki N, et al: Metastatic nodules of hepatocellular carcinoma: Detection with angiography, CT, and US. *Radiology* 1987; 165:61–63.

23. Walter JF, Brookstein JJ, Bouffard EV: Newer angiographic observations in cholangiocarcinoma. *Radiology* 1976; 118:19–23.

24. Whelan JG Jr, Creech JL, Tamburo CH: Angiographic and radionuclide characteristics of hepatic angiosarcoma found in vinyl chloride workers. *Radiology* 1976; 118:549–557.

25. Heaston DK, Chuang VP, Wallace S, et al: Metastatic hepatic neoplasms: Angiographic features of portal vein involvement. *AJR* 1981; 136:897–900.

26. Stack J, Legge D, Behan M: Computed tomographic arteriography in the pre-surgical evaluation of hepatic tumors. *Clin Radiol* 1984; 35:189–192.

27. Lewis E, Bernardino ME, Barnes PA, et al: The fatty liver: Pitfalls in the CT and angiographic evaluation of metastatic disease. *J Comput Assist Tomogr* 1983; 7:235–241.

28. Nakamura H, Tanaka T, Hori S, et al: Transcatheter embolization of hepatocellular carcinoma: Assessment of efficacy in cases of resection following embolization. *Radiology* 1983; 147:401–405.

29. Yamada R, Sato M, Kawabata M, et al: Hepatic artery embolization in 120 patients with unresectable hepatoma. *Radiology* 1983; 148:397–401.

30. Chuang VP, Wallace S: Hepatic artery embolization in the treatment of hepatic neoplasms. *Radiology* 1981; 140:51–58.

31. Mitty HA, Warner RRP, Newman LH, et al: Control of carcinoid syndrome with hepatic artery embolization. *Radiology* 1985; 155:623–626.

32. Rabinowitz JG, Kinkabwala M, Ulreich S: Macro-regenerating nodule in the cirrhotic liver. *AJR* 1974; 121:401–411.

33. Freeny PC: Acute pyogenic hepatitis: Sonographic and angiographic findings. *AJR* 1980; 135:388–391.

34. Baltaxe HA, Fleming RJ: The angiographic appearance of hydatid disease. *Radiology* 1970; 97:599–604.

35. Rösch J, Keller FS: Angiography in diagnosis and therapy of diffuse hepatocellular disease. *Radiologe* 1980; 20:334–342.

36. Casarella WJ, Martin EC: Angiography in the management of abdominal trauma. *Semin Roentgenol* 1984; 19:321–327.

37. Gelin J, Wilms G: Angiodysplasien der Leber und des Gastrointestinaltraktes bei Morbus Rendu-Osler-Weber. *ROFO* 1985; 143:722–724.

38. Danchin N, Thisse JV, Neimann JL, et al: Oster-Weber-Rendu disease with multiple intrahepatic arteriovenous fistulas. *Am Heart J* 1983; 105:856–859.

39. Lyon J, Bookstein JJ, Cartwright CA, et al: Peliosis hepatis: Diagnosis by magnification wedged hepatic venography. *Radiology* 1984; 150:647–649.

40. Tsukamoto Y, Nakata H, Kimoto T, et al: CT and angiography of peliosis hepatis. *AJR* 1984; 142:539–540.

41. Fisher MR, Neiman HL: Periportal sinusoidal dilatation associated with pregnancy. *Cardiovasc Intervent Radiol* 1984; 7:299–302.

42. Cardella JF, Amplatz K: Preoperative angiographic evaluation of prospective liver recipients. *Radiol Clin North Am* 1987; 25:299–308.

43. Cardella JF, Amplatz K: Postoperative angiographic and interventional radiologic evaluation of liver recipients. *Radiol Clin North Am* 1987; 25:309–321.

44. McCarthy PM, van Heerden JA, Adson MA, et al: The Budd-Chiari syndrome: Medical and surgical management of 30 patients. *Arch Surg* 1985; 120:657–662.

45. Maguire R, Doppman JL: Angiographic abnormalities in partial Budd-Chiari syndrome. *Radiology* 1977; 122:629–635.

46. Floyd JL: The radiographic gamut of Budd-Chiari syndrome. *Gastroenterology* 1981; 76:381–387.

47. Uflacker R, Francisconi CF, Rodriguez MP, et al: Percutaneous transluminal angioplasty of the hepatic veins for treatment of Budd-Chiari syndrome. *Radiology* 1984; 153:641–642.

48. Greenwood LH, Yrizarry JM, Hallett JW Jr, et al: Urokinase treatment of Budd-Chiari syndrome. *AJR* 1983; 141:1057–1059.

5 | Portal Venography in Portal Hypertension

KEY CONCEPTS

1. A major indication for portal venography is definition of anatomy and flow prior to surgical shunt placement or hepatic transplantation.
2. Arterial portography does *not* define most esophageal varices, nor does it show active variceal bleeding.
3. Portal hypertension may be caused by presinusoidal or postsinusoidal venous obstruction. Cirrhosis results in an increase in postsinusoidal resistance.
4. Esophageal, colonic, small bowel, and rectal varices represent spontaneous portosystemic shunts, and all can be the source of gastrointestinal bleeding.
5. The direction of portal venous flow reverses with severe cirrhosis.
6. Corrected hepatic vein wedge pressure provides an estimation of hepatic sinusoidal pressure.

INDICATIONS

Portal venography is primarily used for the preoperative evaluation of patients with portal hypertension. The patency and size of portal, splenic, and mesenteric veins are assessed, as well as the direction of flow and presence of portosystemic collaterals. Arterial portography is *not* used to make the diagnosis of esophageal varices or variceal bleeding. Fewer than 25% of patients with endoscopically documented varices will demonstrate these varices by arterial injection of contrast material.[1]

Other indications for angiography of the portal venous system include problems following placement of a portosystemic shunt, suspected mesenteric venous thrombosis, colonic or small bowel varices, and preoperative localization of functioning pancreatic islet cell tumors. Hepatic transplantation is dependent on the patency and size of the portal vein, and portography may be necessary postoperatively as well as preoperatively if complications arise.[2, 3] Portography is occasionally useful for determining if hepatic, pancreatic, or other intra-abdominal tumors are surgically resectable.

Developments in duplex sonography and magnetic resonance imaging may circumscribe the future use of conventional examinations with contrast material.[4, 5] Portography performed for portal hypertension is the subject of this chapter. For islet cell tumor localization and mesenteric thrombosis, please see Chapter 6, Mesenteric Angiography.

ANATOMY

The portal venous system drains blood from the small bowel, stomach, spleen, pancreas, and colon. The confluence of the superior mesenteric vein and splenic vein at the pancreatic head forms the portal vein. The inferior mesenteric vein usually enters the splenic vein several centimeters from the origin of the portal vein.

The portal vein ramifies within liver, defining the segmental anatomy of the organ. Blood flows within the hepatic sinusoids, mixes with hepatic arterial blood, and drains into the hepatic veins at the periphery of the acinus (the basic functional unit of the liver). An older convention defines the hepatic lobule as the tissue surrounding a hepatic vein and places portal triads (portal vein, hepatic artery, and bile duct branch) at the periphery of each lobule.

If normal portal flow is obstructed, blood may pursue various collateral pathways to rejoin the systemic circulation. Potential portosystemic communications include left portal vein–to–paraumbilical veins within the falciform ligament, superior rectal–to–inferior rectal veins, spontaneous splenorenal or other retroperitoneal shunts, and left gastric (coronary vein) or short gastric branches to the azygos system by way of esophageal veins (Fig 5–1). Obstruction of main portal or splenic veins can result in the emergence of gastroepiploic, pancreatic, or biliary/gallbladder venous collaterals.[6] The propensity of gastroesophageal varices to catastrophic rupture is a prime cause of death from portal hypertension.

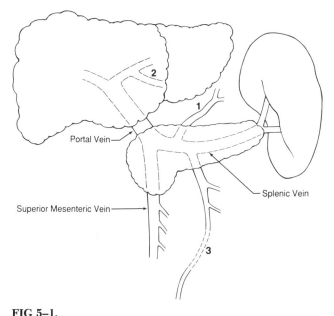

FIG 5–1.
Portosystemic venous channels commonly seen in portal hypertension.
Coronary vein *(1,)* paraumbilical veins *(2)*, inferior mesenteric/rectal
veins *(3)*. (Modified from Reuter SR, Redman HC, Cho KJ: *Gastrointestinal Angiography,* ed 3. Philadelphia, WB Saunders, 1986.)

CAUSES OF PORTAL HYPERTENSION

The major cause of portal hypertension in the U.S. population is
Laennec's (alcoholic) cirrhosis. The condition is characterized by bands
of fibrosis throughout the liver, and relatively small, homogeneous
regenerating nodules. Viral hepatitis and a number of other toxic or
infectious diseases may lead to postnecrotic cirrhosis, which is associated with larger regenerating nodules. Both alcoholic and postnecrotic cirrhosis produce portal venous obstruction predominately at the
postsinusoidal level. Other conditions causing a postsinusoidal increase

in vascular resistance include Budd-Chiari syndrome (hepatic vein thrombosis) and constrictive pericarditis.

Portal hypertension may also arise from *presinusoidal* obstruction, such as tumor invasion or encasement of the portal vein, schistosomiasis, congenital hepatic fibrosis, noncirrhotic idiopathic portal hypertension (Banti syndrome), and portal vein thrombosis. Thrombosis can result from dehydration, trauma, infection, and hypercoagulable states. Splenic vein occlusion, often consequent to pancreatitis or pancreatic malignancy, produces hypertension in a portion of the portal system. The distinction between presinusoidal and postsinusoidal blocks is important in the interpretation of studies with contrast material and in hepatic venous pressure measurements (see "Hemodynamic Measurements," later in this chapter). Rarely, a large arteriovenous shunt produced by trauma or hepatocellular carcinoma can result in a high-output form of portal hypertension.

CLINICAL ASPECTS OF PORTAL HYPERTENSION

As mentioned, a major threat from any form of portal hypertension is exsanguination from ruptured varices. Although rectal, colonic, and other varices may develop, they do not ordinarily pose the same degree of risk that gastroesophageal varices do. Approximately 50% of patients with cirrhosis will develop esophageal varices, and most of these will bleed at some time.[7] Despite aggressive approaches to medical management, the prognosis of such patients is dismal. One third will die during initial hospitalization, many will bleed again within weeks of discharge, and only about one third will survive for 1 year.[7] Bleeding is related to the size of varices, but their size does not correlate well with portal venous pressure.[8]

Other major causes of morbidity are hepatic encephalopathy and liver failure. These are particularly problems in patients with cirrhosis as the cause of portal hypertension, and they may be precipitated by an episode of major hemorrhage. Unfortunately, encephalopathy may occur in some patients after surgical treatment with portosystemic shunt placement for decompression.

Treatment Alternatives

Vasopressin, somatostatin, and propranolol are medications that decrease portal pressure, and they may offer some protection against recurrent bleeding. Intravenous vasopressin has been used successfully to control acute hemorrhage. However, medical treatment has not yet been shown to substantially affect mortality.[8]

In cases of active bleeding, esophageal balloon catheters, such as the Sengstaken-Blakemore tube, are effective in tamponade and are best applied in catastrophic hemorrhage. However, a high rebleeding rate after balloon deflation means that tamponade is merely a temporizing step before definitive treatment.[9]

Because of the deficiencies of more conservative treatment, various surgical portosystemic shunting procedures have been devised. A successful shunt reduces the risk of recurrent hemorrhage to 4% to 7%, and when bleeding occurs it is usually associated with thrombosis or stenosis of the shunt.[10] Emergency surgery carries a high mortality rate, and acute bleeding is best controlled by other means. Alternatives to surgery explored in the past 15 years have included percutaneous transhepatic embolization of varices, and endoscopic injection sclerotherapy.

Transhepatic catheter embolization of varices was investigated at various institutions, but the procedure has generally been abandoned because of a high complication rate and lack of lasting effect (see "Percutaneous Interventions," later in this chapter). More recently endoscopic sclerotherapy has gained wide application; it can effectively stop acute hemorrhage when other measures have failed,[8, 9] but does not prevent or adversely affect later operative portal decompression. Nevertheless, the ultimate role of sclerotherapy in the long-term management of varices is not established, and shunt placement remains standard therapy for portal hypertension.[11, 12]

Surgical Shunts

The type of shunt created operatively depends on patient anatomy, hemodynamics, and the experience of the surgeon. Direct end-to-side and side-to-side portocaval shunts, mesocaval [i.e., superior mesenteric vein–to–inferior vena cava (IVC)], and splenorenal shunts have been employed. A short synthetic graft can be interposed in the placement of portocaval and mesocaval shunts. Splenorenal shunts may either be proximal (Linton) or distal (Warren). In the former, the spleen is removed. In the latter, the splenic vein is divided and drainage from the spleen is directed into the renal vein, essentially isolating the spleen from the rest of the portal circulation.

Although the distal splenorenal shunt is technically more difficult to place, it tends to preserve antegrade (hepatopetal) flow in the portal vein. The development of hepatic encephalopathy has been correlated with loss of normal portal flow.[1] For a patient to be a candidate for such a shunt, antegrade flow must be present preoperatively.

What appears to be more clinically important than flow direction is the rapidity and degree of portal vein flow reversal, for most patients with distal splenorenal shunts eventually develop hepatofugal flow.[13] Use of a flow-limiting graft in portocaval shunt surgery can accomplish the same effect with a low incidence of postoperative hepatic failure or encephalopathy.[4]

Patients with noncirrhotic causes of portal hypertension and good hepatic functional reserve tend to tolerate surgery (as well as variceal hemorrhage) better than those with cirrhosis. Distal splenorenal shunts have produced a mean 5.5-year survival in nonalcoholic patients, as opposed to a mean 4.3-year survival in those with alcoholic cirrhosis.[14] Also, alcoholics are more likely to die of hepatic failure than those having shunt placement for other reasons.

METHODS OF PORTOGRAPHY

Portal angiography can be performed either by direct or indirect means. The portal vein and its branches cannot be catheterized by way of the arterial or systemic venous circulations; consequently, direct portography requires a more invasive approach, such as hepatic or splenic puncture. For this reason most portal studies are performed by injecting mesenteric and splenic arteries with relatively high doses of contrast media. If portal opacification is inadequate, a direct approach can then be employed.

Arterial Portography

Arterial portography requires selective catheterization of the superior mesenteric, splenic, and hepatic arteries. Preliminary biplane aortography may demonstrate stenosis or occlusion of vessel origins, but it also increases the total amount of contrast medium used and is usually unnecessary. A Simmons or sidewinder-shape catheter is quite useful for selective catheterization, and provides stable tip placement. Because of the large amounts of contrast medium injected, stable position is critical. If there are any doubts, one may perform a one-second power test injection at the rate planned, in order to see if the catheter "kicks out" of the selected vessel. A sidehole near the catheter tip makes dissection or inadvertent superselective injection much less likely.

Injection volumes for conventional film runs range from 50 to 80 mL for superior mesenteric artery and splenic injections. Injection rate should be matched to the normal flow of the vessel, as estimated by hand injection and fluoroscopy (usually between 6 and 10 mL/sec).

Filming is rapid during the arterial phase, and slowed after 5 to 6 seconds. Portal opacification normally peaks 15 to 25 seconds after arterial injection, so exposures are best continued to at least 25 seconds. Portal opacification can be enhanced by administration of 50 mg of tolazoline into the artery immediately before injection of contrast material. However, tolazoline is not often needed and it should be used with caution in patients with cardiac disease.

Digital subtraction portography has been performed successfully, and it can substantially diminish the amount of contrast agent used.[15] Patient cooperation, intravenous administration of glucagon, and minimal bowel gas are prerequisites for a satisfactory study.

Hepatic arteriography should be a part of every portal vein examination. The presence of tumor, arterioportal shunting, liver size, and degree of cirrhosis can be assessed. If portal flow is reversed, the portal vein may opacify only after hepatic artery injection. In cases where selective hepatic or splenic catheterization is unsuccessful, a celiac injection may be performed.

Transhepatic Portography

The portal vein can be catheterized through a transhepatic route utilizing a technique similar to that employed for percutaneous biliary drainage. After the lateral costophrenic angle is examined fluoroscopically to prevent puncture of low-lying lung, an 18-gauge sheath needle is directed from the midaxillary line through the right lobe toward the porta hepatis. The trocar is removed, and the sheath is slowly withdrawn until blood can be aspirated. Contrast medium is then injected to confirm that a portal branch (rather than a hepatic vein or hepatic artery) has been entered. If so, a guidewire is introduced, and the sheath is advanced over it into the splenic or superior mesenteric vein. Pressure measurements are made, and contrast material injected for filming. This approach also allows cannulation of the coronary vein and gastroesophageal varices with selective catheters for detailed study or possible embolization.

If no portal branch is entered on the first pass, the needle should not be completely removed from the liver. Instead, redirection from within the tract is advisable in order to minimize the risk of peritoneal hemorrhage or bile leakage. Entry into extrahepatic portal vein must be avoided. At the conclusion of transhepatic portography, the tract is occluded with gelatin sponge (Gelfoam) particles or coils as the sheath is slowly withdrawn through parenchyma. Particular care should be taken to place occlusion devices near the point of entry into the liver capsule.

Transhepatic portography has the advantage of providing excellent opacification of the portal system and direct measurement of portal pressure. Disadvantages include the increased risk of bleeding, as well as possible focal trauma with thrombosis of the portal vein. Needle insertion may be quite difficult in patients with advanced cirrhosis and extensive hepatic fibrosis. Also, many such patients have impaired hemostasis, further increasing risk. Contraindications to transhepatic portography include the presence of hypervascular hepatic lesions (hepatocellular carcinoma or cavernous hemangioma) and portal vein thrombosis.[7]

Splenoportography

This approach shares many of the advantages and disadvantages of transhepatic portography. Unlike the transhepatic approach, selective catheters cannot be safely inserted through the spleen. Splenoportography is performed with an 18-gauge sheath needle placed from a subcostal midaxillary or anterolateral approach.[16] Fluoroscopy or ultrasound can be used to guide the needle into splenic pulp along the long axis of the organ, avoiding possible intervening colon. With proper needle placement, blood should return freely through the sheath. If there is no free return of blood, Brazzini et al. recommend a test injection with contrast material, followed by a rapid injection of 1 to 2 mL of saline to break up the pulp at the sheath tip.[16] Pressures can then be determined, and a contrast study performed with an injection of 4 to 10 mL/sec. At the conclusion of splenoportography, plugs of gelatin sponge (Gelfoam) are deposited in the tract as the sheath is slowly removed. Without tract plugging, the risk of hemorrhage is substantial.

Umbilical Vein Catheterization

An alternative to transparenchymal catheterization of the portal venous system is umbilical vein cannulation.[17] A cut-down is performed in the epigastrium and the umbilical vein remnant isolated. A sound can then be passed through the remnant until the return of blood signals that the left portal vein has been reached. An 8-F sheath is sutured into place, giving access to catheters for selective examination of the portal vein and its tributaries. This approach permits direct portal manometry, while it minimizes the chances of peritoneal hemorrhage. A disadvantage of umbilical vein cannulation is the increased radiation dose to the radiologist from direct exposure to the beam.

HEMODYNAMIC MEASUREMENTS

An integral part of portal vein studies is the determination of portal venous pressure. This is trivial in the case of transhepatic or umbilical cannulation of the portal vein; however, if arterial portography is used, portal pressure is estimated by "corrected" hepatic vein wedge pressure.

Corrected Hepatic (Sinusoidal) Wedge Pressure

An estimation of sinusoidal pressure is obtained by subtracting a "free" hepatic venous pressure (FHVP) from that obtained with a catheter "wedged" into a hepatic vein (WHVP). The latter can be measured with an endhole catheter passed as far as possible into a hepatic vein, such that it occludes outflow through that vein. A somewhat more convenient method of obtaining the same measurement is to employ an occlusion balloon catheter. The balloon is gently inflated, and the efficacy of occlusion is checked fluoroscopically by manual injection of 1 to 2 mL of radiographic contrast medium. After the WHVP is measured, the FHVP is obtained with the balloon deflated.

Some recommend that the pressure of the IVC (rather than FHVP) be subtracted from the WHVP.[18] If the diaphragm exerts a pinchcock effect on the IVC, the FHVP may more closely approximate the pressure within right atrium rather than that of IVC and intra-abdominal systemic veins. The effect of this would be to overestimate the effective portal pressure if the FHVP is used to calculate the corrected WHVP.

If a jugular venipuncture is used, selection of hepatic veins with a catheter is usually uncomplicated. When a femoral approach is used, a deflection wire is required. In this case the catheter is passed to the level of the diaphragm and the deflecting wire advanced until its tip is 2 to 3 cm proximal to the catheter tip (a deflecting wire should never be passed beyond catheter tip, otherwise injury could result). With deflection exerted, the catheter is manipulated until its tip engages the orifice of a hepatic vein. The wire is then held in place with constant deflection as the catheter is fed off of it into the vein. Deflection is gently released, and the wire is removed when satisfactory catheter position has been achieved.

A normal corrected WHVP should be less than 5 mm Hg (1 mm Hg is equivalent to 1.36 cm of water). Corrected wedge pressures reflect portal pressures in cirrhosis and other conditions producing postsinusoidal venous obstruction. However, because the degree of disease may vary from one part of liver to another, at least two separate hepatic veins should be sampled for most accurate results.[18, 19] Corrected wedge pressures will *not* be elevated in cases of portal hypertension caused by presinusoidal venous obstruction.

Other Pressure Measurements and Left Renal Venography

Pressures should also be obtained from the right atrium and from the IVC below its intrahepatic portion. If a splenorenal shunt is under consideration, left renal vein pressure must be measured to exclude left renal vein hypertension, and renal venography is performed to assess size and location of the left renal vein. A circumaortic or retroaortic renal vein makes selective shunt placement more difficult.

Koolpe and Koolpe have found measurement of abdominal aortic diastolic pressure valuable because the ratio of aortic diastolic pressure to hepatic vein wedge pressure has correlated with the type and degree of cirrhosis present.[19] They also recommend that the character of the hepatic venous pressure change occurring with deflation of the occlusion balloon be observed closely (see "Alcoholic Cirrhosis," later in this chapter).

HEPATIC VENOGRAPHY IN PORTAL HYPERTENSION

Although not a standard component of most portal studies, hepatic venography may add useful information. Injections can be made with the catheter either in a free or wedged position. Free hepatic venography is performed with 20 mL of contrast material injected at a rate of 8 to 10 mL/sec, with rapid filming. The appearance of the hepatic vein correlates well with the extent of liver disease. In combination with corrected WHVP measurements, free hepatic venography is valuable in making the diagnosis of cirrhosis in those patients presenting a high risk for biopsy.[18]

Wedge hepatic venography is performed with 4 mL of contrast material injected over 2 seconds. A parenchymal stain results, and reflux of contrast material into the portal vein can be used to assess portal patency and flow. However, this procedure has fallen into disfavor after reports of resultant segmental hepatic infarction and poor correlation of findings with portal hemodynamics.[1, 20]

DIRECT CATHETERIZATION OF SHUNTS

After surgical decompression of the portal venous system, arterial portography may not demonstrate shunt patency well. In such cases, or if direct pressure measurements are desired, direct cannulation of the shunt can be performed. Of course, the type and location of shunt must be known before such a study is attempted.

Any gradient of 5 mm Hg or more across a shunt is abnormal. If a stenosis is present, it may be dilated with balloon catheters. Ruff et al. have successfully treated 20 shunt stenoses as early as 2 weeks postoperatively.[10] However, surgical backup must be arranged for the small, but potentially catastrophic, risk of shunt rupture.

FINDINGS

Varices and Other Collateral Veins

In all forms of portal hypertension, disturbances in normal flow patterns may be found. Although arterial portography is not very sensitive, it may still define gastric or esophageal varices. Colonic, rectal, and small bowel varices are potential causes of gastrointestinal blood loss in patients with portal hypertension. Colonic and small bowel varices are almost always associated with previous abdominal surgery and are presumably related to adhesions. Rectal varices may become more of a problem in those whose esophageal varices have been treated by endoscopic sclerotherapy.[21] Enlargement of superior mesenteric or inferior mesenteric veins with associated retrograde flow should alert one to the possibility of such unusual lesions. Opacification of the IVC after injections of contrast material into the superior mesenteric or inferior mesenteric artery also indicates major portosystemic venous communications.[22] Late venous phase films must be carefully examined to detect small bowel or colorectal varices, and oblique projections may occasionally help confirm the diagnosis.

Portal vein thrombosis commonly stimulates development of a collateral network in the porta hepatis that is often erroneously termed "cavernous transformation of the portal vein." Isolated splenic vein thrombosis tends to produce gastric varices in the absence of esophageal varices. The left hepatic vein may be prominent in portal hypertension, because it serves as a portosystemic collateral through paraumbilical (*not* umbilical) veins in the ligamentum teres.[23]

In interpreting arterial portography, inflow of unopacified blood and reversal of portal flow can be misinterpreted as partial or complete thrombosis of the portal vein. If an examination is equivocal, simultaneous catheterization and injection of the splenic and superior mesenteric arteries may resolve the issue.

Retrograde (Hepatofugal) Portal Flow

Normally, second- and third-order intrahepatic branches of the portal vein are opacified with arterial portography. As portal flow becomes more stagnant with increasing intrahepatic portal obstruction, opacification diminishes. In severe portal hypertension, flow becomes hepatofugal, and the portal vein is not opacified by injection of contrast material into the superior mesenteric or splenic artery. In such cases, hepatic arteriography or wedge hepatic venography may demonstrate reversed portal flow. If the major portosystemic outflow is through the left portal vein and paraumbilical collaterals, the main portal vein will opacify in a normal antegrade manner, even though functionally flow is hepatofugal. In patients with successful creation of decompressive shunts, portal flow generally reverses, even in those with selective splenorenal shunts.[13] In cases where portal vein patency cannot be resolved by angiography, duplex ultrasound, contrast material–enhanced computed tomography and magnetic resonance imaging are noninvasive diagnostic alternatives.

Alcoholic Cirrhosis

In advanced alcoholic cirrhosis the liver may be small, with marked enlargement and "corkscrewing" of hepatic artery branches. Hepatic arterial flow increases to compensate for loss of normal portal flow, and arterioportal shunting may contribute to portal hypertension. It has been asserted that a corrected WHVP of at least 12 mm Hg is necessary for the development of varices in patients with alcoholic cirrhosis.[9] As noted earlier, any corrected wedge pressure ≥ 5 mm Hg represents portal hypertension. Large portosystemic venous collaterals serve to depress the degree of pressure elevation.[22]

Koolpe and Koolpe have found a gradient ≥ 4 mm Hg between the IVC and right atrium to be a reliable indicator of alcoholic cirrhosis.[19] They have also observed that in alcoholic cirrhosis, deflation of the balloon of an hepatic vein occlusion catheter results in a slow and irregular drop in pressure over more than 1 second. Patients with alcoholic cirrhosis will have hepatofugal portal flow if the ratio of aortic diastolic pressure to hepatic vein wedge pressure is ≤ 1.6.[19]

Free hepatic venography shows a normal angle of branch junctions but irregular filling of branches with peripheral occlusions and a "pruned-tree" appearance in alcoholic cirrhosis (see Fig 5–2).[24] Venous anastomoses are rarely defined. Wedge hepatic venography produces an irregular parenchymal stain, and portal branches often opacify.[25]

Postnecrotic Cirrhosis

In postnecrotic cirrhosis, arteriographic and venographic findings are similar to those of alcoholic cirrhosis, but tend to be less severe. Balloon deflation in the hepatic vein produces a rapid (less than 1 second) and smooth drop from WHVP to FHVP. There should be no pressure gradient between abdominal IVC and right atrium.[19] Hepatofugal flow correlates with a ratio of aortic diastolic pressure to WHVP of less than 1.3.[19]

Noncirrhotic Idiopathic Portal Hypertension

This condition, also known as Banti syndrome, is rare in the United States but common in Asia.[24] Hepatic venography shows a distinctive, "weeping willow," appearance: hepatic vein branches join together at acute angles (see Fig 5–2,C). Large communications with other hepatic veins are often evident. Corrected WHVP values and hepatic venograms tend to be normal in the face of multiple occlusions in small and medium-sized portal branches.

PERCUTANEOUS INTERVENTIONS

A percutaneous transhepatic route has been taken to occlude gastroesophageal varices in the past. Various agents including gelatin sponge soaked in Sotradecol (a sclerosing agent), coils, bucrylate, and alcohol have been used to control acute bleeding.[7, 8, 26] However, although such an intervention can control bleeding in 76% of patients not responding to other measures, over two-thirds experience bleeding again within a matter of months.[7] High complication rates and low 1-year survival of patients treated have led to an abandonment of the technique for all but exceptional cases. At Emory University hospitals, transhepatic obliteration of those collaterals diverting flow from the liver has been used to treat patients developing hepatic encephalopathy after portosystemic shunt surgery.[10]

Another intervention attempted in patients with portal hypertension has been the nonsurgical creation of decompressive shunts. Colapinto et al. treated a small number of patients by advancing a modified Ross transjugular biopsy needle into a hepatic vein, through parenchyma, and into a portal branch.[27] After selective catheterization and occlusion

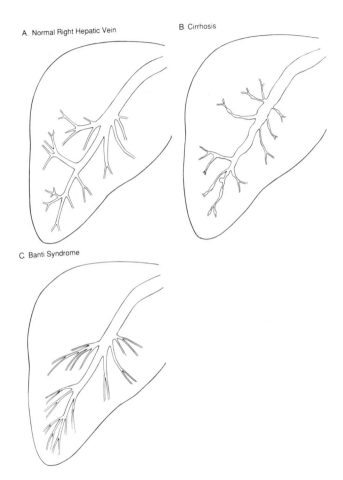

FIG 5–2.
Hepatic venous patterns in healthy (**A**) and diseased (**B** and **C**) livers
A, normal right hepatic vein. **B,** cirrhosis. **C,** Banti syndrome.

of varices through this approach, they dilated the tract with a balloon catheter. Although such shunts tended to remain patent in the short term, patient survival was not improved. It remains to be seen if the use of vascular stents can lead to better results with such catheter-created portosystemic shunts.[28]

REFERENCES

1. Nordlinger BM, Nordlinger DF, Fullenwider JT, et al: Angiography in portal hypertension: Clinical significance in surgery. *Am J Surg* 1980; 139:132–141.
2. Cardella JF, Amplatz K: Preoperative angiographic evaluation of prospective liver recipients. *Radiol Clin North Am* 1987; 25:299–308.
3. Cardella JF, Amplatz K: Postoperative angiographic and interventional radiologic evaluation of liver recipients. *Radiol Clin North Am* 1987; 25:309–321.
4. Helton WS, Montana MA, Dwyer DC, et al: Duplex sonography accurately assesses portocaval shunt patency. *J Vasc Surg* 1988; 8:657–660.
5. Williams DM, Eckhauser FE, Aisen A, et al: Assessment of portosystemic shunt patency and function with magnetic resonance imaging. *Surgery* 1987; 102:602–607.
6. Reuter SR, Redman HC, Cho KJ: Cirrhosis and portal hypertension, in *Gastrointestinal Angiography*, ed 3. Philadelphia, WB Saunders, 1986, pp 382–445.
7. Benner KG, Keefe EB, Keller FS, et al: Clinical outcome after percutaneous transhepatic obliteration of esophageal varices. *Gastroenterology* 1983; 85:146–153.
8. Joffe SN: Non-operative management of variceal bleeding. *Br J Surg* 1984; 71:85–91.
9. Rikkers LF: Variceal hemorrhage. *Gastroenterol Clin North Am* 1988; 17:289–302.
10. Ruff RJ, Chuang VP, Alspaugh JP, et al: Percutaneous vascular intervention after surgical shunting for portal hypertension. *Radiology* 1987; 164:469–474.
11. Levine BA, Gaskill HV, Sirinek KR: Portasystemic shunting remains the procedure of choice for control of variceal hemorrhage. *Arch Surg* 1985; 120:296–300.
12. Botta GC, Contini S: Endoscopic sclerotherapy as emergency and long-term management of esophageal varices: A changing approach? *Int Angiol* 1988; 7:167–171.

13. Robbins AH, Johnson WC, Nabseth DC: Long-term follow-up of distal splenorenal shunts. *Radiology* 1980; 134:341–345.
14. Zeppa R, Lee PA, Hutson DG, et al: Portal hypertension: A fifteen year perspective. *Am J Surg* 1988; 155:6–9.
15. Foley WD, Stewart ET, Milbrath JR, et al: Digital subtraction angiography of the portal venous system. *AJR* 1983; 140:497–499.
16. Brazzini A, Hunter DW, Darcy MD, et al: Safe splenoportography. *Radiology* 1987; 162:607–609.
17. Spigos DG, Tauber JW, Tan WS, et al: Umbilical venous cannulation: A new approach for embolization of esophageal varices. *Radiology* 1983; 146:53–56.
18. Cavaluzzi JA, Sheff R, Harrington DP, et al: Hepatic venography and wedge hepatic vein pressure measurements in diffuse liver disease. *AJR* 1977; 129:441–446.
19. Koolpe HA, Koolpe L: Portal hypertension: Angiographic and hemodynamic evaluation. *Radiol Clin North Am* 1986; 24:369–381.
20. Castañeda-Zuñiga WR, Jauregui H, Rysavy JA, et al: Complications of wedge hepatic venography. *Radiology* 1978; 126:53–56.
21. Foutch PG, Sivak MV Jr: Colonic variceal hemorrhage after endoscopic injection sclerosis of esophageal varices: A report of three cases. *Am J Gastroenterol* 1984; 79:756–760.
22. Burcharth F, Sorenson TIA, Andersen B: Percutaneous transhepatic portography: Relationships between portosystemic collaterals and portal pressure in cirrhosis. *AJR* 1979; 133:1119–1122.
23. Lafortune M, Constantin A, Breton G, et al: The recanalized umbilical vein in portal hypertension: A myth. *AJR* 1985; 144:549–553.
24. Futagawa S, Fukazawa M, Musha H, et al: Hepatic venography in non-cirrhotic idiopathic portal hypertension. *Radiology* 1981; 141:303–309.
25. Heeney DJ, Bookstein JJ, Bell RH, et al: Correlation of hepatic and portal wedged venography with histology in alcoholic cirrhosis and periportal fibrosis. *Radiology* 1982; 142:591–597.
26. Yune HY, O'Connor KW, Klatte EC, et al: Ethanol thrombotherapy of esophageal varices: Further experience. *AJR* 1985; 144:1049–1053.

27. Colapinto RF, Stronell RD, Gildiner M, et al: Formation of intrahepatic portosystemic shunts using balloon dilatation catheter: Preliminary clinical experience. *AJR* 1983; 140:709–714.

28. Palmaz JC, Sibbitt RR, Reuter SR, et al: Expandable intrahepatic portacaval shunt stents: Early experience in the dog. *AJR* 1985; 145:821–825.

6

Mesenteric Angiography

KEY CONCEPTS

1. Angiography is indicated for gastrointestinal bleeding that does not resolve with conservative therapy and cannot be localized or treated endoscopically.
2. Identification of the bleeding source allows directed surgical or nonsurgical intervention and minimizes operative mortality.
3. If the activity of bleeding is in question, radionuclide scans are much more sensitive than arteriography for detection of extravasation. A positive scan increases the diagnostic yield of angiography and limits the amount of contrast medium needed.
4. Selective arterial infusion of vasopressin is most effective for gastric bleeding and colonic diverticular hemorrhage.
5. Biplane aortography is needed to evaluate suspected aortoenteric fistula or intestinal angina.
6. Acute mesenteric ischemia has high mortality and can result from arterial embolus, thrombus, low-flow state, or venous occlusion.

INDICATIONS

Since the aggressive use of endoscopy became widespread in the early 1980s, the role of angiography in the diagnosis of gastrointestinal (GI) bleeding has diminished considerably. Still, arteriography remains a valuable study in those patients with recurrent or continued GI hemorrhage whose source has eluded previous endoscopic, radionuclide, and barium studies, especially those whose bleeding has continued after "blind" laparotomy.[1] Lesions of angiodysplasia [or other arteriovenous malformations (AVMs)] are particularly difficult to identify at surgery, and selective arterial injection of methylene blue can guide resection.[2]

In acute GI bleeding, angiography is most likely to be diagnostic if blood loss exceeds 4 units within 24 hours, if continued intravenous fluids and transfusions are needed to keep the patient stable, and if the nasogastric tube aspirate remains bloody despite lavage with cold saline.[3] When bleeding is intermittent and patient stability permits, radionuclide studies are useful to confirm that extravasation is taking place as well as to guide angiography to the most likely source. Because of their great sensitivity, abnormal red blood cell–labeled scans are associated with angiographic demonstration of active bleeding in over 40% of cases, while a negative scan decreases the likelihood of finding extravasation to about 3%.[4] A not inconsequential number of patients with scans negative for disease *will* have angiographic abnormalities, but catheter studies need not be performed on an emergency basis, and less invasive examinations can be pursued if bleeding has stopped.

Angiography in the face of unrelenting gastric or colonic hemorrhage allows arterial vasopressin infusion or embolization to be performed in selected patients. Operative mortality for emergency surgery is high. Even if focal ischemia results from catheter intervention, the patient's condition may be stabilized prior to any necessary surgery (see Chapter 12, Embolotherapy).

Other indications for mesenteric or celiac angiography include suspected intestinal angina, acute mesenteric ischemia, splenic or other splanchnic artery aneurysm, detection of islet cell tumors not identified by other studies, and trauma. Secreting pancreatic tumors can be localized by transhepatic mesenteric venous sampling. In exceptional cases, angiography may still be requested to determine the resectability of pancreatic carcinoma.

ANATOMIC AND TECHNICAL CONSIDERATIONS

The arterial supply to the stomach comes mainly from the left gastric artery, which usually arises from the celiac trunk and follows the lesser curvature of the stomach to anastomose with the right gastric artery, a much smaller vessel arising from the proper or left hepatic artery. The greater curvature of the stomach is supplied by the gastroepiploic branches of the gastroduodenal and splenic arteries. There are free anastomoses among these various vessels, making it unlikely that occlusion of any single branch will result in ischemia. Short gastric arteries originate along the length of the splenic artery.

The pancreas obtains blood from branches of the gastroduodenal, splenic and superior mesenteric arteries (SMAs). The pancreatico-duodenal arcades supply the head, while blood flow is provided to the body by way of the pancreatica magna from the splenic artery, and dorsal pancreatic artery, which may be a branch of proximal splenic or SMA. These latter vessels join the transverse pancreatic, which follows the axis of the organ. The arterial supply to the pancreas is highly variable, and selective angiography is needed to define anatomy in a given individual precisely.

The SMA provides flow to the entire small intestine beyond the ligament of Treitz. The cecum, right colon, and much of the transverse colon are also supplied by the SMA. The remainder of the colon obtains its blood from the inferior mesenteric artery (IMA). The IMA typically arises from the left anterolateral surface of the aorta at the level of L-3.

The ready availability of large collateral pathways between celiac, SMA, and IMA distributions makes chronic arterial insufficiency highly unlikely in the absence of major occlusive disease affecting at least two of the three vessels. Major communications between celiac artery and SMA are present in the pancreaticoduodenal arcades, and transpancreatic flow over other vessels in the body of the organ can also be seen. In a few patients a large central communication between the celiac artery and SMA persists from early development as the arc of Bühler. Anastomoses between SMA and IMA circulations occur at the junction of middle colic and left colic arteries. The arcade thus formed proximal to the vasa recta at the edge of the colonic mesentery is called the marginal artery of Drummond. A more central pathway between SMA and IMA is the arch of Riolan. These collaterals are commonly filled in elderly patients, because the IMA is frequently occluded by atherosclerosis. In some cases of IMA occlusion the colon may be primarily supplied by the internal iliac arteries over middle rectal–to–superior rectal artery anastomoses.

In performing mesenteric studies, the flow of the vessel should be matched (on the order to 10 mL/sec for celiac, 8 mL/sec for SMA, and 3 mL/sec for IMA) for 6 to 10 seconds. Filming should be rapid during the arterial phase, usually one to two exposures per second, and slowed to a rate of one exposure every 2 to 5 seconds for capillary and venous phases. Imaging should be carried out to at least 30 seconds, to allow appreciation of slow extravasation in cases of GI bleeding and venous filling in patients with portal hypertension or suspected venous occlusion. Digital subtraction techniques are generally of little value in mesenteric studies because of the copious bowel gas often present and the high potential for misinterpreting motion artifact as contrast extravasation. The suspected bleeding vessel should be injected first, and the more selective the catheter placement (e.g., injection of the left gastric artery rather than the celiac axis), the higher the probability that active bleeding will be recognized.

In cases of lower GI hemorrhage, the pelvis should be imaged first (with both IMA and SMA injections), before contrast material accumulates in the bladder. Placement of a Foley catheter prior to angiography is highly recommended. It may be necessary to select each internal iliac artery for a complete study. Because mesenteric angiography often requires multiple injections of large amounts of iodinated contrast medium, the status of the patient's renal function should be known and hydration maintained as well as possible.

Aortography is not routinely needed, except for cases of suspected aortoenteric fistula, intestinal angina, and mesenteric ischemia. Biplane studies are mandatory for these indications. Lateral aortography is also valuable if a major splanchnic vessel cannot be catheterized or identified, a situation which may represent anatomic variation or vascular occlusion.

GASTROINTESTINAL HEMORRHAGE

The great majority of GI bleeding will stop spontaneously, and most episodes do not recur. It is persistent or recurrent massive hemorrhage that accounts for the 8% to 10% mortality of acute GI bleeding.[3] A clinical distinction must be made between upper GI and lower GI hemorrhage. Up to 90% of all GI bleeding is from a lesion proximal to the ligament of Treitz, and most patients will have bloody nasogastric tube aspirates.[5] However, absence of a positive aspirate does not exclude duodenal ulcer as a cause of melena or hematochezia. Fortunately, if it is performed within 6 hours of hospital admission, endoscopy can positively identify bleeding points in all but a few patients with upper GI hemorrhage.[6] Endoscopy is important both for excluding lesions not ordinarily needing angiographic diagnosis or amenable to catheter interventions (such as gastroesophageal varices, nasopharyngeal abnormalities, or anorectal lesions), and for treating upper GI bleeding unresponsive to conservative measures.[3, 7]

Hemorrhage which stops, only to recur at intervals, is a particular problem if endoscopy and barium studies are repeatedly negative. Evaluation of such bleeding can be time-consuming, frustrating, and expensive. Angiography and intraoperative endoscopy are the diagnostic measures most likely to be fruitful in this situation.[1, 8] In some cases computed tomography (CT) and ultrasound may enable one to detect parenchymal abnormalities responsible for GI hemorrhage (such as hepatic aneurysms, arterioportal fistulas, pseudoaneurysms from pancreatitis, or tumors), thus selecting patients for directed arterial study.[9]

Upper Gastrointestinal Bleeding

The most common cause of upper GI blood loss is peptic ulcer disease, followed in frequency by gastroesophageal varices.[5, 7] (For a discussion of the latter, see Chapter 5, Portal Venography in Portal Hypertension.) Stress gastritis was once a major clinical problem, but with improvements in medical therapy and prophylaxis it has virtually disappeared as an indication for emergency gastric surgery.[5] Other less common causes of hemorrhage include various benign and malignant neoplasms, aortoenteric fistula (which should *always* be suspected if the patient has an aortic graft in place), pancreatitis, hemobilia, AVMs, and splanchnic arterial aneurysms. Mesenteric venous thrombosis has produced upper GI hemorrhage in isolated cases.[10]

In an otherwise healthy patient, a standard indication for operative intervention is loss of at least 6 units of blood within 24 hours. Mortality for emergency surgery in cases of upper GI tract bleeding is as high as 23%.[5] Precise identification of the bleeding site and prompt treatment are of paramount importance. Reports of angiographic accuracy have ranged from 60% to 86%, and angiography is indicated when endoscopy fails or is unavailable, as long as the patient is not grossly unstable.[3]

Endoscopic electro-, thermal, or laser coagulation, or mucosal injection of ethanol or epinephrine are techniques capable of controlling many gastric bleeds, but endoscopy cannot reliably treat hemorrhage from vessels larger than 2 mm in diameter.[6] Large bleeding vessels are often amenable to occlusion with embolic particles. Catheter embolization is more effective for gastric bleeding than for duodenal lesions because of patterns of collateral arterial supply (see Chapter 12, Embolotherapy). Smaller vessel bleeding or diffuse gastritis may be treated with selective vasopressin infusion (see "Arterial Infusion of Vasopressin," later in this chapter).

Lower Gastrointestinal Bleeding

Only one patient in ten with GI bleeding will be found to have a source in jejunum, ileum, or colon.[5] An upper GI source must always be excluded before melena or bloody stools are attributed to a more distal abnormality. Patients with lower GI bleeding tend to be more elderly than those with gastric or duodenal lesions.

Acute Hemorrhage

In acute lower GI hemorrhage the role of endoscopy is not well established. If bleeding is massive and continuing, angiography is a primary diagnostic measure. In such a situation, angiographic accuracy is only 40% to 48%, but this is still considerably better than undirected surgical resection, which finds the underlying cause in only 30%.[2, 3] Localization of a bleeding source is of great importance, because limiting the extent of emergency surgery can drop operative mortality from 40% associated with subtotal colectomy to 13% with segmental resection.[2]

Colonic Diverticula

Colonic diverticula are responsible for roughly half of all cases of lower GI hemorrhage, and although most diverticula are found in the left colon, the great majority of bleeding diverticula are in the right colon. Diverticular bleeds are typically abrupt in onset, massive, and painless, and they are unlikely to recur.[11]

Angiodysplasia

Angiodysplasia is another common cause of acute bleeding. Although it is more commonly found in those over 60 years of age, angiodysplasia has been described in much younger patients, as well. These lesions, often multiple, are properly termed telangiectasias, for they represent dilation of preexisting vascular structures. They are vascular tangles of dilated mucosal or submucosal venules most commonly found in cecum or ascending colon. It has been postulated that angiodysplasia is caused by intermittent obstruction of venous outflow, but ischemia or toxic effects of bowel contents have also been implicated in its pathogenesis.[12]

For a positive diagnosis of bleeding from angiodysplasia to be made, actual extravasation must be seen. Up to 27% of all elderly patients may have angiodysplasia, but half of these also have colonic diverticula.[2] If extravasation is not detected, carcinoma or another lesion must always be excluded.

Meckel's Diverticulum

This embryologic remnant, present in distal ileum, has a prevalence of 2% in the adult population.[11] Because ectopic gastric mucosa is found in a large number of Meckel's diverticula, about one fourth of patients eventually present with lower GI hemorrhage. Bleeding is more common in children, but can develop at any age. Technetium pertechnetate radionuclide scans have been used to detect bleeding diverticula with good success.[3] The vitelline artery, a separate branch of the SMA, is a vessel specific to Meckel's diverticulum, but in 80% of diverticula this vessel has involuted and been replaced by a distal ileal branch.[13]

Other Lesions

Endoscopic polypectomy is an iatrogenic cause of acute bleeding. Carcinoma, polyps, or other tumors can bleed spontaneously, but such blood loss is more likely to be chronic and intermittent rather than massive. Inflammatory bowel disease, and small bowel or colonic ulcers must also be considered in the differential diagnosis.

Chronic Lower Gastrointestinal Bleeding

Barium and endoscopic examinations uncover all but a few neoplastic or inflammatory causes of intermittent, recurrent, or low-grade hemorrhage. Even so, about 5% of patients will not have a diagnosis made by conventional means.[14] Small bowel lesions are responsible for many of these cases, with angiodysplasia or AVMs eventually uncovered in 20% to 35%.[14, 15] Angiography is worth performing, and Lau et al. have reported high accuracy for lower GI hemorrhage of obscure origin.[1] Even in patients with initially negative studies, they found that repeat arteriographic examination could reliably establish the bleeding source. Rösch et al. have injected boluses of heparin and streptokinase to unmask bleeding sites in particularly refractory cases, but such intervention is risky and should be avoided.[16]

When a vascular malformation is present, the catheter should be left in place as the patient is taken to the surgery. Arterial injection of methylene blue is a valuable aid in the surgical localization of such lesions. Resected vascular malformations must be marked by a surgical suture because they can be easily overlooked by the pathologist.[8] Ideally, surgical specimens are examined histologically after arterial injection of a silicone or other polymer.

Angiographic Findings in Gastrointestinal Bleeding

The *sine qua non* of bleeding is extravasation of contrast material. This is seen as a puddling or staining that persists beyond the capillary and venous phases. Delayed films sometimes clearly show intestinal folds, if bleeding is brisk enough. Occasionally, the extravasated contrast material may accumulate between two folds and appear radiographically as a "pseudovein." For extravasation to be detectable, bleeding must exceed 0.5 mL/min.[17]

Few lesions responsible for hemorrhage have characteristic angiographic findings. Aneurysms, pseudoaneurysms, and frank AVMs such as those of the Rendu-Osler-Weber syndrome are easily recognized. Angiodysplasia, however, can be quite subtle in appearance. Criteria for diagnosis include vascular tufts appearing in the arterial phase of injection, early opacification of a draining vein, and delayed emptying of a dilated and tortuous intramural vein.[12] Aortoenteric fistula is not often diagnosed angiographically, but observation of a nipple-like projection or a pseudoaneurysm in the presence of a vascular graft is presumptive evidence of a fistula.[3]

Ulcerations, inflammatory lesions, diverticula, and neoplasms usually present nonspecific angiographic findings. Venous phase films must always be inspected for signs of portal or mesenteric vein thrombosis. Unsuspected esophageal or mesenteric varices can be detected on late films. Because bleeding can be very intermittent and spasm can prevent opacification of the extravasating vessel, repeat injection or repeat angiography may be warranted.[1]

Role of Radionuclide Studies

Radionuclide studies are not indicated in patients with massive continuous bleeding. However, if there is question whether bleeding has stopped, or if hemorrhage is intermittent, scans are quite useful. A red blood cell scan or sulfur colloid scan that is negative for disease is highly unlikely to be followed by an arteriogram showing extravasation.[4, 18] Also, a positive scan can guide angiography directly to the area and vessel most likely bleeding, minimizing the amount of contrast medium necessary. In a patient who may have hypotension or renal compromise, contrast dose is an important consideration.

Technetium-labeled sulfur colloid scans can enable one to detect rates of bleeding as low as 0.05 mL/min.[18] Because the tracer is rapidly extracted by liver and spleen, blood pool circulation time is only 12 to 15 minutes (still good in comparison with the few seconds iodinated contrast medium has to demonstrate a bleeding site). Background activity is quickly cleared, but the high number of counts originating in liver and spleen makes the study useless in cases of upper GI hemorrhage, and extravasation at the colonic flexures may be hard to detect.

Technetium-labeled red blood cell studies take more time to prepare, and they detect only greater amounts of bleeding, but they are still more sensitive than angiography.[4] Upper GI hemorrhage can be detected, and blood pool circulation lasts for many hours. At least 85% of positive studies require more than 1 hour of imaging.[4] One potential drawback is that if early imaging does not depict the bleeding site, activity found at a later time may represent blood that has entered bowel and has been propelled away from the bleeding source by peristalsis.

Technetium pertechnetate scanning is used when a Meckel's diverticulum is suspected to be bleeding. Tracer accumulates in gastric mucosa. Because a bleeding Meckel's diverticulum contains ectopic gastric tissue, a technetium scan is positive in up to 75% of cases.[3]

Arterial Infusion of Vasopressin

Direct arterial infusion of vasopressin has been used to treat various causes of GI bleeding. Once employed for variceal hemorrhage, selective infusion has given way to intravenous vasopressin, which is equally effective for acutely bleeding gastroesophageal varices.[19] Vasopressin is particularly effective for bleeding of gastric origin, achieving control with selective left gastric artery infusion in 82% of cases.[20] Infusion of the celiac axis or gastroduodenal arteries is much less likely to produce benefit.

Vasopressin is also useful for colonic diverticular bleeding, with control obtained in the great majority of patients. However, up to half of those controlled acutely are likely to bleed again.[2] Because emergency colonic surgery carries a high mortality, even temporary stabilization of the patient can reduce operative risk. Patients with angiodysplasia or bleeding of small bowel origin are less likely to respond to vasopressin infusion.[21]

When used intra-arterially, vasopressin should be started at 0.2 units/min and the angiographic results checked at 20 minutes. If vasoconstriction is not excessive and bleeding persists, the dose can be increased to 0.4 units/min. After bleeding stops, the infused dose should be tapered gradually over several hours to avoid a rebound hyperemia.[3] Saline may be infused for some time before the catheter is removed, to allow immediate retreatment should bleeding resume.

Vasopressin must be avoided in patients with ischemic heart disease. Therapy is associated with a 9% rate of serious complications, including cardiac, mesenteric, and acral ischemia; hyponatremia; and cerebral edema.[22] If the left gastric artery is being infused, particular attention must be paid to cardiac dysrrhythmias because phrenic branches of the left gastric artery can supply pericardium.[3] Vasopressin must also be avoided after embolotherapy, for bowel infarction is a possible result.

Therapeutic Embolization in Gastrointestinal Hemorrhage

Arterial embolization has been used with success in the gastric and colonic circulations in cases of bleeding not responding to other measures.[22, 23] Embolization has been safely applied to a number of lesions of the small bowel as well.[21] The minimum number of particles needed to stop hemorrhage should be injected as selectively as possible (but not into vasa recta), and embolization in conjunction with vasopressin infusion is not recommended.[22] Gelfoam, a temporary occluder, is the agent of choice. The major risk of bowel embolization is ischemia and stricture formation, but these risks must be balanced against the high mortality rate of emergency surgery.[24] Risk of ischemia is higher in patients with severe atherosclerosis or previous bowel surgery. Embolotherapy is especially useful in the treatment of bleeding pancreatic or hepatic pseudoaneurysms (see Chapter 12, Embolotherapy).[9, 25, 26]

MESENTERIC ISCHEMIA

Inadequate perfusion of the bowel can lead to a variety of symptoms. If there is chronic arterial insufficiency, the syndrome of intestinal angina is produced. Arteriography is an essential part of diagnostic evaluation, as it is in cases of acute arterial ischemia, whether from embolus, in-situ thrombus, or a low-flow state. Mesenteric vein thrombosis is more difficult to diagnose clinically than acute arterial ischemia, and the angiographic findings are usually indirect.

Intestinal Angina

Patients with intestinal angina complain of postprandial pain, nausea, vomiting, and diarrhea. Weight loss is marked, and many patients are cachectic in appearance. Symptoms may be so severe that the patient has a fear of eating. Unfortunately, complaints are not so clear-cut in many cases.

Chronic arterial insufficiency of bowel typically requires occlusion of at least two of the three celiac and mesenteric vessels, and the third must often be stenotic for symptoms to arise. Such extensive occlusive disease must be present because of the rich network of collateral vessels that is normally available. Atherosclerosis is commonly the underlying disease, but Takayasu arteritis, systemic lupus erythematosis, Wegener's granulomatosis, and a variety of other inflammatory conditions have also produced intestinal angina.[27]

Biplane abdominal aortography must be done to evaluate this condition. Injection is made at or above the level of the celiac axis at T-12. The lateral view best demonstrates the origins of the celiac and superior mesenteric arteries. A steep left posterior oblique projection may show the orifice of the IMA if it is not well seen on the biplane study. Note should be made of patterns of filling and development or absence of collaterals. A severe narrowing of the proximal celiac associated with angulation of the vessel (the so-called "median arcuate ligament compression" syndrome) is a common finding in healthy individuals, and its significance in symptomatic patients is questionable.

With proper patient selection, surgical revascularization has a 70% success rate, but a 3% to 8% mortality.[28] Odurny et al. have used balloon angioplasty to produce relief of symptoms for up to 2 years, but lesions recurred in many of their patients.[28] Median arcuate ligament compression and ostial lesions are not amenable to dilatation.

Acute Arterial Ischemia

Symptoms of acute ischemia include severe pain, nausea, vomiting, and diarrhea (with or without gross blood). Although acute mesenteric ischemia is an uncommon cause of hospital admission, mortality is quite high: 70% to 90%.[27] Angiography should be done early if the diagnosis is suspected and symptoms have persisted at least 2 hours. Most have arterial occlusion from thrombus or embolus (onset of pain is usually very abrupt in the latter case), and surgical intervention is indicated. However, 20% of patients will be found to have non-occlusive mesenteric ischemia.[10]

Non-occlusive ischemia is associated with low cardiac output; with use of digitalis, ergot alkaloids, propranolol, or vasoconstricting drugs; and with hypovolemia.[27] Patients are typically quite elderly. Angiographic findings consist of slow flow and prominent arterial spasm. Catheter infusion of papaverine directly into the SMA (30 to 60 mg/hr) has provided some benefit, but patients must be closely observed for signs of peritonitis. Surgery should be reserved for those developing bowel necrosis.

Mesenteric Vein Thrombosis

Mesenteric ischemia of venous origin is much less common than acute arterial ischemia. It also tends to have a more gradual onset and better prognosis, with mortality under 40%.[10] As mentioned earlier, upper GI bleeding may be seen, a symptom not found in ischemia of arterial cause. Patients with a predisposing condition for thrombosis, such as use of oral contraceptives, antithrombin III or protein C deficiency, neoplasm, or cirrhosis, generally do not do as well as those with no underlying disease.

Angiography may be nondiagnostic. Slow arterial flow, spasm, poor filling, and failure of venous opacification are typical findings.[27] Computed tomography or duplex ultrasound can be useful when the diagnosis is in doubt.[29]

Veins extending from the marginal mesenteric arcades are usually involved, so thrombectomy is of little value. Standard treatment consists of bowel resection and anticoagulation. There should be wide margins in the resection, because early recurrent ischemia is a common postoperative problem.[10]

PANCREATIC ANGIOGRAPHY

Computed tomography has proved quite accurate in the staging of pancreatic carcinomas, and angiography need not be performed in the great majority of cases.[30] Encasement of major arterial or portal venous branches makes a tumor unresectable. Care should be taken that focal spasm not be misinterpreted as encasement.

Angiography is still used for evaluation of patients with suspected islet cell tumors. These neoplasms tend to be slow growing and small. Aside from insulinomas, most endocrine pancreatic tumors are malignant.[31] Although many lesions can be identified by means of CT and ultrasound, their accuracy is limited by the small tumor size. Arteriography is most valuable when other imaging has been negative or equivocal, and arteriography can enable detection of multiple lesions, hepatic metastases, and vascular encasement.

Islet cell tumors tend to be circumscribed and have a dense, homogeneous capillary blush.[31] Technique is critical in their demonstration, and superselective catheterization and distention of the stomach with gas are recommended to improve accuracy. Intestinal mucosal blushes and accessory spleens are potential sources of confusion. Oblique projections can be helpful; CT-angiography can provide additional information when angiographic findings are equivocal. Even so, it has been found that all but a few tumors detected angiographically can be palpated at surgical exploration.[32] Intraoperative ultrasound may be able to localize lesions that cannot be palpated, so the application of angiography for islet cell tumors may diminish in the future. Preoperative localization may be restricted to patients submitting to reexploration.

MESENTERIC VENOUS SAMPLING FOR ISLET CELL TUMORS

An invasive examination that in some hands has been even more accurate than arteriography for islet cell tumor localization is selective mesenteric venous sampling for hormonal assay.[32] The approach is identical to transhepatic cannulation of the portal vein for direct portography, but the catheter tip is advanced into the splenic vein. Blood samples are obtained at 10- to 15-mm intervals as the catheter is withdrawn toward the liver. Accuracy is greatest if small tributary veins from the pancreas are sampled and the hormone levels are carefully mapped.[31] With the advent of intraoperative ultrasound, such invasive sampling is presently rarely needed for localization.

MISCELLANEOUS CONDITIONS

Splenic Artery Aneurysms

These are the most common non-aortoiliac aneurysms in the abdomen, but incidence is less than 1 in 1,000 in the general population.[33] They are relatively more common in younger women than other aneurysms. Splenic artery aneurysms may rupture, embolize, or thrombose. Because of the high mortality from rupture, all symptomatic lesions should be resected, and asymptomatic aneurysms exceeding 5 mm diameter in patients under 60 years of age should also be treated. Patients at great risk for surgery may be managed by catheter embolization.[34] Aneurysms in pregnant women or women of childbearing age pose particular hazards for rupture and should be aggressively treated.

Splenic Trauma

Angiography is reserved for suspected cases of primary vascular injury. The spleen often shows a mottled parenchymal blush, which should not be confused with contusion or laceration. Barely half of lesions found at surgical pathologic examination are detected by splenic angiography.[35] However, catheterization allows bleeding to be controlled by embolotherapy if extravasation is found. Embolization should be very selective in such instances, with as much splenic pulp preserved as possible.

REFERENCES

1. Lau WY, Ngan H, Chu KW, et al: Repeat selective visceral angiography in patients with gastrointestinal bleeding of obscure origin. *Br J Surg* 1989; 76:226–229.
2. Leitman IM, Paull DE, Shires GT: Evaluation and management of massive lower gastrointestinal hemorrhage. *Ann Surg* 1989; 209:175–180.

3. Kadir S, Ernest CB: Current concepts in angiographic management of gastrointestinal bleeding. *Curr Probl Surg* 1983; 20:281–343.

4. Winzelberg GG, McKusik KA, Froelick JW, et al: Detection of gastrointestinal bleeding with 99m Tc-labeled red blood cells. *Semin Nucl Med* 1982; 12:139–146.

5. Greenburg AG, Saik RP, Bell RH, et al: Changing patterns of gastrointestinal bleeding. *Arch Surg* 1985; 120:341–344.

6. Gostout CJ: Acute gastrointestinal bleeding: A common problem revisited. *Mayo Clin Proc* 1988; 63:596–604.

7. Petrini JL Jr: Endoscopic therapy for gastrointestinal bleeding. *Postgrad Med* 1988; 84:239–244.

8. Schwartz RW, Hagihara PF, Griffin WO Jr: Intraoperative endoscopy for recurrent gastrointestinal bleeding. *South Med J* 1988; 81:1106–1108.

9. Savastano S, Feltrin GP, Miotto D, et al: Vascular parenchymal sources of upper gastrointestinal bleeding. *Acta Radiol* 1989; 30:39–43, Fasc 1.

10. Clavien P-A, Dürig M, Harder F: Venous mesenteric infarction: A particular entity. *Br J Surg* 1988; 75:252–255.

11. Lawrence MA, Hooks VH, Bowden TA Jr: Lower gastrointestinal bleeding: A sytematic approach to classification and management. *Postgrad Med* 1989; 85:89–100.

12. Hemingway AP: Angiodysplasia: Current concepts. *Postgrad Med J* 1988; 64:259–263.

13. Bree RL, Reuter SR: Angiographic demonstration of a bleeding Meckel's diverticulum. *Radiology* 1973; 108:287–288.

14. Thompson JN, Hemingway AP, McPherson GAD, et al: Obscure gastrointestinal hemorrhage of small-bowel origin. *Br Med J* 1984; 288:1663–1665.

15. Monk JE, Smith BA, O'Leary JP: Arteriovenous malformations of the small intestine. *South Med J* 1989; 82:18–22.

16. Rösch J, Feller FS, Wawrukiewicz AS, et al: Pharmacoangiography in the diagnosis of recurrent massive lower gastrointestinal bleeding. *Radiology* 1982; 145:615–619.

17. Peterson WL: Obscure gastrointestinal bleeding. *Med Clin North Am* 1988; 72:1169–1176.

18. Alavi A: Detection of gastrointestinal bleeding with 99m Tc-sulfur colloid. *Semin Nucl Med* 1982; 12:126–138.

19. Joffe SN: Non-operative management of variceal bleeding. *Br J Surg* 1984; 71:85–91.

20. Eckstein MR, Kelemouridis V, Athanasoulis CA, et al: Gastric bleeding: Therapy with intraarterial vasopressin and transcatheter embolization. *Radiology* 1984; 152:643–646.

21. Palmaz JC, Walter JF, Cho KJ: Therapeutic embolization of the small-bowel arteries. *Radiology* 1984; 152:377–382.

22. Uflacker R: Transcatheter embolization for treatment of acute lower gastrointestinal bleeding. *Acta Radiol* 1987; 28:425–430, Fasc 4.

23. Rösch J, Keller FS, Kozak B, et al: Gelfoam powder embolization of the left gastric artery in treatment of massive small-vessel gastric bleeding. *Radiology* 1984; 151:365–370.

24. Mitty HA, Efremidis S, Keller RJ: Colonic stricture after transcatheter embolization for diverticular bleeding. *AJR* 1979; 133:519–521.

25. Tegtmeyer CJ, Bezirdjian DR, Ferguson WW, et al: Transcatheter embolic control of iatrogenic hematobilia. *Cardiovasc Intervent Radiol* 1981; 4:88–92.

26. Huizinga WKJ, Kalideen JM, Bryer JV, et al: Control of major hemorrhage associated with pancreatic pseudocysts by transcatheter arterial embolization. *Br J Surg* 1984; 71:133–136.

27. Hunter GC, Guernsey JM: Mesenteric ischemia. *Med Clin North Am* 1988; 72:1091–1115.

28. Odurny A, Sniderman KW, Colapinto RF: Intestinal angina: Percutaneous transluminal angioplasty of the celiac and superior mesenteric arteries. *Radiology* 1988; 167:59–62.

29. Mildenberger P, Jenny E, Schild H: Nachweis einer thrombose der vena mesenterica superior nach splenektomie mittels real-time- und dopplersonographie. *Radiologe* 1988; 28:395–398.

30. Jafri SZH, Aisen AM, Glazer GM, et al: Comparison of CT and angiography in assessing resectability of pancreatic carcinoma. *AJR* 1984; 142:525–529.

31. Rossi P, Allison DJ, Bezzi M, et al: Endocrine tumors of the pancreas. *Radiol Clin North Am* 1989; 27:129–161.

32. Proye C, Boissel P: Preoperative imaging versus intraoperative localization of tumors in adult surgical patients with hyperinsulinemia: A multicenter study of 338 patients. *World J Surg* 1988; 12:685–689.

33. Greene DR, Gorey TF, Tanner WA, et al: The diagnosis and management of splenic artery aneurysms. *J R Soc Med* 1988; 81:387–388.

34. Baker KS, Tisnado J, Cho S-R, et al: Splanchnic artery aneurysms and pseudoaneurysms: Transcatheter embolization. *Radiology* 1987; 163:135–139.
35. Casarella WJ, Martin EC: Angiography in the management of abdominal trauma. *Semin Roentgenol* 1984; 19:321–327.

7 | Thoracic Aortography and Bronchial Angiography

KEY CONCEPTS

1. Aortic laceration should be suspected in anyone with major deceleration injury and mediastinal widening.
2. The most common sites of aortic injury are the aortic isthmus, ascending aorta, and diaphragmatic hiatus.
3. Dissection is the most common acute primary aortic disease, with an incidence two to three times that of ruptured abdominal aortic aneurysm.
4. In 13% of aortic dissections no false lumen or intimal flap can be identified on aortography. Flattening of the true lumen and wall thickening are indirect signs.
5. Atherosclerotic aneurysms less than 5 cm diameter are unlikely to rupture, but patients with Marfan's syndrome, mycotic aneurysms, or luetic aneurysms should have elective repair of even small, asymptomatic lesions.

INDICATIONS

Thoracic aortography remains the definitive diagnostic study for the detection of acute and chronic aortic injury, as well as for the preoperative delineation of aortic aneurysms and dissections. It defines the relationship of aortic abnormalities to the heart and branch vessels as no cross-sectional imaging technique can. Thoracic angiography is also used to clarify the nature of para-aortic masses which might represent aortic diverticula, pulmonary sequestrations, or in rare cases, invading neoplasms.

Bronchial/intercostal arteriography is mainly reserved for determining the source of recurrent and massive hemoptysis, with embolization as a treatment option. Occasionally it is employed in the preoperative evaluation of chronic pulmonary embolism.

AORTIC TRAUMA

Clinical Aspects

The thoracic aorta is susceptible to injury by rapid deceleration or crushing, and aortic tears are a common cause of death in high-speed motor vehicle accidents.[1, 2] The majority of those with aortic rupture do not survive to reach medical care. However, those patients that are admitted with acute injury are at a 40% to 50% risk of exsanguination within 24 hours if the diagnosis is not made and surgical treatment begun.[1, 3] Risk of death is as high as 90% by 1 month after injury.[4] Although some may survive hospitalization and be discharged without aortic injury suspected, they are prone to developing chronic pseudoaneurysms, which may enlarge and rupture as long as 49 years later.[5, 6]

Aortic disruption may be limited to the intima or it may involve the entire thickness of the vessel wall. Extent of injury ranges from only a discrete portion of the aortic circumference to complete transection. The site of tear most likely to be diagnosed angiographically is at the aortic isthmus, where the ligamentum arteriosus joins the pulmonary artery and aorta. Here the relatively mobile aortic arch joins the descending aorta, which is fixed by investing fascia. With sudden deceleration, great shear stress can be generated at this junction. Of all aortic disruptions diagnosed, 80% to 86% are found at the isthmus.[2, 4, 7] The next most likely site of angiographically diagnosed disruption is the aortic root, with an incidence of about 9%.[7] The number of traumatic injuries to the ascending aorta is actually higher, but such injuries are more likely to be fatal, and fewer patients come to aortography. A small number of tears occur at the level of the diaphragm, and imaging studies should always extend to this area.

Plain Radiographic Findings

A large number of signs have been described as suggesting aortic injury on chest radiographs. Chief among these is mediastinal widening, measured as a ratio of superior mediastinal to chest wall width.[3] A ratio of 0.25 to 0.28 on a chest radiograph with the patient supine should raise the suspicion of aortic injury. Greater ratios are more specific (but less sensitive) for aortic rupture. Care must be taken in interpreting films with any degree of patient rotation. Many apparent cases of mediastinal widening will resolve on an anteroposterior view of the chest with the patient upright and leaning several degrees forward, but such a study is not always possible to obtain. It should be noted that only a minority of mediastinal hematomas are actually related to aortic laceration.

Other plain radiographic signs reported include shift of a nasogastric tube to the right, depression of the left bronchus, tracheal deviation, and presence of an apical cap. Pneumothorax, hemothorax, pulmonary contusion, and rib or clavicular fractures are indicative of major trauma, but do not have specific value in predicting aortic injury.[3] None of the signs listed has high accuracy alone, but in 49 confirmed cases of disruption, Stark et al. observed that not a single chest study was within normal limits.[2] The consequences of a missed diagnosis are so grave that if any chest abnormality is seen or if clinical suspicion is high enough, aortography is justified. One should expect to perform eight or nine negative aortograms for every positive arteriogram; otherwise, some patients may remain undiagnosed.[2, 3]

Arteriographic Technique

Thoracic aortography should be performed with a high-flow pigtail catheter, 6 F or larger in size, placed within 1 to 2 cm of the aortic valve. Injection should be of a sufficient rate and volume to ensure that the valve cusps are well defined on filming (30 mL/sec for 2 seconds is adequate for most adults). Filming must be rapid (two to three exposures per second) and should be carried out well into the "washout" phase to detect persistent puddling or extravasation. Biplane filming is advantageous, because *at least two views of the aorta are mandatory* before a laceration can be excluded. If there is any question of intimal irregularity or transection, other views must be obtained. Although rapid-frame digital subtraction angiography has been successfully applied in cases of aortic injury,[1] the large field of view, maximum spatial resolution, and capability for biplane study make conventional film arteriography the standard for diagnosis.

If a femoral approach is used for catheter placement, any resistance to advancement of catheter or wire should raise concern, and in the absence of other disease such resistance is diagnostic of aortic disruption.[8] Deaths have been reported after power injection of contrast material directly into traumatic pseudoaneurysms.

Findings of aortic trauma range from minimal wall irregularity representing intimal tear to complete transection with contained or uncontained extravasation from the lumen. Bronchial artery origins and atherosclerotic or other mural irregularities sometimes pose diagnostic problems. Usually these can be resolved angiographically, but at times thoracic exploration is unavoidable. A ductus bulge or diverticulum has a fairly characteristic appearance and should not be a cause for confusion (see the following subsections). When a true tear is discovered, all film studies must be carefully examined. Multiple arterial lacerations are not rare, especially if the aortic injury is at a site other than the isthmus.[4]

Chronic Aortic Pseudoaneurysms

Patients with this condition may present with pain, hoarseness, or dyspnea months to years after injury.[5] Many chronic aortic pseudoaneurysms are discovered incidentally on chest radiographs. A focal bulge from the ventral aspect of the proximal descending thoracic aorta is characteristic, but some pseudoaneurysms can affect the ascending aorta.[7] Wall calcification is common, but mural thrombus is not.[5] Review of radiographs from the time of trauma usually shows mediastinal widening in retrospect. Because of the distinct tendency of these lesions to rupture and the low operative mortality for elective repair, it is generally recommended that all chronic pseudoaneurysms of the thoracic aorta be treated surgically.

AORTIC DIVERTICULA

There are various forms of aortic diverticula. One consists of a bulbous widening to the origin of an aberrant right subclavian artery, which represents incomplete regression of the primitive right aortic arch.[9] A much rarer diverticulum is the mirror image of this variant in a right aortic arch. The most common aortic diverticulum is thought by most to be a remnant of the ductus arteriosus, seen as a bulge at the anteromedial aspect of the aortic isthmus. It is a variant of normal in infants, regressing with time so that a focal bulge at the isthmus persists into adulthood in only 9% of the general population.[10] Ductus diverticula are distinguished from aortic lacerations and pseudoaneurysms by their smooth margins and lack of a delay in washout of iodinated contrast material.

The differential diagnosis of masses in the aortopulmonary window must include ductus arteriosus aneurysm.[11] Ductus aneurysms are likely to contain mural thrombus, a feature distinguishing them from the chronic pseudoaneurysms of trauma.[12] Like chronic pseudoaneurysms, they often show peripheral calcification. Diagnosis is important, for untreated ductus aneurysms incur a high mortality.

AORTIC DISSECTION

Predisposing Factors and Clinical Presentation

Aortic dissection is the most common acute disease affecting the aorta, and its incidence is two to three times that of ruptured abdominal aortic aneurysm.[13] In fact, up to 5% of patients admitted with a clinical diagnosis of acute myocardial infarction may actually be suffering aortic dissection. The great majority of patients have a history of hypertension. Those rare cases presenting under age 40 tend to have Marfan's syndrome, coarctation of the aorta, or involve pregnant women.

Cystic medial necrosis, a degenerative condition of the aortic wall, was long thought responsible for dissection. However, Wilson and Hutchins reviewed the autopsies of over 204 individuals with aortic dissection and found only 10% had evidence of cystic medial necrosis.[14] Others have done case control studies, finding that those with aortic dissection are no more likely than patients dying from other causes to have cystic medial necrosis. Pathogenesis of dissection is no doubt complex and multifactorial. Biomechanical stress in those with hypertension or iatrogenic trauma can cause intimal disruption. Abnormal connective tissue in conditions such as Marfan's syndrome has been implicated. More than 10% of cases may not have *any* tear found at autopsy, and dissection may possibly be initiated by bleeding into vasa vasorum in these instances.

Typical symptoms include severe chest pain of abrupt, rather than gradual, onset. However, pain may be localized to the back, abdomen, or neck. Patients may complain of dyspnea, hoarseness, nausea and vomiting, or anuria. Those with extension of dissection into the aortic root may have systolic murmurs of aortic insufficiency. Peripheral pulses can be unequal, damped, or absent. Death results from pericardiac tamponade, secondary myocardial or cerebral infarction, rupture, or from mesenteric or renal ischemia.

Dissection vs. Aneurysm vs. Pseudoaneurysm

The widely used term "dissecting aneurysm" is a misnomer, because most acute dissections occur in an aorta of normal size (i.e., one that is *not* aneurysmal). Dissection is just that—a dissection or separation of arterial wall layers. Separation of aortic wall layers is most likely to occur in the outer third of the media.[13] Pusatile blood works to extend a dissection.

An aneurysm is focal dilatation of a vessel. All three layers of aortic wall (intima, media, and adventitia) are intact in an uncomplicated aortic aneurysm (Fig 7–1). A pseudoaneurysm is a disruption of the

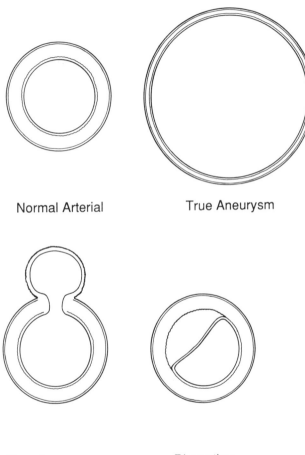

Normal Arterial   True Aneurysm

Pseudoaneurysm   Dissection

FIG 7–1.
Differentiation of aneurysm, pseudoaneurysm, and dissection.

arterial wall. This may be limited to intima (without the layer separation found in dissection) with bulging of the other layers, or it may involve all layers. In the latter case, the pseudoaneurysm can also be termed a contained hematoma.

Classification and Treatment of Dissections

The traditional classification of aortic dissection was proposed by deBakey[13]: type I, both ascending and descending thoracic aorta involved; type II, confined to ascending aorta and arch; type III, only descending or abdominal aorta affected. A more practical modification was proposed by Daley in 1970: type A, dissections involving ascending aorta; type B, all other aortic dissections[13] (Fig 7–2). Daley's classification is more relevant to treatment, because proximal dissections are rapidly fatal if not treated surgically, usually by graft placement. Those with uncomplicated type B dissections respond much better to medical management with propranolol and other antihypertensive medications, and 80% now survive to be discharged from the hospital.[15] With the drop in surgical mortality resulting from increased experience and improved technique, some advocate surgery even for type B patients, because long-term conservative treatment can result in aneurysm, rupture, or redissection.

The great majority of dissections arise from a tear in the ascending aorta or arch. However, in defining the type of dissection present *location of an intimal tear is irrelevant*. For example, a dissection can propagate in a retrograde fashion from a tear in the descending aorta to involve the aortic root or great vessel origins. It is the extension of a dissection into the great vessels, coronary arteries, or pericardium that determines prognosis. Resection or repair of a tear does not play a critical role in surgical treatment.[13]

Noninvasive Imaging

The most common plain film finding in aortic dissection is mediastinal widening. There may be an increased diameter to the aortic knob, "double density," or deviation of the trachea.[16] Pleural effusion or cardiac enlargement may represent rupture into the pleural space or pericardium. Separation of intimal calcification from the margin of the aortic knob by more than 4 mm is a sign suggestive of, but not entirely specific for, dissection. An unequivocally normal chest film makes the diagnosis less likely but does not exclude it. About 6% of cases show no abnormality on plain radiographs.[17]

Echocardiography can assist one in making the diagnosis of dissection when it involves the aortic root, and in many patients a length of the descending aorta can also be evaluated.[18, 19] Still, spurious echoes

Aortic Dissection

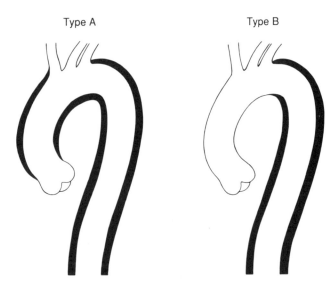

Type A   Type B

FIG 7–2.
Simplest clinically relevant classification of aortic dissections. *Type A*
involves ascending aorta or arch, whether or not the dissection extends
into the descending aorta. *Type B* involves descending aorta *only*.

can be misinterpreted as intimal flaps.[18, 20] Computed tomography (CT)
is a more reliable noninvasive study, allowing more complete depiction
of the aorta than sonography. The protocol of Thorsen et al. specifies
rapid injection of intravenous contrast medium and repeated scanning
(angio-CT) at three levels: aortic root, mid-ascending aorta, and aortic
arch.[21] They have found CT to be more accurate than angiography in
the diagnosis of dissection. Even so, if a patient with suspected dissection is a surgical candidate, angiography should take precedence,
with CT reserved for those cases with equivocal angiographic findings.[15, 22] Angiography remains the best procedure for demonstrating
occlusion of branch vessels and aortic regurgitation, both important
for planning surgery.

Angiographic Findings and Accuracy

Aortography can be performed by a femoral route in patients with preserved femoral pulses. The pigtail catheter must be positioned above the aortic valve to demonstrate any possible regurgitation or coronary artery occlusions consequent to dissection. Rapid injection of contrast material with rapid biplane filming remains the standard for diagnosis. Filming should be carried out at least 12 to 15 seconds after injection to allow for delayed filling or slow washout of either true or false lumen. If a test injection shows slow flow, the power injection rate should be modified accordingly. Multiple projections are mandatory, for an intimal flap or multiple lumina may not be recognized in one projection. Digital subtraction arteriography is more appropriate for postoperative or long-term follow-up of patients with known dissection.[23] If thoracic aortography fails to confirm the presence of a dissection, biplane abdominal study should still be performed, because primary abdominal dissections do occur.

The *sine qua non* of aortic dissection is the demonstration of an intimal flap, separating a true and false lumen. The lumina may opacify sequentially or simultaneously. In the review of Earnest et al., the false lumen opacified in 87% of aortograms.[16] Even when the false lumen fails to fill, if the true lumen is compressed or if widening of the extraluminal border exceeds 10 mm, the diagnosis can be made with reasonable certainty. However, an aneurysm with mural thrombus or with leakage into the mediastinum can have a similar appearance. Abnormal position of the angiographic catheter (which should follow the outside curve of the aorta) may suggest aortic dissection. Dissections tend to follow the outer curve of the aorta, and they commonly extend into the left renal artery when the abdominal aorta is involved. Care must be taken that prolonged washout of dependent iodinated contrast media in the descending aorta of a patient with isolated aortic insufficiency is not misinterpreted.[24] At times a dissection may have the appearance of a mural "ulcer."[24, 25]

False lumen injection will not opacify the sinuses of Valsalva in a type A dissection. Power injection into a false lumen does not appear to present any extraordinary risk, but as noted earlier, the injection rate should be modified to match observed flow. Precise differentiation of false lumen as opposed to true lumen generally has little bearing on diagnosis or treatment. The false lumen often provides blood to kidneys or other viscera. In this regard it should be noted that therapy is directed at prevention of extension rather than obliteration of the false lumen. Repeated observations have shown that a distal false channel commonly persists after proximal surgical intervention.[6, 26]

Accuracy of arteriography for acute dissection has been cited as 95% to 99%.[15] In their review of 55 aortograms determined to be negative, Eagle et al. found that many patients were ultimately determined to have primary myocardial infarction, aneurysm, aortic valve disease, or pericarditis.[27] However, four patients were found to have dissections, and the lower limit of angiographic false negative studies was calculated to be 2% to 4%. The diagnosis of chronic dissection can be problematic if the false lumen has thrombosed, and many cases may be angiographically indistinguishable from aneurysms. Whether the condition is acute or chronic, if angiographic findings are equivocal, angio-CT or magnetic resonance imaging can often resolve the issue.[28]

ANEURYSMS OF THE THORACIC AORTA

Atherosclerotic

Intrathoracic aneurysms most often affect the descending aorta, and most true aneurysms are atherosclerotic in origin. As with lesions in other arteries, thoracic aneurysms may produce symptoms by compression of adjacent structures, distal embolization of mural thrombus, occlusion, or rupture. Among the various clinical presentations are pain, stridor, dysphagia, superior vena cava syndrome, hoarseness, or aortic insufficiency.[6] In rare cases, hemoptysis may be due to aneurysm-induced aortobronchopulmonary fistula.[29] Risk of rupture is related to size in atherosclerotic lesions, almost negligible in those aneurysms smaller than 5 cm, but over 40% if the aneurysm exceeds 10 cm in diameter.[6] The size and stability of thoracic aneurysms can be economically and noninvasively assessed by serial chest radiographs.

Thoracic or thoracoabdominal aneurysms are much more difficult to treat than abdominal aortic aneurysms, and surgical mortality remains high. One of the greatest risks in elective surgery for descending thoracic lesions (whether they are aneurysms, dissections, or traumatic lesions) is spinal cord ischemia and paraplegia. The artery of Adamkiewicz usually arises between the T-8 and L-1 levels. Reattachment of large intercostal or lumber vessels during graft placement can decrease the incidence of paraplegia to about 16% in extensive resections.[30] However, preoperative angiographic identification of the spinal artery and its origin has not been shown to be a useful measure.[31]

Mycotic, Luetic, and Other Aneurysms

Mycotic aneurysms are typically eccentric saccular lesions that often involve the aortic root and have a high tendency to rupture. They usually represent a complication of bacterial endocarditis or other dis-

tant infection. Organisms most commonly implicated are staphylococ-
cal, streptococcal, or enterococcal species.[32] Syphilitic aneurysms, much
more common in the preantibiotic era, tend to affect the ascending
aorta and arch. Because aneurysms due to syphilis and other micro-
organisms may enlarge suddenly and rupture, prompt elective repair
is indicated, even if the lesions are small and asymptomatic.

Marfan's syndrome not only predisposes to aortic dissection but also
to true aneurysms of the ascending aorta. Surgery is recommended as
soon as dilation becomes evident. Sinus of Valsalva aneurysms are
associated with a subendocardial cushion defect and high ventriculo-
septal defect.[6] They most often arise in the right sinus and must be
distinguished from mycotic lesions.

TAKAYASU AORTITIS

Patients with Takayasu arteritis are relatively young, more often
women, and are likely to have symptoms and serologic abnormalities
suggesting inflammatory disease.[33] The condition is a panarteritis of
unknown cause, affecting the vasa vasorum of elastic arteries. Although
it is one of the most common vascular diseases of the Orient, its incidence in North America and Europe is low.

About one third of patients have irregular lesions difficult to distinguish angiographically from atherosclerosis. More typically, long,
smooth, concentric, segmental stenoses of the thoracic aorta are present. Proximal portions of the brachiocephalic vessels are also usually
involved, as are the abdominal aorta, superior mesenteric artery, and
renal arteries.[34] The pulmonary circulation is often affected by the
disease, and many patients demonstrate abnormalities on perfusion
scans.

Aside from stenoses, mild ectasia and small cystic protrusions from
the arterial wall can be seen. Renovascular hypertension is a common
complication of Takayasu arteritis, and it poses the greatest risk from
the disease. Remarkably, cerebral and mesenteric ischemia are rare.[34]

PULMONARY SEQUESTRATION

Pulmonary sequestration is a developmental anomaly in which a
portion of the lung is partially or completely isolated from the pulmonary circulation (with or without a separate pleural investment).
The most common location for a sequestration is on the left side in the
vicinity of the posterior basal segment.[35] Pulmonary sequestration can

present as a asymptomatic para-aortic mass, rarely containing calcification.[36] Symptomatic lesions are commonly infected, and resection is indicated for these cases.

Arterial supply generally comes from the aorta, sometimes from below the diaphragm. Multiple arteries may be seen in 20%, and the vessels usually traverse the pulmonary ligament.[37] Aortography is recommended both to confirm the diagnosis and to delineate the supplying and draining vessels.

BRONCHIAL AND INTERCOSTAL ANGIOGRAPHY

Bronchial and intercostal arteriography are most useful in the diagnosis and embolic treatment of refractory massive hemoptysis (see Chapter 12, Embolotherapy). Tuberculosis, sarcoidosis, aspergilloma, and bronchiectasis are among the disorders responsible for such hemoptysis. Systemic arterial branches can also enlarge to provide collateral blood supply to areas of lung affected by chronic pulmonary embolism, and angiography is useful prior to surgical treatment of such disease.[38]

Bronchial arteries have a highly variable distribution. Most commonly there is a single right intercostobronchial trunk, and one or two left bronchial arteries, but this pattern is present in fewer than half of patients.[39] The bronchial arteries usually originate from the descending thoracic aorta near the level of T-5 or T-6, and they are not normally seen peripheral to the hilus.[40] Cobra, Simmons, or Mikaelsson catheter configurations are most useful for selective catheterization of intercostal or bronchial arteries.

Low-osmolar contrast media and careful injections of small volumes are advised in the study of these vessels. Although the risk of spinal cord injury is low, the anterior spinal artery can originate from bronchial or intercostal arteries. Paraplegia remains a possible complication.

REFERENCES

1. Mirvis SE, Pais SO, Gens DR: Thoracic aortic rupture: Advantages of intraarterial digital subtraction angiography. *AJR* 1986; 146:987–991.
2. Stark P, Cook M, Vincent A, et al: Traumatic rupture of the thoracic aorta: A review of 49 cases. *Radiologe* 1987; 27:402–406.

3. Mirvis SE, Bidwell JK, Buddemeyer EU, et al: Imaging diagnosis of traumatic aortic rupture: A review and experience at a major trauma center. *Invest Radiol* 1987; 22:187–196.
4. Fleckenstein JL, Schultz SM, Miller RH: Serial aortography assesses stability of "atypical" aortic arch ruptures. *Cardiovasc Intervent Radiol* 1987; 10:194–197.
5. Heystraten FM, Rosenbusch G, Kingma LM, et al: Chronic posttraumatic aneurysm of the thoracic aorta: Surgically correctable occult threat. *AJR* 1986; 146:303–308.
6. Althaus U, Marincek B: Thorakale aortenaneurysmen. *Schweiz Med Wochenschr* 1984; 114:1547–1559.
7. Lundell CJ, Quinn MF, Finck EJ: Traumatic laceration of the ascending aorta: Angiographic assessment. *AJR* 1985; 145:715–719.
8. LeBerge JM, Jeffrey RB: Aortic lacerations: Fatal complications of thoracic aortography. *Radiology* 1987; 165:367–369.
9. Salomonowitz E, Edwards JE, Hunter DW, et al: The three types of aortic diverticula. *AJR* 1984; 142:673–679.
10. Goodman PC, Jeffrey RB, Minagi H, et al: Angiographic evaluation of the ductus diverticulum. *Cardiovasc Intervent Radiol* 1982; 5:1–4.
11. Traughber PD, Wojtowycz M, Karwande SV, et al: Roentgenologic CPC: Enlarging mediastinal mass. *Invest Radiol* 1987; 22:240–243.
12. Mitchell RS, Seifert FC, Miller DC, et al: Aneurysm of the diverticulum of the ductus arteriosus in the adult. *J Thorac Cardiovasc Surg* 1983; 86:400–408.
13. Anagnostopoulos CE: *Acute Aortic Dissections*. Baltimore, University Park Press, 1975, pp 1–249.
14. Wilson SK, Hutchins GM: Aortic dissecting aneurysms: Causative factors in 204 subjects. *Arch Pathol Lab Med* 1982; 106:175–180.
15. DeSanctis RW, Doroghazi RM, Austen WG, et al: Aortic dissection. *N Engl J Med* 1987; 317:1060–1067.
16. Earnest F, Muhm JR, Sheedy PF Jr: Roentgenographic findings in thoracic aortic dissection. *Mayo Clin Proc* 1979; 54:43–50.
17. Kaufman SL, White RI Jr: Aortic dissection with "normal" chest roentgenogram. *Cardiovasc Intervent Radiol* 1980; 3:103–106.

18. Victor MF, Mintz GS, Kotler MN, et al: Two dimensional echocardiographic diagnosis of aortic dissection. *Am J Cardiol* 1981; 48:1155–1159.

19. Come PC: Improved cross-sectional echocardiographic technique for visualization of the retrocardiac descending aorta in its long axis. *Am J Cardiol* 1983; 51:1029–1032.

20. Kolettis M, Toutouzas P, Avgoustakis D: False echocardiographic diagnosis of aortic root dissection in case of abdominal aortic dissection. *Br Heart J* 1981; 45:602–604.

21. Thorsen MK, SanDretto MA, Lawson TL, et al: Dissecting aortic aneurysms: Accuracy of computed tomographic diagnosis. *Radiology* 1983; 148:773–777.

22. Dee P, Martin R, Oudkerk M, et al: The diagnosis of aortic dissection. *Curr Probl Diagn Radiol* 1983; 12:1–55.

23. Guthaner DF, Miller DC: Digital subtraction angiography of aortic dissection. *AJR* 1983; 141:157–161.

24. Shuford WH, Sybers RG, Weens HS: Problems in the aortographic diagnosis of dissecting aneurysm of the aorta. *N Engl J Med* 1969; 280:225–231.

25. Hekali P, Velt P, Gutierrez O, et al: Radiology of aortic dissection: Pitfalls in diagnosis. *Eur J Radiol* 1986; 6:314–318.

26. Pinet F, Froment JC, Guillot M, et al: Prognostic factors and indications for surgical treatment of acute aortic dissections: A report based on 191 observations. *Cardiovasc Intervent Radiol* 1984; 7:257–266.

27. Eagle KA, Quertermous T, Kritzer GA, et al: Spectrum of conditions initially suggesting acute aortic dissection but with negative aortograms. *Am J Cardiol* 1986; 57:322–326.

28. Akins EW, Carmichael MJ, Hill JA, et al: Preoperative evaluation of the thoracic aorta using MRI and angiography. *Ann Thorac Surg* 1987; 44:499–507.

29. Coblentz CL, Sallee DS, Chiles C: Aortobronchopulmonary fistula complicating aortic aneurysm: Diagnosis in four cases. *AJR* 1988; 150:535–538.

30. Pokela R, Karkola P, Tarkka M, et al: Surgery of thoracoabdominal aortic aneurysms. *Scand J Thorac Cardiovasc Surg* 1984; 18:179–189.

31. Fereshetian A, Kadir S, Kaufman SL, et al: Digital subtraction spinal cord angiography in patients undergoing thoracic aneurysm surgery. *Cardiovasc Intervent Radiol* 1989; 12:7–9.

32. Castañeda-Zuñiga WR, Nath PH, Zollikofer C, et al: Mycotic aneurysm of the aorta. *Cardiovasc Intervent Radiol* 1980; 3:144–149.

33. Yamato M, Lecky JW, Hiramatsu K, et al: Takayasu arteritis: Radiographic and angiographic findings in 59 patients. *Radiology* 1986; 161:329–334.

34. Liu YQ: Radiology of aortoarteritis. *Radiol Clin North Am* 1985; 23:671–688.

35. Khalil KG, Kilman JW: Pulmonary sequestration. *J Thorac Cardiovasc Surg* 1975; 70:928–937.

36. Wojtowycz M, Gould HR, Atwell DT, et al: Calcified bronchopulmonary sequestration. *J Comput Tomogr* 1984; 8:171–173.

37. Bolman RM, Wolfe WG: Bronchiectasis and bronchopulmonary sequestration. *Surg Clin North Am* 1980; 60:867–882.

38. Mills SR, Jackson DC, Sullivan DC, et al: Angiographic evaluation of chronic pulmonary embolism. *Radiology* 1980; 136:301–308.

39. Cadotte R, Leger C, Harel C, et al: Bronchial angiography: A report of 21 patients. *J Can Assoc Radiol* 1986; 37:22–24.

40. North LB, Boushy SF, Houk VN: Bronchial and intercostal arteriography in non-neoplastic pulmonary disease. *AJR* 1969; 107:328–342.

8

Upper Extremity Angiography

KEY CONCEPTS

1. Manifestations of upper extremity arterial disease include claudication, ulcers, and Raynaud's phenomenon.
2. Raynaud's phenomenon without fixed vascular occlusion does not pose the threat of tissue loss.
3. Unilateral Raynaud's phenomenon signals an underlying arterial lesion.
4. Embolization from proximal plaques or aneurysms can lead to amputation.
5. Arterial spasm in the hand is countered by extremity warming and direct injection of vasodilators; the pain of contrast injection is minimized by low-osmolar agents and digital subtraction angiography.
6. Thoracic outlet syndrome is responsible for most cases of intermittent venous obstruction and spontaneous thrombosis.

INDICATIONS

Angiography of the upper extremities is used to evaluate arterial insufficiency, aneurysms, arteriovenous malformations, venous occlusion, and traumatic, thermal, or electrical injuries. Arteriography is much less often needed in the upper than in the lower extremities, for only about 5% of all cases of limb ischemia involve the arms.[1-3] A tendency for spasm to affect the vessels of the hand makes Raynaud's phenomenon a prominent feature in many cases of ischemia, and spasm calls for special techniques of examination. Claudication and gangrene can also be features of arterial occlusive disease in the arms.

Primary distal arterial disease must be differentiated from that produced by more proximal lesions. Noninvasive studies may fail to uncover an aneurysm or nonocclusive plaque giving rise to small emboli, and a majority of patients with digital ischemia will actually have a proximal lesion responsible for their symptoms.[4] Thoracic outlet compression of arteries or veins, and involvement of hand vessels by connective tissue disorders and systemic vasculitis are among the factors that distinguish upper extremity disease from common lower extremity disorders.[3, 5–7] Unlike the foot, the hand is prone to ischemic complications from repetitive trauma.[8, 9] Because of the often confusing clinical presentation of upper extremity vascular problems, angiography is valuable both for diagnosis and treatment planning.

ARTERIAL EXAMINATION TECHNIQUE

As a rule, the arterial supply to the extremity should be delineated completely, from the origin of the innominate or subclavian artery through the digital vessels. This is best done from a femoral approach, but in cases of severe proximal disease or systemic atherosclerosis axillary or brachial artery puncture may be needed. In relatively young patients without tortuous vessels an H1H configuration catheter is well adapted for the catheterization of brachiocephalic vessels from the femoral artery. A sidewinder or Simmons catheter is a better choice for most older individuals. Stenoses involving the origins of vessels arising from the aortic arch may be missed on a single projection, and attention must be paid to the presence of collateral arterial channels or reversed flow in the vertebral artery.

Once the intrathoracic arteries are examined, the catheter is advanced over a guidewire into the axillary artery for more distal injections. If possible, the tip is placed into the distal brachial artery for examination of the hand. Up to 15% of people have a high origin to the radial artery, which sometimes bifurcates from the axillary artery, a potential source of error if proximal imaging is neglected.[4, 10]

The injection of iodinated contrast medium into upper extremity arteries can be extremely painful. Fortunately, the availability of low-osmolar agents and digital subtraction angiography has improved the tolerability of arteriography considerably. Lidocaine should not be mixed with contrast agent because of the possibility of reflux into the vertebral artery.

Methods of Countering Spasm

Because of the difficulties posed by spasm, various schemes for vasodilatation have been tested to provide optimal filling of palmar and digital vessels. Although it has been suggested that contrast media cause a vasodilatation that can be used to improve flow for an early second injection, this has not been confirmed in practice.[11, 12] Oral administration of alcohol, and brachial plexus block with lidocaine yield unpredictable results.[10, 12] General anesthesia produces rapid and complete fill of digital vessels, but it is not commonly employed or desired for angiography in the United States. Tolazoline (15 to 25 mg), phentolamine (2 to 4 mg), and reserpine (0.5 mg in 10 mL of saline injected slowly) are drugs that have been injected intra-arterially to increase flow.[4, 10, 13] With tolazoline or phentolamine, filming should be performed within 30 seconds to several minutes after injection. Reserpine produces its greatest effect 40 minutes after administration.[10]

Effects of Temperature

Temperature is a prime factor affecting vascular tone, and normal digital blood flow correlates with a fingertip temperature above 31°C.[10, 11] The extremity may be warmed directly, or a large heating pad can be used to increase body core temperature. A brachial artery catheter–to–wrist flow time of 4 seconds or less indicates effective ablation of any vasospasm. Arneklo-Nobin et al. recommend body warming and phentolamine in combination as the best routine preparation for hand arteriography.[10]

ANATOMIC CONSIDERATIONS

Proximal arterial occlusions alone rarely lead to tissue necrosis because of rich collateral networks available around the shoulder and elbow. The forearm and hand are supplied by the radial, ulnar, and interosseous arteries. The radial artery continues as the deep (proximal) palmar arch (Fig 8–1). The ulnar artery becomes the superficial palmar arch, which provides the primary vascular supply to the fingers. If there is a major anastomosis between the ulnar and radial circulations, a palmar arch is called complete. The deep palmar arch is complete in 95% of the population, while the superficial arch is complete in only about 80%.[4] The variations in palmar anatomy are responsible for the individual differences in the ability to tolerate arterial occlusions in the forearm and wrist. A clinical assessment of the adequacy of collateral circulation can be made by the Allen test: compression of radial and ulnar arteries at the wrist with observation of the reversal of finger blanching by the sequential release of compression.

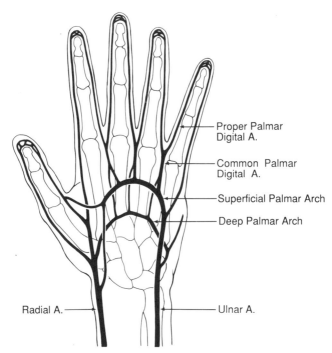

FIG 8–1.
Arterial anatomy of the hand and wrist.

ARTERIAL DISORDERS

Raynaud's Phenomenon

The symptoms first described by Raynaud in the late 19th century are precipitated by cold or other sympathetic stimuli: the affected digits become blanched, cold, and numb, later turning cyanotic and ruborous with an attendant burning, throbbing pain. Although the lower extremities may be affected, symptoms are more common and more pronounced in the fingers and hands. The syndrome may be primary

(Raynaud's disease) if no underlying vascular lesion or systemic disease can be found to account the symptoms. If an underlying condition is established, the clinical manifestations are called secondary Raynaud's phenomenon.

Primary Raynaud's Phenomenon

Besides absence of other illness to account for arterial insufficiency, the following conditions must be met to make the diagnosis of primary Raynaud's phenomenon: bilateral episodes, absence of gangrene, and presence of symptoms for at least 2 years.[4] Leppert et al. have found the prevalence of Raynaud's phenomenon to be 16% among Swedish women, which is in line with other similar surveys.[14] The prevalence among men appears to be less than half this figure. There is a small, but significant association with smoking, and patients with extremity symptoms are more likely to also suffer recurrent headaches, muscle pains, and joint pain.[14]

Although vasospasm is strongly implicated, the reasons for heightened vascular sensitivity to stimuli in primary Raynaud's phenomenon remain controversial.[3, 15, 16] Surgical sympathectomy and numerous medical treatments have been tried with only limited success, but there is evidence that nifedipine may decrease the incidence and severity of attacks.[16]

Because primary Raynaud's phenomenon is unlikely to result in tissue loss, arteriography is rarely indicated and should be reserved for those with evidence of arterial occlusion. Noninvasive studies such as segmental blood pressure measurements, Doppler, and plethysmography are excellent screens for fixed obstructions down to the level of the fingertips.[10] Those with uncomplicated primary Raynaud's phenomenon have normal resting digital arterial pressures.

Secondary Raynaud's Phenomenon

Raynaud's phenomenon is often the first manifestation of a variety of disorders, and connective tissue diseases may become evident only years later.[2, 17] Fixed digital artery occlusions are responsible for ulcers and gangrene, but tissue breakdown seldom occurs before other systemic signs of scleroderma, rheumatoid arthritis, systemic lupus erythematosis, or mixed connective tissue disease appear.[4] Raynaud's phenomenon has also been observed in such diverse conditions as chronic renal failure, hypothyroidism, and occult malignant disease.[16] Angiography is not particularly useful for making the diagnosis of collagen vascular disease or for distinguishing among its various forms. Hematologic, pulmonary, and gastrointestinal studies are more productive in this regard.

Arteriography has greater value in evaluating *unilateral* Raynaud's symptoms. These may be consequent to thoracic outlet compression, aneurysm, radiation therapy, chronic or repetitive trauma, iatrogenic brachial artery injury, or emboli from a cardiac source. In such cases surgical vascular reconstruction and/or cervical sympathectomy can prevent tissue loss.[5]

Atherosclerosis

Atherosclerosis predominately involves proximal vessels, and many innominate or subclavian artery lesions remain asymptomatic unless they produce emboli. If flow is reversed in the vertebral artery (subclavian steal), symptoms, if present, are more likely to relate to central nervous system ischemia rather than to the arm. When ischemia does complicate upper extremity atherosclerosis, amputation is the ultimate result in up to 22%.[7] If the only forearm vessel intact is the interosseous artery, the prognosis is poor.[18] As noted, proximal plaques producing emboli can be difficult to detect. Multiple views and pull-back pressure measurements improve diagnostic yield.[13]

Some subclavian or axillary artery aneurysms are caused by extrinsic compression at the thoracic outlet and development of poststenotic dilatation (see "Thoracic Outlet Syndrome," later in this chapter). However, most true aneurysms at this location are caused by atherosclerosis, and many patients will also have abdominal aortic, femoral, or other peripheral aneurysms.[19] These lesions characteristically contain mural thrombus, and although they can rupture, the greatest threat they pose is from distal embolization.

Thromboangiitis Obliterans (Buerger's Disease)

This disease often produces painful ischemic ulcers of the fingertips, as well as in the feet. At times the lesions of thromboangiitis obliterans are difficult to distinguish angiographically from those of atherosclerosis. However, they are generally more distal, with concomitant forearm and hand arterial occlusions.[7] In their most typical appearance, occlusions are long and are bridged by paralleling "corkscrew" collaterals. Arterial spasm is not a prominent finding. Patients are relatively young, almost invariably male, and there is a strong association with tobacco use (see Chapter 2, Angiography of the Abdominal Aorta, Pelvis, and Lower Extremities).

Embolization From a Cardiac Source

Most emboli lodging in the upper extremity come from the heart, and atrial fibrillation is a predisposing factor. They are more likely to occlude the axillary and brachial arteries than vessels of the forearm and hand.[20] Onset of symptoms is typically abrupt, and a filling defect or sharp cutoff is found at angiography. However, a sharp arterial cutoff is not specific to embolization, and other causes of occlusion cannot be excluded by angiographic criteria alone.

Vasculitides

A great number of systemic diseases affect upper extremity vessels, and only a few are addressed specifically here.

Connective Tissue Disorders

Raynaud's phenomenon accompanies many cases of scleroderma, systemic lupus erythematosis, rheumatoid arthritis, and related diseases, and—as noted earlier—frequently constitutes the initial clinical presentation. Occlusive disease is typically distal and bilaterally symmetric in its distribution (an exception is giant cell arteritis, which affects larger vessels such as the axillary artery). Collateral vessels are usually inadequately developed, and filling beyond the obstructed segments is poor.[7]

Surgical sympathectomy has been advocated as a treatment for distal occlusive disease, but results of standard cervical or thoracic procedures have been poor in patients with scleroderma and related illnesses.[2, 21, 22] Alternative interventions proposed include digital sympathectomy and intra-arterial injection of reserpine, but symptomatic relief obtained is more often than not incomplete and temporary.[15, 21, 23] In connective tissue disorders the activity of disease in the hands usually parallels disease progression or regression seen elsewhere. Recurrent tissue loss and ulceration are more likely in those with connective tissue diseases than in patients with small vessel occlusions from other causes.[1]

Takayasu Arteritis

This vasculitis primarily affects the aorta, its major branches, and the pulmonary arteries. It is a giant-cell necrotizing arteritis, and it commonly compromises the subclavian arteries.[24] Stenotic lesions are characteristically smooth, with normal appearance to those arterial segments unaffected by the disease. Most patients have symptoms of fever and malaise, suggesting an active inflammatory process.

Hypersensitivity and Toxic Vasculitis

Hypersensitivity angiitis is characterized by rapid onset of small vessel occlusions. The process is usually self-limited, and ulcers respond to local care.[1] In many cases an inciting agent is never identified.

Use of ergotamine (an alpha-adrenergic blocking agent) for migraine headaches can sometimes elicit pronounced peripheral arterial constriction and occlusion.[25] Symptoms are usually acute in onset, and both upper and lower extremities can be affected. Arteriography shows long segments of severe narrowing.

Chronic occupational exposure to vinyl chloride has been associated with arterial wall thickening, perivascular inflammation, and Raynaud's phenomenon.[26] Stenoses and occlusions are limited to digital arteries, sparing the palmar arches. Angiography can confirm the diagnosis before the bony changes of acro-osteolysis become evident.

Arterial Injury

As in the lower extremities, blunt or penetrating trauma can produce occlusion, hemorrhage, pseudoaneurysm, or arteriovenous fistula. Examination technique and findings in acute trauma are similar (see Chapter 2).

Repetitive Trauma

The hand and wrist are particularly sensitive to recurrent vibration or occupational trauma from tools. Motions involving twisting or striking with the palm, even apparently trivial, have been implicated in vascular occlusion or aneurysm formation. The ulnar artery is particularly susceptible to trauma where it passes over the hamate bone, superficial to the carpal ligament.[9] The clinical presentation of hypothenar hammer syndrome is usually insidious, with ischemia typically affecting the fourth and fifth fingers of the dominant hand.

Because of its relatively protected position and the complete formation of the deep palmar arch in most people, the radial artery is less likely to suffer trauma from pressure on the thenar eminence. Still, cases of occlusion have been described.[8] One must be aware of these effects of repetitive trauma in order to elicit the necessary history, which may not be volunteered or considered relevant by the patient.

Long-term use of axillary-support crutches poses a risk for axillary artery aneurym.[27] Presenting symptoms of acute or chronic ischemia are the result of distal embolization. Early detection and aneurysm repair can prevent limb loss in this situation.

Iatrogenic Trauma

Widespread use of the brachial artery for catheterization, particularly for cardiac angiography, has currently made iatrogenic trauma one of the most common causes of upper extremity arterial insufficiency. In a survey of over 12,000 patients having cardiac studies performed with 8-F catheters in a single institution, McCollum and Mavor found that 0.9% needed surgical repair of the puncture site.[28] A more recent report of patients undergoing angiography with 4- and 5-F catheters from a brachial approach noted a thrombectomy rate of

4%, despite the routine use of heparin during angiography.[29] No doubt the tendency of the brachial artery to go into spasm predisposes it to postcatheterization complications.

Other Injuries

Intra-arterial injection of illicit drugs can produce multiple digital artery occlusions, mycotic aneurysms, and arteriovenous fistulas. Severe electrical injuries are associated with segmental stenoses or occlusions, and angiography is instrumental in determining the level of any necessary amputation.[7] Angiography serves the same function in cases of frostbite.[30]

THORACIC OUTLET SYNDROME

The narrow costoclavicular space, bounded inferiorly by the first rib, superiorly by clavicle, posteriorly by the scalenus medius muscle, and anteriorly by the costoclavicular ligament, forms a functional bottleneck for the vessels to the upper extremity. Anomalous ribs, muscles, and fascial slips can compress the brachial plexus, subclavian artery, or subclavian vein. Local radiation therapy, clavicular fractures, and osteomyelitis can evoke perivascular fibrosis and cause an acquired thoracic outlet syndrome.[5] It has been postulated that some cases are caused by a cervical stretch injury followed by prolonged muscle contraction and hypertrophy. The great majority of thoracic outlet problems arise from brachial plexus compression, and only about 3% of cases have signs of arterial or venous obstruction.[31]

Most patients with arterial compression have an associated fusiform aneurysm, but the presence of mural thrombus may obscure the abnormality on arteriography. Machleder recommends examination with arm abduction and external rotation to elicit compression.[4] However, this challenge and others that have been recommended are nonspecific, and many people without clinical symptoms can obliterate their brachial pulses by such maneuvers.[32]

Venous compression (or the "thoracic inlet syndrome") can produce intermittent symptoms or acute thrombosis. The most common surgical treatment of arterial or venous compression is resection of the first rib or clavicle, but in cases of venous obstruction recurrent symptoms can arise from postoperative scarring.[31]

UPPER EXTREMITY VENOGRAPHY

Venography is used to diagnose venous obstruction in patients suffering arm swelling and pain. Thrombosis must be distinguished from lymphedema, hemorrhage, and other conditions that present in a similar fashion. Venous thrombosis may be a primary condition, or it may be secondary to central venous catheterization (for a discussion of the latter, see Chapter 9, Dialysis Access, Central Venous Catheters, and Other Central Venous Problems). When thrombosis arises spontaneously, a hypercoagulable condition or central obstructing tumor must be excluded diagnostically.

Primary axillary-subclavian vein thrombosis (Paget-Schroetter syndrome) typically affects young men, usually in the dominant arm. A recent history of vigorous or unusual exercise can often be obtained.[33] If the clot is detected early, thrombolysis or thrombectomy is recommended, for long-term morbidity is a common consequence of conservative treatment.[33, 34] Thoracic outlet syndrome is implicated in 80% of cases of spontaneous venous occlusion.[4]

The majority of patients with signs of venous obstruction from thoracic outlet compression have intermittent symptoms relieved by rest, and do not demonstrate thrombosis.[35] Hyperabduction (to 180°) may produce occlusion on venography, but this sign is not very specific. Schubart et al. have used venous pressure measurements to assess the significance of obstruction induced by abduction and external rotation, finding normal resting pressure to be 5 mm Hg and a rise of $\geq$ 10 mm Hg to indicate disease.[35]

THERAPEUTIC VASCULAR INTERVENTIONS

In selected cases thrombolytic therapy is quite helpful in distal arterial occlusions of the upper extremity.[36] As in experience with other vessels, best results are obtained with relatively acute thromboses. Thrombolysis of primary venous thrombosis has also been successful, but any underlying external compression must be treated surgically.[31] Balloon angioplasty has been applied to subclavian artery stenoses with good results (see Chapter 10, Percutaneous Angioplasty, Recanalization, and Vascular Stents). Experience with transluminal angioplasty in the more distal arteries of the upper extremity has been sparsely reported.

In the past, angiographers have been called upon to treat Raynaud's phenomenon with intra-arterial injection of reserpine. Slow injection of 1.0 mg into the brachial artery has produced objective and subjective signs of improvement for up to 13 months in some patients.[15, 21] However, injectible reserpine is no longer available in the United States.

REFERENCES

1. Mills JL, Friedman EI, Taylor LM Jr, et al: Upper extremity ischemia caused by small artery disease. *Ann Surg* 1987; 206:521–527.

2. Welling RE, Cranley JJ, Krause RJ, et al: Obliterative arterial disease of the upper extremity. *Arch Surg* 1981; 116:1593–1595.

3. Kobinia GS, Olbert F, Russe OJ, et al: Chronic vascular disease of the upper extremity: Radiologic and clinical features. *Cardiovasc Intervent Radiol* 1980; 3:25–41.

4. Machleder H: Vaso-occlusive disorders of the upper extremity. *Curr Probl Surg* 1988; 25:7–67.

5. Bouthoutsos J, Morris T, Martin P: Unilateral Raynaud's phenomenon in the hand and its significance. *Surgery* 1977; 82:547–551.

6. Adler J, Hooshmand A: The angiographic spectrum of the thoracic outlet syndrome with emphasis on mural thrombosis and emboli and congenital vascular anomalies. *Clin Radiol* 1973; 24:35–42.

7. Erlandson EE, Forrest ME, Shields JJ, et al: Discriminant arteriographic criteria in the management of forearm and hand ischemia. *Surgery* 1981; 90:1025–1036.

8. Wandtke JC, Spitzer RM, Olsson HE, et al: Traumatic thenar ischemia. *AJR* 1976; 127:569–571.

9. Pineda CJ, Weisman MH, Bookstein JJ, et al: Hypothenar hammer syndrome: Form of reversible Raynaud's phenomenon. *Am J Med* 1985; 79:561–570.

10. Arneklo-Nobin B, Albrechtsson U, Eklöf B, et al: Indications for angiography and its optimal performance in patients with Raynaud's phenomenon. *Cardiovasc Intervent Radiol* 1985; 8:174–179.

11. Rösch JF, Antonovic R, Porter JM: The importance of temperature in angiography of the hand. *Radiology* 1977; 123:323–326.

12. Viehweger G, Plötz J: Vergleichende angiographische untersuchungen in lokal-, regional- und allgemein-anästhesie an der oberen extremität. *ROFO* 1974; 121:303–310.

13. Maiman MH, Bookstein JJ, Bernstein EF: Digital ischemia: Angiographic differentiation of embolism from primary arterial disease. *AJR* 1981; 137:1183–1187.

14. Leppert J, Åberg H, Ringqvist I, et al: Raynaud's phenomenon in a female population: Prevalence and association with

other conditions. *Angiology* 1987; 38:871–877.

15. Romeo SG, Whalen RE, Tindall JP: Intra-arterial administration of reserpine. *Arch Intern Med* 1970; 125:825–829.

16. Smith CR, Rodeheffer RJ: Treatment of Raynaud's phenomenon with calcium channel blockers. *Am J Med* 1985; 78(suppl 2B):39–42.

17. Dale WA, Lewis MR: Management of ischemia of the hand and fingers. *Surgery* 1970; 67:62–79.

18. Gross WS, Flanigan DP, Kraft RO, et al: Chronic upper extremity arterial insufficiency. *Arch Surg* 1978; 113:419–423.

19. Pairolero PC, Walls JT, Payne WS, et al: Subclavian-axillary artery aneurysms. *Surgery* 1981; 90:757–762.

20. Janevski B: Arterial embolism of the upper extremities. *ROFO* 1986; 145:431–434.

21. Nilsen KH, Jayson MI: Cutaneous microcirculation in systemic sclerosis and response to intra-arterial reserpine. *Br Med J* 1980; 280:1408–1411.

22. van de Wal HJ, Skotnicki SH, Wijn PF, et al: Thoracic sympathectomy as a therapy for upper extremity ischemia: A long-term follow-up study. *Thorac Cardiovasc Surg* 1985; 33:181–187.

23. Flatt AE: Digital artery sympathectomy. *J Hand Surg* 1980; 5:550–556.

24. Sano K, Aiba T, Saito I: Angiography in pulseless disease. *Radiology* 1970; 94:69–74.

25. Bagby RJ, Cooper RD: Angiography in ergotism: Report of two cases and a review of the literature. *AJR* 1972; 116:179–186.

26. Falappa P, Magnavita N, Bergamaschi A, et al: Angiographic study of digital arteries in workers exposed to vinyl chloride. *Br J Ind Med* 1982; 39:169–172.

27. Abbott WM, Darling RC: Axillary artery aneurysms secondary to crutch trauma. *Am J Surg* 1973; 125:515–520.

28. McCollum CH, Mavor E: Brachial artery injury after cardiac catheterization. *J Vasc Surg* 1986; 4:355–359.

29. Grollman JH Jr, Marcus R: Transbrachial arteriography: Techniques and complications. *Cardiovasc Intervent Radiol* 1988; 11:32–35.

30. Gralino BJ, Porter JM, Rösch J: Angiography in the diagnosis and therapy of frostbite. *Radiology* 1976; 119:301–305.

31. Roos DB: Thoracic outlet syndromes: Update 1987. *Am J Surg* 1987; 154:568–573.
32. Warrens AN, Heaton J: Thoracic outlet compression syndrome: The lack of reliability of its clinical assessment. *Ann R Coll Surg* 1987; 69:203–204.
33. Becker GJ, Holden RW, Rabe FE, et al: Thrombolytic therapy for subclavian and axillary vein thrombosis. *Radiology* 1983; 149:419–423.
34. Smith-Behn J, Althar R, Katz W: Primary thrombosis of the axillary/subclavian vein. *South Med J* 1986; 79:1176–1178.
35. Schubart PJ, Haeberlin JR, Porter JM: Intermittent subclavian venous obstruction: Utility of venous pressure gradients. *Surgery* 1986; 99:365–367.
36. Tisnado J, Bartol DT, Cho S-R, et al: Low-dose fibrinolytic therapy in hand ischemia. *Radiology* 1984; 150:375–382.

9 | Dialysis Access, Central Venous Catheters, and Other Central Venous Problems

KEY CONCEPTS

1. Hemodialysis access fistulas and grafts have a limited lifetime, usually no more than 2 years.
2. Early detection and treatment of stenoses can prolong the use of shunts by 6 months or more.
3. Catheter-related thrombosis is the most common complication of prolonged central venous catheterization.
4. Many cases of catheter-induced central thrombosis go unrecognized, but the incidence of pulmonary embolism may be as high as 12%.
5. Embolized fragments of venous catheters, pacer wires, and other foreign bodies can cause fatal complications if not removed, but most can be snared percutaneously.

With the increasing use of hemodialysis in the treatment of chronic renal failure, and the increasing reliance on central venous catheters in administration of long-term cancer chemotherapy, antibiotics, and parenteral alimentation, radiologists are seeing a growing number of patients with problems arising from such vascular access devices. Although catheters and fistulas have a limited lifetime, a number of interventions can be performed percutaneously to prolong their usefulness or to guide the surgical correction of complications. For those patients who may need treatment for an indefinite period of time, each loss of vascular access is a major setback, making future care more difficult. Therefore, the early recognition of dysfunction, often remediable by relatively minor radiologic or surgical procedures, is of great clinical importance.

DIALYSIS FISTULAS AND SHUNTS

The number of patients undergoing hemodialysis for chronic renal failure in the United States increased dramatically in the 1980s, reaching over 90,000 persons in 1987.[1, 2] Maintenance of vascular access in these patients presents a major problem requiring a great commitment of resources.

Permanent peripheral hemodialysis access can be created by direct anastomosis of a vein to the side of an artery, or by placement of a graft bridging artery and vein. The introduction of the Brescia-Cimino forearm fistula in the 1960s was a breakthrough in the care of dialysis patients. It remains the procedure of choice, because of its better long-term patency and lower complication rate when compared with dialysis grafts.[2] However, thrombosis of antecubital veins after repeated venipuncture, or atherosclerotic disease involving the upper extremity, limit the application of surgical fistulas. In such cases, synthetic or bovine vein grafts can be placed in straight or looped configurations. Bovine grafts have not proved very suitable, and most dialysis grafts are now made of polyfluorotetraethylene. Although arteriovenous (AV) shunts have been created in the lower extremities, they predispose a patient to sepsis, and upper extremity dialysis access is the generally preferred route.[3] Mean patency of AV grafts is 1.7 years vs. 2.8 years for forearm fistulas.[2]

Complications of Dialysis Access

Potential problems with grafts and fistulas include failure of maturation, thrombosis, stenosis, infection, pseudoaneurysm formation, and vascular "steal."

In a fistula, 4 to 6 weeks are needed for the vein to respond to increased flow by enlargement and mural thickening. A fistula cannot be used for hemodialysis until it has matured (shunts formed by grafts can be used earlier, and no maturation interval is necessary). The maturation of a dialysis fistula is impaired if flow is diminished by anastomotic or arterial stenosis, or if a single unimpeded outflow vein is absent.

Difficulties acquired after maturation and use often manifest themselves prior to thrombosis by poor flow, increased venous resistance (in outflow problems), or generation of negative pressures during dialysis (in arterial inflow problems). Venous anastomotic or more proximal (central) stenoses are *much* more common than arterial lesions, with an incidence ratio of about 15:1.[1] Neointimal hyperplasia and fibrosis account for most stenotic lesions.

Occlusions may result from stenoses, or they may occur with prolonged compression after removal of dialysis needles, hypovolemia, other low-flow states, infection, or hypercoagulability. Pseudoaneurysms and infection are rare in Brescia-Cimino fistulas (present in 1% to 2%), but are fairly common complications of synthetic grafts: with 10% and 15% occurrences, respectively.[2]

The diversion of blood flow from the distal extremity through a shunt can cause pain and ischemia. This vascular "steal" commonly results from too large an anastomosis. The incidence of this problem has diminished with the increasing experience of vascular surgeons and introduction of grafts tapered to 4 mm at their arterial ends.[1]

Preoperative Venography

In most cases clinical examination can determine if an extremity is suitable for AV fistula or graft. However, there have been patients in whom prior use of a subclavian venous catheter has resulted in clinically inapparent stenosis or occlusion, which becomes symptomatic only after the placement of a dialysis shunt.[4, 5] If there is any indication of upper extremity venous insufficiency or history of prolonged central venous catheterization, preliminary venography may be indicated. If a fistula or graft has failed, the surgeon may request a venogram to define remaining veins and to exclude a central abnormality.

Venography is performed through a needle placed in a hand or distal forearm vein, with a proximal tourniquet or blood pressure cuff inflated above venous pressure. The study is best performed with large-format spot films (two- or three-on-one multiexposure format), with the patient's forearm placed in various positions to obtain multiple projections of the veins when filled. Filling should be documented centrally into the superior vena cava (SVC) after the tourniquet is removed. The central portion of the examination is facilitated by digital subtraction imaging (DSA). Inflow of unopacified blood from jugular veins or contralateral innominate vein must not be confused with clot. Filling of collateral veins is a very reliable sign of obstruction.

Imaging and Evaluation of Dialysis Shunts

Duplex Sonography

Improvements in vascular sonography have led to its use in evaluating dialysis access. Duplex ultrasound can be used to identify mechanical kinks, stenoses, and thromboses with high accuracy.[6, 7] However, the examination procedure is more involved and time-consuming, and requires more technical expertise than conventional studies with contrast material. Ultrasound does not present a "global" view of the fistula or graft, and acceptance of scanning results by the referring clinician may be limited. Because conventional studies are relatively simple, inexpensive, and minimally invasive, ultrasound can be reserved for conditions in which it may be more accurate than angiography, such as suspected pseudoaneurysm or vascular steal.[7] Duplex sonography can also be applied to patients with a history of hypersensitivity to contrast material or in whom angiography reveals an equivocal abnormality.

Contrast Media Examinations

Early approaches to examinations with contrast material involved selective arteriography or brachial artery puncture.[8, 9] Time and experience have shown, however, that direct puncture of the shunt presents extremely low risk and provides good diagnostic information with a much less cumbersome technique.

Angiography should be performed at the first sign of dysfunction, for cannulation of an occluded fistula will not uncover an underlying abnormality unless thrombolysis is pursued. Performance and interpretation are helped by knowledge of the type of fistula or graft in place, including location of anastomoses. The graft or venous limb of a fistula can be examined through a dialysis needle left in place, or by puncture with a 21-gauge butterfly infusion needle. From a diagnostic standpoint, it makes no difference if the needle is directed retrograde or antegrade. Free return of blood indicates proper position, but this is confirmed by a test injection of contrast material under fluoroscopy.

Digital subtraction angiography, 100-mm spot films, or cut-film angiography can all be used, but DSA permits rapid evaluation of a shunt with dilute contrast medium. Initial injections are made without cuff occlusion of venous outflow, and these define flow, venous anastomosis, and proximal veins. If the patient has clinical signs of outflow obstruction (diffuse edema and high venous resistance during dialysis), and no abnormality is discovered, additional projections must be obtained. Should a distal obstruction to outflow be found, the examination should not be concluded before more central flow through subclavian vein, innominate vein, and SVC is defined.

Examination of the graft or vessel "upstream" from the needle, as well as the arterial anastomosis, calls for occlusion of outflow by a blood pressure cuff. The cuff is inflated above arterial pressure immediately before filming, and the pressure is released immediately after the injected contrast material fills both the arterial and venous sides of the graft or fistula. Again, use of DSA simplifies evaluation by allowing immediate assessment of the degree of filling attained. Also, because reflux of contrast material into the arteries of the upper extremity can be extremely painful, DSA adds to patient comfort by allowing use of very dilute contrast media. Definition of the arterial anastomosis is particularly important in cases of suspected steal or if poor flow or pressures during dialysis indicate obstructed inflow. Rarely is standard catheter arteriography needed to rule out a proximal arterial lesion.

All injections of iodinated contrast media may be performed manually or mechanically. The volume and rate chosen depend on the flow observed during test injections. Generally, volumes of 10 to 25 mL administered at 1–5 mL/sec are adequate for each field imaged. An exposure rate of at least one frame per second should be used, except in central studies of poor venous outflow. A rapid frame rate is essential for appreciating the pattern of arm and forearm collateral venous filling in the presence focal outflow obstruction.

Interventions

Because thrombectomy and surgical revision can extend dialysis access lifetime by an average 4 to 8 months, prolongation of graft or AV fistula use by at least 6 months can be considered a successful intervention.[1] By this criterion, combined use of thrombolysis and/or balloon angioplasty has succeeded in treatment of graft thrombosis or poor flow in 53% to 76% of such attempts.[9–12] The importance of early intervention is underscored by the considerably better 6-month patency attained by angioplasty of stenotic, but still patent, shunts: 76% to 82% vs. 36% to 62% for occlusions treated in these same reports.

Percutaneous Angioplasty

Percutaneous transluminal angioplasty (PTA) can proceed either through the graft, through the venous limb of the fistula, or in cases of proximal (central) venous stenoses, through a femoral venous approach. Risk of pseudoaneurysm from placement of a balloon through graft appears to be minimal, and puncture site compression with surgical collagen sponge promotes hemostasis.[10]

As mentioned previously, most stenoses in dialysis access shunts result from intimal hyperplasia, felt to be incited by the turbulence and shear stresses resulting from high blood flow.[12] Intimal hyperplasia seems to be particularly severe at venous anastomoses, and successful dilation often requires prolonged inflation (minutes at a time) with high-pressure balloons. Especially refractory lesions have been opened with coaxial Teflon fascial dilators.[9] Because dilatation can be quite painful in these patients, infiltration of lidocaine about the site of the lesion before PTA is advisable.[9, 10]

Some have found that central venous stenoses in the presence of dialysis shunts are quite readily treated by PTA.[5, 12] Lesions in central vessels are dilated with larger balloons, up to 20 mm in diameter. Arm and forearm stenosis with grafts or veins usually require balloons 6 to 10 mm in diameter. Arterial inflow lesions may be treated with 4-mm balloons. As in any PTA procedure, balloon size is predicated on the size of the vessel to be treated, allowing for a small amount of over-dilation.

Although early response tends to be good, most vessels that stay open for 6 months will have recurrent stenosis within 1 year. Still, dialysis fistula or graft PTA can be repeated with good results.[12] Some patients apparently have a predisposition to intimal hyperplasia and have early restenosis whether they are treated percutaneously or by surgical revision.[11] Intravascular stents have been promoted as a possible solution to the problem of restenosis, but early reports indicate that new stenoses can occur within expandable stents.[13]

Stenoses in fistulas that have failed to mature are not amenable to PTA, and they may be prone to rupture.[9, 12, 14] A narrowing that extends over more than 4 cm in length is also unlikely to respond to balloon dilatation.[10, 12]

Thrombolysis

Earlier experience with graft declotting by infusion of streptokinase was disappointing, and treatment was often complicated by bleeding through the graft or hemodialysis puncture sites.[3, 15] More rapid administration of the fibrinolytic agent, mechanical disruption of the clot, and use of urokinase have allowed better control of lysis and greater success.[3, 11] Even if a lesion not treatable by angioplasty is uncovered, surgical revision can often be more limited than it would have been without lytic therapy.

The technique described by Davis et al. involves puncture of occluded grafts near arterial and venous anastomoses, with the placement of crossing hook-tipped catheters.[11] Entry into the lumen is indicated by the return of a small amount of serosanguinous fluid or easy insertion of a soft-tipped guidewire. Each catheter is slowly rotated and withdrawn from the respective anastomosis, while a concentrated dose (25,000 units/mL) of urokinase is injected into the clot. The total dose so administered is 100,000 to 150,000 units. A heparin bolus of 2,000 to 3,000 units is also given, followed by an intravenous heparin drip. One catheter tip is then positioned near the arterial anastomosis and the other left in midgraft, each infusing 2,000 units/min until pulse or bleeding from the puncture sites is noted. With flow reestablished, an angioplasty balloon is used to compress residual thrombus and to dilate any stenosis found.

By this method, Davis and associates achieved lysis in 90% of patients in an average of 90 minutes with a mean total dose of less than 400,000 units of urokinase.[11] Patency was preserved for 6 to 18 months in 16 of 26 patients with lysis. The short duration of infusion by this technique allows the patient to be observed in the angiography suite throughout the procedure. Any bleeding that arises can be immediately controlled by direct compression.

CENTRAL VENOUS CATHETERS

Large-bore central venous catheters have been applied increasingly for long-term administration of cancer chemotherapy, antibiotics, hyperalimentation, dialysis, and blood withdrawal. These catheters may be partially or totally implanted, and have a longer useful lifetime than conventional smaller bore central catheters. Catheters with subcutaneous infusion chambers have been reported functional for a mean of 6 months or more, in comparison with subcutaneously tunneled Hickman catheters, which have a mean lifetime of 40 to 110 days.[16, 17] Patency is maintained by regular flushing with heparinized saline when the lines are not in use. Serious complications with these devices are quite rare, with sepsis, endocarditis, catheter fracture, and embolization representing the major life-threatening problems.

Fibrin Sheaths and Mechanical Problems With Catheters

There is more than a trivial incidence of mechanical malfunctioning of catheters or of central venous thrombosis requiring radiologic evaluation and possible intervention. In several recent series, 22% to 54% of patients with various infusion catheters experienced problems at some point.[16-18] Most commonly a fibrin sheath forms about the catheter, making blood withdrawal difficult or impossible. If the sheath is complete, it may cause infusate to track back along the catheter into the soft tissues of the chest wall. This must be differentiated from catheter breakage or disconnection causing extravasation. Fibrin deposits occur on virtually all long-term catheters, but when difficulties arise, the majority of catheters can be reopened by direct instillation of low-dose thrombolytic agents.[18] The same holds true for many clotted catheters.

Fibrin sheaths may be seen as thin membranes about the catheter, with or without associated pericatheter clot. On a normal catheter injection contrast material should flow freely from the port being injected, with rapid flow of unopacified blood diluting the contrast stream.

Retrograde flow is abnormal, and it may be the only appreciable sign of a fibrin sheath.

Among the other mechanical problems that have been described are catheter malposition, catheter tip abutting the wall of the SVC and preventing aspiration, tight sutures causing occlusion, back-migration of catheter into the subcutaneous tract, and local infection. Catheter malposition can sometimes be corrected by percutaneous transvenous manipulation similar to that used to remove embolized catheter fragments (see "Percutaneous Removal of Intravascular Foreign Bodies," later in this chapter).

Totally implanted systems, which have access through a subcutaneous silicone membrane, must be punctured *only* by specially designed needles[17]; otherwise, the membranes may be damaged, leading to premature failure of the catheter. When the proper needles are used, many cases of extravasation are the result of poor needle placement rather than to a problem intrinsic to the infusion port.

Catheter-Related Central Venous Thrombosis

Thrombosis of central veins as a result of long-term infusion catheters has been said to be a rare complication, presenting clinically in fewer than 1% of patients, according to some.[4, 19] However, others quote an occult rate of thrombosis between 17% and 40%.[19–22] It is clear that such occlusions usually do not present acute or long-term problems, and venous insufficiency or claudication is less likely to result from secondary subclavian vein thrombosis than it is with primary venous occlusion.[19] Still, occlusion must be recognized and treated in those patients who are entirely dependent on venous hyperalimentation for nonmalignant conditions, or who need long-term hemodialysis. Progressive loss of central venous access in such patients is a serious problem.

One underappreciated consequence of catheter-related central venous thrombosis is pulmonary embolism, which may be fatal. Multiple reports indicate an incidence of about 12% of this complication in patients with catheter-related clots.[20, 21]

When symptoms of venous occlusion do arise, standard therapy has been removal of the catheter and anticoagulation by heparin.[21] Before any such measures are taken venous thrombosis should be confirmed radiographically, because hematoma, trauma, tumor, and lymphedema can present similar symptoms. Not only should the catheter be injected with iodinated contrast material, but an upper extremity venogram must also be acquired to define the extent of thrombosis. Venous Doppler systems are not sensitive as a screening measure because of flow through collateral veins, and most Doppler studies do not permit detection of occlusion.[19, 21] It remains to be seen if duplex sonography will play a useful role.

Although catheter removal and anticoagulation are generally advised for thrombosis, full anticoagulation alone has resolved symptoms in patients without signs of infection.[20] It is unclear how many patients with clinical improvement actually have recanalization of their veins. No doubt a portion of them, perhaps the majority, will have chronic thrombosis and development of adequate flow through collaterals. A more aggressive approach involves the use of thrombolysis.

Fraschini et al. have described 81% success with local urokinase infusion of clots present for less than 1 week by patient history.[22] In most cases they did not find catheter removal necessary. Success was highly correlated with *direct* infusion of clot; infusion of a peripheral vein in the arm was almost uniformly unproductive. They recommend a urokinase dose of 1,000 to 2,000 units/kg/hr, and have found fibrinogen levels to be the best test for predicting puncture site bleeding complications (which correlate with serum fibrinogen levels of less than 100 mg/dL). Those with residual thrombus after recanalization are prone to reocclusion, and these patients may best be treated with long-term anticoagulation. Otherwise, a short course of intravenous heparin after successful lysis may be all that is necessary for a good result. Patients with active phlebitis should not undergo infusion until their inflammatory symptoms subside.[22]

Radiologic Catheter Placement

Most long-term infusion catheters can be placed under local anesthesia, and many procedures have been performed at bedside. Selby et al. have used initial bilateral upper extremity venography to guide Seldinger puncture of the subclavian vein for insertion of 11-F double-lumen dialysis catheters.[23] After blood return from the needle, and confirmation of appropriate entry by contrast material injection, the tract is widened with Teflon dilators that are advanced over a standard angiographic guidewire. A DSA "roadmapping" technique would be a useful adjunct to the method they describe.

Hickman catheters, originally implanted by way of cephalic vein cut-down, have been placed percutaneously under fluoroscopic guidance in 51 patients by Robertson et al.[24] They give their patients 1.0 g of cefoxitin intravenously 1 hour prior to catheter placement. In the procedure they describe, the skin inferior to the junction of the mid and distal thirds of the clavicle is sterilely prepared, and a 21-gauge needle attached to an aspirating syringe is advanced from this spot, taking care to avoid the periostium of rib or clavicle. Once the subclavian vein is entered, an 0.018-inch mandril guidewire is passed, and a transition dilator is used to introduce a larger wire. The tract chosen to tunnel the catheter is liberally anesthetized with lidocaine, and the tunneling device supplied with the catheter is inserted through a second incision toward the site through which the needle has entered the vein. The Hickman catheter is attached to the tunneling tool and pulled back through the tract and out of the second incision. The Dacron cuff of the catheter (important for prevention of bacterial colonization and infection) should lie about 3 cm within the tunnel. Before the intravenous portion of the catheter is inserted through a peel-away sheath, that end of the catheter is trimmed to an appropriate length, ideally allowing the tip to lie at the junction of the SVC and right atrium. After proper positioning is confirmed by chest radiography, the remaining exposed catheter is buried and the incisions sutured. Robertson et al. have had 100% success in catheter placement and a low complication rate with this technique.[24]

In patients with progressive loss of vascular access, subclavian and iliac veins may not prove suitable for catheter insertion. Successful translumbar insertion into the inferior vena cava (IVC) has been reported in such exceptional cases.[24] At the University of Wisconsin–Madison, we have placed a hyperalimentation catheter transhepatically, through a hepatic vein into the right atrium. This radical approach was necessary after surgical implantation into the azygos vein was complicated by infection, and the IVC was not available, having been surgically occluded at an earlier time.

VENAE CAVAE RECANALIZATION

Superior vena cava syndrome (swelling of the neck and upper extremities from venous congestion) may result from tumor, granulomatous disease, or central venous catheter–induced thrombosis. Tumor-induced obstruction of the SVC has caused pulmonary embolism when anticoagulation has not been instituted, and high mortality has been reported for SVC occlusion.[25, 26] Still, symptoms often resolve with heparin/Coumadin (warfarin) therapy and treatment of any underlying disorder. In isolated cases more aggressive therapy has included systemic or local thrombolysis and balloon angioplasty.[26–28] However, stenoses of the SVC tend to resist dilatation, and Putnam and associates have found insertion of intravascular stents necessary.[28] It remains to be seen if such an approach to SVC syndrome offers substantial advantages in comparison to standard anticoagulation and supportive therapy.

Occlusion of the IVC can cause similar problems, and may manifest itself as the Budd-Chiari syndrome. Obstruction may be caused by tumor, hypercoagulable state (e.g., paroxysmal nocturnal hemoglobinuria, protein C deficiency, antithrombin III deficiency), or idiopathic segmental occlusion.

The last condition named is relatively frequent in Japan, and has been treated by careful advancement of a stiff guidewire–cannula–catheter system from a femoral approach under biplane fluoroscopic guidance.[29] The procedure is completed by dilatation of the obstruction with two or more balloons. More recently Furui et al. have employed a ceramic-capped thermal laser probe to perform the recanalization.[30] We, as well as others, have successfully treated cases of IVC thrombosis not associated with obstructing tumor by the local infusion of thrombolytic agents.[31] Surgical approaches to IVC thrombosis may have prohibitive morbidity and mortality, and percutaneous interventions are preferable, when feasible.

PERCUTANEOUS REMOVAL OF INTRAVASCULAR FOREIGN BODIES

Percutaneous intervention is also well suited for retrieval of embolized catheter fragments, cardiac pacemaker wires, and other intravascular foreign bodies. Successful retrieval often takes but a matter of minutes, and thoracotomy with an open heart procedure can be averted in many instances.

Catheter and other fragments frequently lodge in the right side of the heart or in major pulmonary artery branches. Myocardial perforation, pericardiac tamponade, dysrrhythmias, endocarditis, and pulmonary embolli are possible consequences of failure to remove such fragments. Removal is mandatory, for mortality has been reported between 27% and 61%.[32] If one end is free within the IVC, right atrium, or a major pulmonary artery, an intravascular retrieval basket, or a simple snare consisting of a long, thin guidewire (such as a 200-cm, 0.025-inch wire) doubled over and passed through a 7- or 8-F nontapered catheter, can be used to grasp the fragment. Retrievals should be performed through a sheath. A large fragment may need simultaneous removal of the sheath or open femoral venotomy for complete extraction, once it has been captured and withdrawn to the femoral vein.

An SVC or jugular thrombus may bury the severed end of a cardiac pacing wire, or an embolized fragment not recognized early (as in asymptomatic breakage of a ventriculoatrial shunt for hydrocephalus) may become adherent to vessel wall. In such cases it is necessary to free one end by hooking a pigtail catheter or a curved catheter/deflecting wire combination about the body of the embolized catheter or wire. Sustained traction may loosen the tip of the foreign body, allowing it to be grasped or snared. At times, intravascular forceps must be employed.[33] Grabenwoeger et al. have used a curved catheter to pass a wire over the foreign body, snaring the tip of the wire with a basket introduced from the opposite femoral vein.[32] With both ends of the guidewire in hand, a considerable amount of traction can be exerted on the embedded fragment.

With all intravascular retrieval devices great care must be exercised; otherwise, vessel perforation or damage to papillary muscles or valves can result. If a severe arrhythmia is evoked by retrieval manipulations, the foreign body is best removed surgically.

REFERENCES

1. Bell DD, Rosental JJ: Arteriovenous graft life in chronic hemodialysis. *Arch Surg* 1988; 123:1169–1172.
2. Zibari GB, Rohr MS, Landreneau MD, et al: Complications from permanent hemodialysis vascular access. *Surgery* 1988; 104:681–686.
3. Zeit RM, Cope C: Failed hemodialysis shunts: One year of experience with aggressive treatment. *Radiology* 1985; 154:353–356.
4. Davis D, Petersen J, Feldman R, et al: Subclavian vein stenosis: A complication of subclavian dialysis. *JAMA* 1984; 252:3404–3406.
5. Ingram TL, Reid SH, Tisnado J, et al: Percutaneous transluminal angioplasty of brachiocephalic vein stenoses in patients with dialysis shunts. *Radiology* 1988; 166:45–47.
6. Pieterman H, Tordoir JHM: Non-invasive evaluation of prosthetic dialysis shunt in asymptomatic patients. *ROFO* 1986; 145:541–546.
7. Middleton WD, Picus DD, Marx MV, et al: Color Doppler sonography of hemodialysis vascular access: Comparison with angiography. *AJR* 1989; 152:633–639.

8. Glanz S, Bahist B, Gordon DH, et al: Angiography of upper extremity access fistulas for dialysis. *Radiology* 1982; 143:45–52.

9. Hunter DW, Castañeda-Zuñiga WR, Coleman CC, et al: Failing arteriovenous dialysis fistulas: Evaluation and treatment. *Radiology* 1984; 152:631–635.

10. Glanz S, Gordon D, Butt KMH, et al: Dialysis access fistulas: Treatment of stenoses by transluminal angioplasty. *Radiology* 1984; 152:637–642.

11. Davis GB, Dowd CF, Bookstein JJ, et al: Thrombosed dialysis grafts: Efficacy of intrathrombotic deposition of concentrated urokinase, clot maceration, and angioplasty. *AJR* 1987; 149:177–181.

12. Saeed M, Newman GE, McCann RL, et al: Stenoses in dialysis fistulas: Treatment with percutaneous angioplasty. *Radiology* 1987; 164:693–697.

13. Günther RW, Vorwerk D, Bohndorf K, et al: Venous stenoses in dialysis shunts: Treatment with self-expanding metallic stents. *Radiology* 1989; 170:401–405.

14. Bourne EE: Late venous rupture after angioplasty of an arteriovenous dialysis fistula. *AJR* 1988; 150:797–798.

15. Young AT, Hunter DW, Castañeda-Zuñiga WR, et al: Thrombosed synthetic hemodialysis access fistulas: Failure of fibrinolytic therapy. *Radiology* 1985; 154:639–642.

16. Repelaer van Driel OJ, Kuin CM, van de Velde CJH: Surgically implanted subcutaneous venous access devices in cancer patients. *Neth J Surg* 1988; 40:97–99.

17. Lorenz M, Hottenrott C, Seufert RM, et al: Total implantierbarer dauerhafter zentralvenöser zugang: Langzeiterfahrung mit subcutanen infusionskammern. *Langenbecks Arch Chir* 1988; 373:302–309.

18. Cassidy PF Jr, Zajko AB, Bron KM, et al: Noninfectious complications of long-term central venous catheters: Radiologic evaluation and management. *AJR* 1987; 149:671–675.

19. Smith VC, Hallett JW Jr: Subclavian vein thrombosis during prolonged catheterization for parenteral nutrition. *South Med J* 1983; 76:603–606.

20. Anderson AJ, Krasnow SH, Boyer MW, et al: Thrombosis: The major Hickman catheter complication in patients with solid tumor. *Chest* 1989; 95:71–75.

21. Horattas MC, Wright DJ, Fenton AH, et al: Changing concepts of deep venous thrombosis of the upper extremity: Report of a series and review of the literature. *Surgery* 1988; 104:561–567.
22. Fraschini G, Jadeja J, Lawson M, et al: Local infusion of urokinase for the lysis of thrombosis associated with central venous catheters in cancer patients. *J Clin Oncol* 1987; 5:672–678.
23. Selby JB, Tegtmeyer CJ, Amodeo C, et al: Insertion of subclavian hemodialysis catheters in difficult cases: Value of fluoroscopy and angiographic techniques. *AJR* 1989; 152:641–643.
24. Robertson LJ, Mauro MA, Jaques PF: Radiologic placement of Hickman catheters. *Radiology* 1989; 170:1007–1009.
25. Adelstein DJ, Hines JD, Carter SG, et al: Thromboembolic events in patients with malignant superior vena cava syndrome and the role of anticoagulation. *Cancer* 1988; 62:2258–2262.
26. Smith NL, Ravo B, Soroff HS, et al: Successful fibrinolytic therapy for superior vena cava thrombosis secondary to long-term total parenteral nutrition. *JPEN* 1985; 9:55–57.
27. Ali MK, Ewer MS, Balakrishnan PV, et al: Balloon angioplasty for superior vena cava obstruction. *Ann Intern Med* 1987; 107:856–857.
28. Putnam JS, Uchida BT, Antonovic R, et al: Superior vena cava syndrome associated with massive thrombosis: Treatment with expandable wire stents. *Radiology* 1988; 167:727–728.
29. Yamada R, Sato M, Kawabata M, et al: Segmental obstruction of the hepatic inferior vena cava treated by transluminal angioplasty. *Radiology* 1983; 149:91–96.
30. Furui S, Yamauchi T, Ohtomo K, et al: Hepatic inferior vena cava obstructions: Clinical results of treatment with percutaneous transluminal laser-assisted angioplasty. *Radiology* 1988; 166:673–677.
31. Sholar PW, Bell WR: Thrombolytic therapy of inferior vena cava thrombosis in paroxysmal nocturnal hemoglobinuria. *Ann Intern Med* 1985; 103:539–541.
32. Grabenwoeger F, Dock W, Pinterits F, et al: Fixed intravascular foreign bodies: A new method for removal. *Radiology* 1988; 167:555–556.
33. Mocellin R: Transluminale extraktion intrakardial embolisierter katheterfragmente bei kindern. *Z Kinderchir* 1987; 42:343–345.

10 | Percutaneous Angioplasty, Recanalization, and Vascular Stents

---

KEY CONCEPTS

1. Balloon angioplasty (PTA) is a safe and effective treatment for short arterial stenoses and occlusions.
2. PTA enlarges the vessel lumen by disrupting plaque, intima, and media in a controlled manner.
3. For prevention of early thrombosis, platelet aggregation inhibitors must be started prior to PTA. Nitroglycerine, nifedipine, and heparin are other essential adjunct medications.
4. A guidewire should be maintained across a lesion until the dilation procedure is deemed to be completed.
5. Failure or complication of PTA very rarely adversely affects a patient's clinical status or surgical options.

In 1964 Dotter and Judkins introduced a new method of treating vascular occlusive disease, a technique which has become known as percutaneous transluminal angioplasty (PTA).[1] Initially, a system of stiff coaxial catheters was used to open stenoses and occlusions. Unfortunately, the resulting arteriotomy matched the diameter of the vessel lumen produced. The large catheters also caused extensive endothelial trauma, leading to high rates of early thrombosis, restenosis, and generally poor long-term results.

Various investigators experimented with balloon catheters, but Grüntzig and Hopff succeeded in developing a balloon that would reliably inflate to a predetermined diameter and not beyond.[2] The Grüntzig balloon permitted the widespread application of PTA not only to the peripheral arteries but also to renal, coronary, and other central vessels. Since the dissemination of the technique in the late 1970s, much progress has been made in balloon and catheter materials, guidewires, and adjuvant medical therapy. As a result, PTA has become widely accepted as a safe alternative or complement to surgical intervention for many patients with vascular disease. Early skepticism and the reluctance of many vascular surgeons to refer patients are slowly being overcome by data, such as that from the Veterans Administration Cooperative study, confirming that if PTA is initially successful, clinical results are as good and as durable as those of bypass surgery.[3] Moreover, failed or complicated PTA is unlikely to worsen the patient's condition or hinder attempts at reconstructive surgery.[3, 4]

Nevertheless, the limitations of balloon angioplasty have prompted the investigation of novel recanalization devices and percutaneously placed vascular stents. Laser-assisted angioplasty, ultrasonic, and mechanical atherectomy devices are designed to traverse long occlusions; remove plaque; and, along with stents, prevent restenosis.[5–12] Although preliminary results are becoming available, the ultimate role of these devices remains to be defined.

INDICATIONS FOR BALLOON ANGIOPLASTY

Leaving aside the use of PTA in the coronary circulation and the heart, the prime indication for balloon angioplasty is lower extremity occlusive disease. Limiting claudication, pain when at rest, or gangrene indicate the need for revascularization. After history, clinical evaluation, and noninvasive studies confirm the presence of vascular occlusive disease, angiography is mandatory to determine its level and extent.

Although PTA is best suited for treating relatively isolated disease, it can be applied successfully to more extensive lesions in limb salvage situations, particularly in those patients with high operative risk.[13] If the length of stenosis does not exceed 7 cm, if an occlusion is no longer than 10 to 12 cm, and if the lesion can be safely approached percutaneously, PTA should be considered a primary treatment option. Best results are obtained for short iliac or aortic bifurcation obstructions, with femoropopliteal disease somewhat less likely to respond. However, technical advances have made the effective treatment of even tibial vessels possible.[14] A more controversial suggested indication for balloon angioplasty is the "blue toe" syndrome.[15, 16]

In the abdomen, PTA has become recognized as the treatment of choice for renal fibromuscular dysplasia and renal transplant arterial stenosis, even by the most skeptical surgeons.[17] It is also appropriate for many patients with atherosclerotic stenoses implicated in renovascular hypertension or azotemia, including those patients with solitary kidneys.[18–20]

Balloon angioplasty has also been employed in upper extremity ischemia, "subclavian steal" syndrome, mesenteric ischemia, dialysis access stenoses, and central venous obstructions.[21–24] Clinical efficacy has ranged from very good to poor, depending on the indication. However, the low morbidity of balloon angioplasty has led to its application in situations where long-term success is not likely to be good, because in some individuals a major surgical procedure may be postponed or avoided altogether.

PATIENT PREPARATION

The relationship of a patient's complaints to vascular compromise must be established by history, physical examination, and such noninvasive studies as segmental Doppler blood pressure measurements, pulse volume recordings, or duplex ultrasound. The presence and relative strength of peripheral pulses must be assessed and recorded. Note should be made of hypertension, coagulation status, cardiac and cerebrovascular disease, diabetes, renal insufficiency, and patient medications. A bed must be available on a ward where nursing and other support personnel are experienced in monitoring and treating patients after PTA. In many cases, such as after renal artery dilatation, scheduled admission to an intensive care unit must be arranged. Although outpatient angioplasty has been described, its use has not been widely accepted.[25]

After the patient is examined, the procedure is explained. Attendant risks (including limb loss and death) are presented, and a realistic appraisal of potential benefits is given. Alternative treatment, whether surgical or conservative, is described. The necessity of strict bedrest after PTA with 12- to 18-hour immobilization of the extremity through which the catheter is placed is made clear. If a recent diagnostic arteriogram is not available, the patient is informed that a treatment decision may be made at the conclusion of angiography in consultation with the referring physician and the patient.

Standard preangiography orders are written, including intravenous hydration and restriction of oral intake to clear fluids after midnight prior to the procedure. Unless there is a contraindication, platelet aggregation inhibitors (125 to 325 mg of aspirin orally, with or without dipyridamole 75 mg three times daily) should be given prior to PTA and continued for 3 to 6 months afterward. There is compelling evidence that lack of antiplatelet premedication increases the rates of platelet deposition at angioplasty sites and therefore of early thrombosis.[26–28] For renal, popliteal, and trifurcation PTA, nifedipine 10 mg may be given orally the evening before and the morning of the procedure, in order to prevent spasm. Ideally, angioplasty procedures are performed early in the day, to ensure optimal monitoring and patient care, as well as to allow diabetics to resume their diets as soon as possible.

A good working relationship with one's vascular surgical colleagues is absolutely necessary for the radiologist performing percutaneous angioplasty. Many patients with "angioplastiable" lesions may be treated at the conclusion of diagnostic angiography, and the referring surgeon can be called to review the films before the diagnostic catheter is removed. If the surgeon concurs that PTA is appropriate, the patient's hospitalization can be shortened, and the additional expense of a separate catheter procedure may be avoided. It is for this reason that we routinely include "possible percutaneous angioplasty" in the consent obtained from patients undergoing runoff angiography. Otherwise, the administration of sedatives during the diagnostic examination interferes with the ability of the patient to provide informed consent.

At our institution, we insist that all patients having angiography for peripheral vascular disease be evaluated by a vascular surgeon, to ensure that a given patient is a candidate for revascularization. Surgical back-up is particularly important for aortic, iliac, subclavian, and renal artery PTA. Although major complications are uncommon, the potential consequences of arterial rupture or occlusion in central vessels demand the availability of prompt operative intervention. For patients with transplanted or other solitary kidneys, a surgical team and operating room on standby may be warranted.[4]

RISKS

The hazards of PTA include all the risks of selective angiography, including hypersensitivity reaction to iodinated contrast media, renal failure, and vascular injury related to catheterization. In addition to these basic risks, PTA, by the directed focal trauma it induces as well as by the larger hole needed at the percutaneous entry site and concurrent anticoagulation, presents a greater chance of bleeding at the entry site, acute arterial dissection, thrombosis, or distal embolization. Rarely, the artery dilated may rupture.[29, 30] Death from the procedure, albeit rare, is most commonly due to rupture or hypotension with myocardial infarction.

The risk of complication is dependent on the patient's general condition, the difficulty of approaching and crossing the lesion in question, and the experience of the operator. According to three large series from Boston, Toronto, and Freiburg in which data from 352 to 4,380 patients undergoing a variety of PTA procedures were analyzed, hospital mortality associated with angioplasty ranges from 0.07% to 0.4%.[29, 31, 32] Limb loss attributable to PTA has occurred in 0.4% of procedures.[32] Complications needing prompt surgical attention or more extensive surgery than would have been performed otherwise are cited in 1% to 4% of balloon angioplasties.[29, 31, 32] Major problems are most often uncontrollable bleeding at puncture site, acute arterial occlusion, or distal embolization. Local thrombolysis can often be used to treat the last two of these conditions. Total morbidity of PTA varies from 5% to 19%, depending on the vessels treated.[29] Most complications are minor and do not require any directed therapy or prolongation of hospitalization. Risks specific to certain treated sites are addressed in greater detail later.

To place the hazards of PTA in perspective, one should be aware of the morbidity and mortality of surgery for similar indications. Several recent surgical series have reported mortality of 0.8% to 3.1% for lower extremity bypass grafts, early occlusion or surgical reexploration in 3% to 11%, and wound infection complicating 13% to 19% of in situ saphenous vein grafts.[3, 33, 34]

BASIC TECHNIQUE AND AVAILABLE MATERIALS

Vascular Access

One of the most important technical factors in determining the success and safety of a planned percutaneous angioplasty is vascular access. This usually means a patent common femoral artery providing an antegrade or retrograde approach to the lesion. The procedure can be performed through axillary artery catheterization, and this may be the only practical approach for some stenoses, particularly in mesenteric or renal arteries. However, the greater difficulty in controlling postprocedural puncture site bleeding and the possible neurologic consequences of major hematoma in the arm argue against the axillary approach. The popliteal artery has been used by some for percutaneous recanalization of superficial femoral arteries, but this route is quite unconventional.[35] It should be kept in mind that smaller arteries are more prone to spasm and occlusion.

Vascular Sheaths

Sheaths are invaluable for PTA, because they limit the amount of puncture site trauma from manipulation, make the procedure less uncomfortable for the patient, and permit the reinsertion of a smaller angiographic catheter after balloon removal for postangioplasty arteriography. When antegrade puncture is performed, progress of the procedure may be monitored through intermittent injection of contrast media through the sideport of the sheath. Also, many high-pressure balloons form stiff "wings" when they are forcefully deflated, projections that could lacerate the puncture site during removal. Use of a sheath removes this theoretical hazard. Before a sheath is inserted, one should check that the balloon to be used will pass easily through the sheath and its valve. The size of sheath needed is usually 1.0- to 1.5-F greater than the shaft of the balloon catheter.

Passing the Obstruction

Guidewires

After obtaining vascular access, passage of a guidewire beyond the obstruction is the most critical step in PTA. Rarely, a stenosis may be crossed by the catheter alone, with intermittent injections of small amounts of contrast medium during advancement[36]; however, standard practice is to probe the stenosis or occlusion gently with a soft-tipped guidewire. For very narrow stenoses, a straight or tight (1.5-mm curve) J-tip wire is preferable. A Bentson wire, which is extremely floppy, is often helpful. If the tip of this wire engages a plaque, a loop can be formed by further advancement of the wire. The loop often seeks the lumen of the vessel and secures passage of the obstruction. Other wires that are quite useful include the various torque-control floppy-tipped guidewires. The new hydrophilic polymer-coated guidewires, by their extremely slippery nature when wet, have proved invaluable in difficult lesions.

Digital "Roadmap"

When probing the lesion, a digital vascular "roadmap," a feature available on many digital subtraction angiographic units, can be used to direct the manipulations. A "roadmap" is obtained by contrast medium injection and storage of the resulting image for real-time subtraction during later fluoroscopy. This feature is helpful both for wire placement and for balloon introduction and inflation.[37] If no "roadmap" is available, forceps or towel clips placed adjacent to lesion can be used as reference points during PTA.

Predilatation

A slight curve or hook to the catheter tip, such as present in Berenstein or H1H catheters, helps engage eccentric stenoses. Once a wire traverses the obstruction, the catheter can be advanced through it. This maneuver serves to "predilate" the lesion, and allows the operator to safely exchange the wire for a heavier, stiffer one, as necessary. In some particularly resistant or calcified lesions, passage of a tapered Teflon van Andel catheter can open the lumen to a 7-F size, easing placement of the dilatation balloon.

Dissection or Perforation

If at any point during the manipulations subintimal dissection or perforation is suspected, a sidearm Y-adapter can be attached to the catheter hub, and its O-ring tightened about the guidewire. Contrast medium can be injected through the sidearm, although the pressure needed for injection around the guidewire may be great. If the tip of the catheter is at the suspected perforation or beyond, the wire may be removed and the catheter hub checked for backflow of blood. A small amount of contrast material is injected during fluoroscopy to clarify the situation. If a sheath has been used in an antegrade puncture, the sidearm can be injected to check intraluminal placement of catheter or wire.

When an intimal flap has been raised, the lesion can rarely be traversed; if a guidewire or catheter perforation is found, the procedure must be terminated. In both cases, it is best to stop and try at another date, if indicated. Small catheter or wire perforations in arteries of an extremity are rarely of clinical consequence as long as any anticoagulation is promptly reversed. Patients with perforations involving central vessels (renal, iliac, subclavian) must be closely monitored for bleeding. Because of the possibility of perforation, it is best to employ a 5-F catheter during initial manipulations until the guidewire is securely placed beyond the obstruction.

Use of Medications

As previously noted, patients should be given platelet aggregation inhibitors prior to PTA. Once the lesion has been successfully crossed, heparin should be administered (5,000 units, intravenously or intra-arterially). Heparin need not be continued after a successful procedure, but long-term use of aspirin is strongly recommended.

In vessels prone to spasm, such as distal popliteal, tibial, and renal arteries, direct arterial injection of nitroglycerin (50 to 100 µg) after wire passage and intermittently during subsequent manipulations is advisable. Sublingual nifedipine (10 to 20 mg) may be given before or during PTA, and its combination with injected nitroglycerin may have an additive effect, countering spasm.[38] In any case, vascular spasm can be more easily prevented than treated, once present. If acute thrombosis occurs during the procedure, a thrombolytic agent infused locally often restores flow.

Choosing a Balloon

The appropriate size of balloon can be gauged by the diameter of the relatively normal artery adjacent to the lesion being treated. Conventional film angiography provides for direct measurement of a slightly magnified vessel (approximately 120% the true diameter of the lumen). Alternatively, the size of the vessel may be compared with the diameter of the catheter (or calibration markers on some catheters), if both are included on the same digital subtraction frame or 100-mm spot film.

Most angiographers choose a balloon that matches the diameter of the slightly magnified image, while some prefer a balloon 1 to 2 mm larger.[38] In experimental models, a dilated artery is unlikely to rupture unless the balloon diameter exceeds 150% of the normal true diameter of the vessel.[39] Although it has been asserted that deliberate overdilation improves results, a prospective randomized study conducted by Roubin et al. found no significant difference in restenosis rates between moderate overdilation and underdilation of coronary artery lesions.[40] The length of the balloon should not greatly exceed the length of the lesion, otherwise endothelial trauma will unnecessarily extend into nonstenosed portions of the vessel. Excessive balloon length increases the surface area denuded of endothelium and the risk of acute thrombosis after PTA.

Balloon technology has improved greatly since Grüntzig's first catheters. High-pressure polyethylene balloons have enhanced the effectiveness of PTA in tough, calcified lesions. Low-profile balloons are more easily positioned into tight stenoses. Catheters with short, soft tips are less likely to traumatize vessels when balloon inflation straightens the curve of a vessel. Many balloons are now available on 5-F catheter shafts (rather than 7 F), diminishing the likelihood of puncture site bleeding. Balloons have even been manufactured directly on standard-size angiographic guidewires.[41] The Fogarty-Chin dilatation balloon is an evaginating device purported to decrease shear stress at the lesion.[42] It must be inserted through a sheath, for it cannot be placed over a wire.

Balloon Inflation, End Points, and Predictors of Success

If a high-pressure balloon is used, a balloon pressure gauge is not required, as long as hand inflation is performed with a syringe of 10 mL or greater volume. Smaller syringes can be used to generate higher pressures, and such inflations are best monitored by one of the various gauges available. The same holds true for balloons or other PTA catheters not designed to tolerate more than 7 to 8 atmospheres of pressure. When high inflation pressure is desired, the LaVeen syringe, a device with a threaded screw attached to the plunger, is useful. The manufacturer's recommended pressure limits should not be exceeded. Balloon rupture (ideally a longitudinal tear) generally does not cause arterial injury, but can make catheter removal quite difficult, and cases of separation and embolization of balloon material have occurred.

Inflation proceeds with injection of 30% contrast material monitored fluoroscopically. More concentrated contrast material is viscous and slows the inflation-deflation process. A balloon should not be test-inflated prior to insertion into the patient. The profile of the device is disturbed by such a test, and introduction can be made quite difficult. As the balloon is inflated, the indentation or "waist" formed by the stenosis is noted. If it remains evident during deflation, the balloon is reexpanded until residual stenosis disappears. If a lesion is refractory to initial dilatation, a larger balloon may be tried.

Balloon angioplasty produces controlled injury, irreversibly stretching the vessel wall, and cracking plaque, intima, and media.[43] Successful angioplasty produces intimal clefts normally, and these become endothelialized and remodeled to some extent by healing over 4 to 6 weeks. Only if clefts are seen to extend substantially beyond the PTA site should dissection be considered a complication. Because of the splits produced, a guidewire must be maintained through the lesion at all times after initial inflation until the decision is made to terminate the procedure. Otherwise, there is risk of producing an occluding dissection if a guidewire is reintroduced into the lesion. Also, should vessel rupture occur (a rare, but potentially fatal complication), an occlusion balloon can be introduced to tamponade hemorrhage while the patient is taken to surgery.

The duration of balloon inflation optimal for PTA has not been well established. Outside of the coronary and brachiocephalic circulations, duration of inflation is rather arbitrary, and the balloon may be left expanded for over 60 seconds at a time. An inflation of 20 to 40 seconds is commonly used, and multiple intermittent inflations are the rule. If there is a persistent "waist" to the balloon at the stenosis, longer inflations may result in more irreversible stretching of fibrous and elastic elements in the vessel wall. On the other hand, if the lesion is seen to "pop" open to full vessel diameter on the initial balloon dilatation, multiple inflations may be superfluous.

By the nature of the angioplasty mechanism, concentric stenoses are more likely to respond than very eccentric lesions. In the latter, the plaque itself may not be disrupted, and pressure will simply expand the relatively normal elastic wall of the vessel. Heavily calcified lesions or fibrous stenoses (such as anastomoses) can be difficult to open.

A technically successful PTA is measured by absence of residual stenosis greater than 30%, no residual systolic pressure gradient above 10 mm Hg, and good blood flow through the vessel. Pressure measurements made before and after angioplasty are often helpful in assessing the results of treatment in iliac and renal artery lesions. Direct measurements are rarely obtained in vessels distal to the inguinal ligament. Additional lesions not likely to be obstructing blood flow are left alone in order to minimize the extent of endothelial trauma.

Restenosis

Postangioplasty restenosis has been a particular problem in the coronary vessels, with rates exceeding 40% in some series.[40] In the renal arteries, recurrence is less likely, with 13% representing a typical restenosis rate at 2 years of follow-up.[44] Similar figures can be expected in the peripheral circulation, and most recurrences are evident within 1 year.[45, 46] If a patient develops new symptoms after 2 years, a new lesion at a separate site is usually responsible. Despite the non-negligible rates of recurrence, PTA remains a valuable option, particularly because most recurrent lesions can be successfully dilated a second time without any increased likelihood of subsequent recurrence.[21, 31, 34, 47, 48] It should also be noted that previous surgical endarterectomy does not present a contraindication to balloon angioplasty.[49]

LOWER EXTREMITY ANGIOPLASTY

In selecting patients for PTA in a lower extremity, it is important to correlate the location, extent, and morphology of lesions with the patient's symptoms. For limiting claudication, PTA produces good results in those with focal, isolated lesions and intact distal vessels. It is not good practice to dilate more than five discrete lesions in an extremity; surgery is a better choice. However, if a patient is suffering tissue necrosis or pain while at rest, balloon angioplasty can be applied to more extensive lesions or in the face of diffuse disease, for such patients tend to do poorly with surgery.[13]

Diagnostic arteriography is best done from the side opposite the area of major symptoms. If PTA is to follow immediately, many lesions in the iliac, common femoral, or proximal superficial femoral arteries can be catheterized antegrade by way of the aortic bifurcation. Should this fail, the retrograde approach can be used for pelvic lesions, or antegrade puncture employed to reach femoral or more distal obstructions. Whether the artery is entered in an antegrade or retrograde fashion, special care must be taken to avoid arterial entry above the inguinal ligament, for massive bleeding can result.

Gross obesity is a relative contraindication to femoropopliteal PTA, because adequate postprocedural compression of the puncture site is difficult, and such patients are prone to developing large hematomas. Introducing wires, catheters, and sheaths through a deep layer of fat can also be quite problematic. If safe vascular access is unavailable, surgical arteriotomy must be considered.[50]

Aortic Procedures

Most focal occlusive disease of the abdominal aorta involves the bifurcation. However, purely aortic stenoses may arise in women with a history of smoking. These lesions are quite amenable to transluminal angioplasty, often with a single balloon because of the small size of the aorta (less than 12 mm diameter) in most patients. Results are comparable to those of surgical bypass.[51, 52] However, because of the greater stress placed on the walls of larger vessels, the aorta can rupture after PTA, particularly if heavy calcification is present.[53]

Aortic bifurcation lesions are best treated by placement of balloons from each femoral artery with simultaneous inflation ("kissing balloon" technique). This maneuver prevents embolization to or occlusion of the contralateral iliac artery. Bifurcation stenoses can be opened in over 90% of cases, with 90% patency at 1 year.[54]

Iliac Procedures

If only technically successful PTA procedures are considered, the only large-scale prospective randomized study available indicates there is no difference in 3-year cumulative patency between patients treated with balloon angioplasty or those who underwent surgical bypass.[3] For stenotic lesions, immediate success can be expected in over 90%, and 5-year patency rates between 63% and 90% have been recorded.[31, 34, 45, 55]

There was early hesitancy to refer patients to transluminal angioplasty for iliac occlusion, because of fears of vessel perforation or embolization. Still, some investigators have shown that recanalization can be performed safely and effectively, especially if a lesion is no longer than 5 cm.[56, 57] Although a straight wire and catheter can be used, Colapinto et al. prefer to advance a catheter shaped like a hockey stick into the thrombus and inject small amounts of contrast material, finding that the lumen is often defined in this manner.[57] Technical success can be expected in 67% to 78% of patients, the lower percentage for long occlusions. However, long-term patency is good for long occlusions if they are successfully reopened.[57] Acute occlusions should not be treated by angioplasty alone, but PTA can follow successful thrombolysis (see Chapter 11, Local Thrombolytic Infusion).

An advantage of aortoiliac balloon angioplasty vis-à-vis surgical bypass in men is that sexual potency may be preserved. In fact, PTA has been used specifically in the treatment of vasogenic impotence in isolated cases.[58]

Femoropopliteal Procedures

Long-term results of PTA in the superficial femoral or popliteal arteries are comparable to those of surgical bypass if one acknowledges the lower initial technical success rate of PTA.[3, 34] Effective dilation of stenoses can be expected in 85% to 96% of cases, but recanalization of occlusions is considerably less successful depending on the length of the lesion.[3, 46, 47] Still, occlusions up to 3 cm in length can be opened as readily as focal stenoses.[46]

If one includes all procedures as a base (including unsuccessful PTAs), long-term (2- to 5-year) patency rates in patients treated for claudication are on the order of 53% to 58%.[31, 34, 59] Those treated for limb salvage have a somewhat worse prognosis (see "Limb Salvage," later in this chapter). It is clear that lesions longer than 3 to 7 cm are both more difficult to recanalize and have higher reocclusion rates.[46, 47, 59]

Procedures Below the Knee

The use of coronary-type 0.016- and 0.018-inch guidewires, small-shaft (4.5 F) balloon catheters, and drugs such as nifedipine, nitroglycerin, and verapamil has greatly improved results of balloon angioplasty in the distal popliteal artery and trifurcation vessels. Schwarten and Cutcliff have achieved a remarkable 97% technical success rate for PTA in 98 patients, with limb salvage at 2 years in 86%.[14] In small-vessel PTA, it is reasonable to continue heparin use for several days after intervention. Surgical bypass (in situ or reversed saphenous vein grafts) in patients with distal lower extremity vascular disease has a 2-year patency of 73%, but if prosthetic material must be used, only 30% of grafts remain open for that period.[33] If a vein is not available for graft placement, PTA should be attempted in patients with disease in the calf arteries, and it may be pursued in others with limited tibial-peroneal lesions, in order to preserve the saphenous vein for later cardiac or other revascularization surgery.

Limb Salvage

When a patient is treated for rest pain or gangrene, results of revascularization, whether surgical or by balloon angioplasty, are predictably worse than for other indications for PTA. Complications of the procedure are twice as common in such patients as in those treated for claudication only.[55] Even so, it is difficult to predict who will have a good long-term outcome, and late occlusion does not necessarily mean clinical failure, because many ischemic ulcers may be healed and will not recur.[13] The presence of occlusion, long-segment disease, or poor distal runoff does not preclude a good clinical result. Therefore, because many patients with rest pain or tissue loss have multiple medical problems, and because many do not have donor veins for grafting, PTA can be considered a first-line treatment for limb salvage.

Patency at hospital discharge can be expected in about 70% of all patients treated by PTA, with long-term patency of approximately 40% at 2 to 5 years.[13, 34, 55] Limb salvage rates exceed long-term patency.[60] If a dilatation is initially successful, clinical benefit is as likely to be durable as that provided by bypass, and failed PTA is unlikely to worsen the patient's clinical status or operative risk.[13] On the other hand, failure of a femoropopliteal graft in a patient treated for threatened lower extremity is more likely to result in major amputation than a failed PTA.[61]

Bypass Grafts

There has been relatively little information presented on the use of balloon angioplasty alone in failing bypass grafts, but it does appear that both venous and prosthetic grafts can be dilated with excellent long-term results.[62] Nonanastomotic stenoses tend to be more difficult to open, and may need serial dilatation.[63] The more common situation has been graft thrombosis, which requires thrombolysis followed by PTA or graft revision (see Chapter 11, Local Thrombolytic Infusion). However, more careful noninvasive monitoring of bypass grafts, particularly by regular duplex ultrasound, makes it likely that more patients with failing (nonoccluded) grafts will present for evaluation and treatment in the future.

Blue Toe Syndrome and Balloon Angioplasty

The blue toe syndrome is the sudden unilateral or bilateral appearance of digital ulcerations and severe pain, arising from a shower of emboli from a proximal source. Most are due to platelet thrombi, and treatment is needed to prevent recurrence and possible amputation.[16] Anticoagulation may not be an adequate measure, and two recent reports[15, 16] have stressed the usefulness of PTA if an iliac or femoropopliteal lesion is felt to be the source of unilateral emboli. Brewer et al.[16] have advocated 6 to 12 weeks of anticoagulation prior to elective angioplasty, while Kumpe et al.[15] have dilated lesions immediately at the time of discovery. More experience is needed to determine the role of PTA in this clinical setting.

RENAL ARTERY ANGIOPLASTY

Renal PTA has allowed control or resolution of renovascular hypertension in many patients, who thus avoid the mortality and morbidity of surgical revascularization or nephrectomy. Lateralized elevation of renin values in selectively obtained venous samples helps predict a beneficial result, but absence of lateralization should not be seen as a contraindication, because about 50% of such patients will still improve following PTA. Balloon angioplasty has also been of value in selected patients with severe renal vascular disease and renal insufficiency. This includes individuals with renal transplants and others with solitary functioning kidneys.

Patients undergoing renal artery angioplasty should have continual monitoring afterward, preferably in an intensive care unit, because blood pressure can change rapidly. If blood pressure drops, it is likely to do so in the first 1 to 2 days after PTA. However, transient early hypertension can be noted in 30% of patients within 2 hours of the procedure and may persist up to 24 hours, presumably a renin "washout" phenomenon.[64] It is advisable not to discontinue antihypertensive medications prior to PTA, and angiotensin converting enzyme inhibitors help prevent a sudden blood pressure drop after the dilation.[38]

Patients with extremely atherosclerotic "shaggy" aortas should have PTA only after very careful deliberation, for they are more likely to experience cholesterol embolization, a cause of renal loss or death in isolated cases.[19, 65] Simmons catheters are effective in catheterizing most renal arteries, but difficult stenoses may require changing the degree of inspiration or injection of contrast material during catheter advancement.[38] In some cases a coaxial catheter balloon system is useful in catheterizing stenoses, particularly in branch renal vessels.

The guidewire placed after initial catheterization of the lesion must be long enough to allow exchange for the dilating catheter. In practice, a Rosen 1.5-mm J-tip wire (180 cm long) is a good choice for most exchanges. However, care must be taken to prevent traumatization of small branch vessels, otherwise a use of a straight-tip exchange wire is advised. During balloon catheter introduction, the position of the wire must be followed carefully by fluoroscopy to prevent accidental withdrawal or excessive advancement of wire. Spasm can be a difficult problem in renal angioplasty; premedication with nifedipine and frequent intra-arterial injections of nitroglycerin (100 to 200 μg) are strongly recommended.

Hydration is particularly important, because many patients already have functional impairment of the kidneys prior to PTA. Martin et al. have found that administration of mannitol (25%) over 8 hours at 50 mL/hr in those deemed to be at high risk for contrast material–related renal failure has dropped the incidence of this complication from 11% to 5%.[36]

Hypertension

Technical success for renal artery balloon angioplasty has been reported as between 81% and 97%.[36, 38, 44, 65, 66] Major complications are seen in 11% to 13% of patients, and risks of permanent impairment of renal function, nephrectomy, and death are each under 1%.[65, 66] Use of undersized balloons and residual stenosis greater than 30% are factors associated with increased recurrence.[38] Bilateral lesions can be treated during one session without any increase in morbidity. However, if any difficulty is encountered in the treatment of one kidney, deferring dilation of the opposite side is prudent.

Fibromuscular Dysplasia

Clinical results in hypertension (cure if diastolic blood pressure is less than 90 mm Hg without medications; improvement if diastolic pressure significantly drops on the same or fewer medications) are best for patients with fibromuscular dysplasia (FMD). At 6-month to 2-year follow-up, there is clinical benefit in 81% to 100% of those with successful dilations.[17, 44, 65, 66–68] Cumulative 5-year patency has been 89% in FMD patients among those with initial response, underscoring the durability of results.[65] Restenosis rates for FMD patients are low. Contray to common perception, atypical fibromuscular lesions unlikely to represent medial fibroplasia *can* be effectively dilated.[69]

Atherosclerosis

Renovascular hypertension from atherosclerotic disease responds to PTA in a less predictable manner, and few patients are likely to be cured of their high blood pressure. While 47% to 65% of patients will have at least some improvement 6 to 16 months after dilatation of unilateral renal artery lesions, only 14% to 46% of those treated for bilateral lesions have had benefit persist for that long a time.[66, 68] Sos et al. attributed their very poor clinical results in bilateral atheromatous disease to an inability to achieve a satisfactory technical result.[68] A more recent study by Klinge et al., however, reported treatment of 42 patients with bilateral lesions with good initial clinical response in 38 (90%).[65] In the same study, of all 133 patients in whom dilatation of atherosclerotic lesions was attempted, 77% had cure or improvement in their hypertension at 6 months, and cumulative 5-year response stayed high. The difference in these reports might be explained by improvement in technique, equipment, and patient selection over time.

Dilation of Ostial Stenoses

Many atherosclerotic lesions are ostial: that is, the stenosis involves the main renal artery near its origin. Ostial stenosis dilatation procedures have generally had poor technical and clinical results, ostensibly because it is aortic plaque that is actually subjected to longitudinal displacement, rather than renal plaque undergoing the desired radial disruption. Which particular lesion actually is ostial depends on the thickness of plaques involving the aorta. In practice, this can only be estimated, but narrowings in the first 5 to 10 mm of the renal artery probably involve its origin. Although success is not as good as for nonostial stenoses, PTA is still worth pursuing in selected patients. Klinge et al. have been able to achieve a satisfactory technical result in 82% of their ostial stenosis patients, with subsequent recurrent hypertension in 18%.[65] Martin et al. have found that response to PTA performed to improve renal function did not depend on the ostial vs. nonostial location of stenosis.[19]

Special Situations in Renal Angioplasty

Renal Transplants

The incidence of renal artery stenosis after transplantation varies widely, but can be as high as 16%.[70] End-to-end anastomoses to the internal iliac artery are best approached by catheterization from the contralateral femoral artery; end-to-side anastomoses to the external iliac should be catheterized from the ipsilateral approach. Most stenoses involve the anastomosis or the main renal artery, and balloon dilatation can be performed with greater than 80% technical success.[70] Improvement in function and hypertension is maintained in 50% to 54% of patients at 1 year or more.[17, 70] As in other vessels, restenosis can usually be treated by a second dilatation procedure.

Azotemia

Patients with azotemia and renal vascular disease may have a single functioning kidney, two kidneys with bilateral lesions, or two kidneys but only unilateral arterial obstruction. Hypertension is a common accompanying problem. Persons suffering severe and long-standing renal insufficiency are not likely to benefit by an attempted balloon angioplasty, but roughly half of those not yet on dialysis will have improved function at 1 year of follow-up.[19, 71, 72] Balanced against this potential benefit is the possibility of at least transient deterioration of renal status in about one in four such patients.[17, 71] Treatment can be justified by the high mortality and cost of chronic hemodialysis in those who are not candidates for transplantation.

Recanalization of Occlusions

Kadir et al. have used angioplasty to treat acute arterial occlusion in five patients with solitary kidneys, achieving return of function in three.[20] Others have approached chronically occluded vessels with balloons and have produced mixed results.[38, 65] Such attempts are best reserved for extraordinary situations.

Surgery for Failure or Complication of Balloon Angioplasty

Although balloon angioplasty may fail or may be responsible for acute arterial occlusion or rupture in a small number of cases, prompt surgical intervention can usually salvage the kidney. Beebe et al. found that their surgical approach was altered by attempted PTA in only 1 of 9 patients, and all but 1 have a good result at a mean of 3 years of follow-up.[17] Revascularization was successful in 10 of 13 kidneys treated by McCann et al., including 4 of 5 patients experiencing a complication of angioplasty.[5]

SUBCLAVIAN BALLOON ANGIOPLASTY

Despite initial fears that balloon angioplasty in the brachiocephalic vessels would be complicated by stroke, subclavian PTA has proved quite safe, even with antegrade flow in a vertebral artery originating distal to the treated lesion.[73] Balloon angioplasty has been used effectively at a number of centers in patients suffering subclavian steal syndrome.[21, 73, 74] In all cases of subclavian angioplasty, it is strongly suggested that a full four-vessel cerebral study be obtained to assess the pattern of intracranial circulation before an intervention is attempted. Although the retrograde brachial or axillary approach has been employed, it is not as likely to be successful as a femoral approach. Subclavian arteries have been recanalized in a small number of cases, but results have been poor for occlusions.[21, 74] Lesions involving the segment from which a patent vertebral artery arises are best not dilated. If a stenosis is near (but not involving) the vertebral, a balloon can be inflated across the vertebral artery origin with little danger of adverse consequences.[73]

OTHER APPLICATIONS

Aside from its application in coronary artery disease, and in instances of cerebrovascular disease, transluminal angioplasty has also been applied to mesenteric ischemia. With the exception of ostial stenosis and median arcuate ligament compression of the celiac exis, mesenteric PTA has had good technical success, but a high incidence of restenosis.[23] Long-term results with central venous and dialysis access stenosis PTA have been disappointing. Still, the relatively short life of a dialysis shunt and the difficulty of treating central venous strictures by others means, as well as the low morbidity of a second PTA procedure, support the use of transluminal dilatation for these clinical problems.[22, 75] The use of stents may play a significant role in treating venous obstructions (see Chapter 9, Dialysis Access, Central Venous Catheters, and Other Central Venous Problems).

SALVAGE OF PROCEDURES COMPLICATED BY THROMBOSIS OR EMBOLISM

If catheterization or dilatation of a lesion results in acute arterial thrombosis, all is not necessarily lost. Immediate local infusion of urokinase or other thrombolytic agent can reestablish flow in the great majority of cases[76] (see Chapter 11, Local Thrombolytic Infusion). However, distal embolization of plaque fragments will cause occlusion unresponsive to thrombolysis.

Distal embolization (as well as thrombotic occlusion) can be treated by passage of a Fogarty embolectomy catheter into the affected vessel by way of a sheath.[77] If this maneuver is chosen, the sheath should be cut or the cap of the sheath removed after the Fogarty catheter is withdrawn, in order to allow any retained clot or debris to be flushed out of the sheath by back-bleeding.

Starck et al. have used percutaneous aspiration thromboembolectomy (PAT) in similar situations.[78] The PAT procedure involves use of a special long 8-F sheath with a removable hemostasis valve. A minimally tapered thin-wall catheter is passed to the clot or obstruction, and strong suction is applied with a large syringe. The catheter is removed, the debris expelled, and the procedure repeated until distal runoff is reestablished.

COST EFFECTIVENESS OF PERCUTANEOUS ANGIOPLASTY

Considering the low morbidity and high effectiveness of PTA in properly selected patients, it is an underutilized treatment of peripheral vascular disease. Up to 40% of patients presenting with lesions amenable to surgical revascularization are candidates for balloon angioplasty.[79] Hospital stay and treatment charges may be less than one third those of surgically treated patients.[79] Although no large prospective cost-comparison studies are at hand, it is reasonable to assume that even if the costs of complications and technical failures are factored in, PTA remains clearly less expensive than surgical bypass. Average hospital stay is but a fraction of that for surgical patients.[34] However, PTA need not be considered competition for vascular surgery, because its availability may actually greatly increase the total number of patients referred to an institution for treatment of peripheral vascular disease![34]

NEW DIRECTIONS

Because of the lower success rates of PTA in recanalization of long occlusions and in patients with diffuse disease, and because of continuing problems with restenosis, a number of laser, thermal, ultrasonic, and mechanical atherectomy devices have been developed and are being tested.

Laser-Assisted Angioplasty

Early attempts to use laser light from bare fibers were complicated by numerous vessel perforations. Widespread application of lasers to the treatment of atherosclerotic disease only followed development of capped catheters in which laser light merely heated the metallic tip.[6] Such "hot-tip" lasers recanalize vessels by a combination of mechanical means and plaque vaporization. However, hot-tip catheters tend to follow the path of least resistance.[80] By deflecting off of calcified plaques, catheters can produce medial dissection. Perforation rates of 4% to 14% have been reported.[6, 80, 81] Newer modifications of the original hot-tip catheters include the capability of using a guidewire, and placement of a small sapphire cap to allow some egress of laser light from the tip.

In another approach, laser light is completely dispersed from the catheter tip by a lens, with plaque ablation primarily due to light absorption.[7, 81] The tips of such catheters are cooled by perfusion of saline, which also clears blood and debris from the field of treatment. One lens-tip laser system uses a distal balloon for centering, but this modification does not completely eliminate the risk of perforation.[7] Other "smart" systems under investigation obtain feedback by spectral analysis of reflected laser light, or by intravascular ultrasound to prevent the exposure of normal structures to laser energy.

Disadvantages of laser recanalization devices include the high cost compared with balloon catheters, the need for repeated removal and cleaning of debris from the catheter tip in long lesions, and the small size of lumen produced. Because of the last factor cited, balloon dilatation must be performed as an adjunct to laser recanalization in the great majority of cases. Rates of technical success are high for stenoses or short (<3 cm) occlusions, but they can be as low as 57% to 66% for hot-tip laser catheters in long occlusions.[6, 80] A reasonable overall recanalization rate to be expected with lasers seems to be 80%.[6, 7, 80–82] To date there is no conclusive evidence that use of a laser improves clinical outcome, restenosis rates, or complication rates of percutaneous angioplasty. What lasers *do* provide is an extension of PTA to longer or otherwise impassable lesions. Further technical modifications, increased operator experience, and refined indications can be expected to improve clinical results.

Mechanical and Other Atherectomy Devices

Transluminal systems for mechanical removal or disruption of plaque have already entered clinical use. The Simpson atherectomy catheter, 7 F to 9 F in diameter, can be inserted through a percutaneous sheath. A motor-driven cutting blade is housed in a cylindrical metal capsule. The aperture of the capsule is forced against a plaque by a low-pressure balloon inflated on the side of the capsule opposite the cutting aperture. Although a small wire is appended to the leading end of the capsule, the catheter is not guided *over* a wire. The Simpson catheter must work through an established lumen; an occluded vessel must be partially reopened to allow its introduction. Also, because only a finite amount of material can be held by the capsule, the catheter must be repeatedly removed and emptied for extensive atherectomy procedures. Very early results indicate Simpson catheters are quite effective for treating isolated lesions, but restenosis is not completely eliminated, occuring in 7% of patients by 6 months.[10]

The Kensey catheter uses a rotating cam to bore through and pulverize occlusive lesions. Early modifications have encountered problems with perforation.[83] Also, embolic particles from fragmented plaque have produced small cerebral and renal infarcts in animal models, but peripheral embolization does not appear to be a clinical problem in the lower extremities.[11, 83, 84] As with laser catheters, balloon dilatation must be performed at the conclusion of most recanalizations.

The Auth burr is a rotating, football-shaped brass device containing embedded diamond chips. Due to its relative lack of compliance, calcified plaque is abraded preferentially by the burr, most resulting particles being <10 nm in diameter.[9] Because of their size (up to 4.5 mm in diameter), larger burrs must be introduced through a surgical arteriotomy. The recanalization of occlusions is limited to lesions that can be traversed prior to atherectomy, for the Auth device is advanced over a guidewire. Other alternatives for plaque ablation being explored currently include the application of electrothermal and ultrasonic energy.[8, 85]

Endovascular Stents

Vascular stents are metallic meshes of stainless steel, tantalum, or titanium alloys that are introduced by means of sheaths and fixed into place by inflation of a balloon carrier-catheter or by properties of self expansion.[12, 86, 87] Stents are meant to improve the results of PTA by resisting any elastic elements in vascular segments not opened completely by balloon alone, and to tack down intimal flaps. It is hoped they will diminish rates of restenosis as well. Isolated reports support the use of stents in patients with superior vena caval syndrome.[88] However, it is too early to judge the overall utility of catheter-placed stents.

Which of the various devices described (or of others being developed) will find wide use depends on the safety, efficacy, and economy demonstrated in clinical trials currently under way. A device that greatly adds to cost while only modestly improving results is not likely to be accepted. The procedure of balloon angioplasty will not be rendered obsolete in the near future. On the contrary, the indications for PTA may be considerably expanded, if the promise of new revascularization devices is borne out.

REFERENCES

1. Dotter CT, Judkins MP: Transluminal treatment of arteriosclerotic obstruction: Description of a new technique and preliminary report of its application. *Circulation* 1964; 30:654–670.

2. Grüntzig A, Hopff H: Perkutane rekanalisation chronischer arterieller verschlüsse mit einem neuen dilatationskatheter: Modifikation der Dotter-technik. *Dtsch Med Wochenschr* 1974; 99:2502–2505.

3. Wilson SE, Wolf GL, Cross AP, et al: Percutaneous transluminal angioplasty versus operation for peripheral arteriosclerosis. *J Vasc Surg* 1989; 9:1–9.

4. McCann RL, Bollinger RR, Newman GE: Surgical renal artery reconstruction after percutaneous transluminal angioplasty. *J Vasc Surg* 1988; 8:389–394.

5. White RA, White GH: Laser thermal probe recanalization of occluded arteries. *J Vasc Surg* 1989; 9:594–604.

6. Sanborn TA, Cumberland DC, Greenfield AJ, et al: Percutaneous laser thermal angioplasty: Initial results and 1-year follow-up in 129 femoropopliteal lesions. *Radiology* 1988; 168:121–125.

7. Nordstrom LA, Young EG: Direct laser recanalization of occluded superficial femoral and iliac arteries: Primary success and low complication rate. *Semin Intervent Radiol* 1988; 5:277–280.

8. Siegel RJ, Fishbein MC, Forrester J, et al: Ultrasonic plaque ablation: A new method for recanalization of partially or totally occluded arteries. *Circulation* 1988; 78:1443–1448.

9. Ahn SS, Auth D, Marcus DR, et al: Removal of focal atheromatous lesions by angioscopically guided high-speed atherectomy: Preliminary experimental observations. *J Vasc Surg* 1988; 7:292–300.

10. Höfling B, Pölnitz AV, Backa D, et al: Percutaneous removal of atheromatous plaques in peripheral arteries. *Lancet* 1988; 1(8582):384–386.

11. Zeitler E, Kensey K: First own results with dynamic angioplasty with the Kensey-catheter. *Ann Radiol* 1988; 31:77–81.

12. Palmaz JC: Balloon-expandable intravascular stent. *AJR* 1988; 150:1263–1269.

13. Milford MA, Weaver FA, Lundell CJ, et al: Femoropopliteal percutaneous transluminal angioplasty for limb salvage. *J Vasc Surg* 1988; 8:292–299.

14. Schwarten DE, Cutcliff WB: Arterial occlusive disease below the knee: Treatment with percutaneous transluminal angioplasty performed with low-profile catheters and steerable guide wires. *Radiology* 1988; 169:71–74.

15. Kumpe DA, Zwerdlinger S, Griffin DJ: Blue digit syndrome: Treatment with percutaneous transluminal angioplasty. *Radiology* 1988; 166:37–44.

16. Brewer ML, Kinnison ML, Perler BA, et al: Blue toe syndrome: Treatment with anticoagulants and delayed percutaneous transluminal angioplasty. *Radiology* 1988; 166:31–36.

17. Beebe HG, Chesebro K, Merchant F, et al: Results of renal artery balloon angioplasty limit its indications. *J Vasc Surg* 1988; 8:300–306.

18. Working Group on Renovascular Hypertension: Detection, evaluation, and treatment of renovascular hypertension. *Arch Intern Med* 1987; 147:820–829.

19. Martin LG, Casarella WJ, Gaylord GM: Azotemia caused by renal artery stenosis: Treatment by percutaneous angioplasty. *AJR* 1988; 150:839–844.

20. Kadir S, Watson A, Burrow C: Percutaneous transcatheter recanalization in the management of acute renal failure due to sudden occlusion of the renal artery to a solitary kidney. *Am J Nephrol* 1987; 7:445–449.

21. Erbstein RA, Wholey MH, Smoot S: Subclavian artery steal syndrome: Treatment by percutaneous transluminal angioplasty. *AJR* 1988; 151:291–294.

22. Glanz S, Gordon DH, Lipkowitz GS, et al: Axillary and subclavian vein stenoses: Percutaneous angioplasty. *Radiology* 1988; 168:371–373.

23. Odurny A, Sniderman KW, Colapinto RF: Intestinal angina: Percutaneous transluminal angioplasty of the celiac and superior mesenteric arteries. *Radiology* 1988; 167:59–62.

24. Ali MK, Ewer MS, Balakrishnan PV, et al: Balloon angioplasty for superior vena cava obstruction. *Ann Intern Med* 1987; 107:856–857.

25. Lemarbre L, Hudon G, Coche G, et al: Outpatient peripheral angioplasty: Survey of complications and patients' perceptions. *AJR* 1987; 148:1239–1240.

26. Richter E-I, Zeitler E: Percutaneous transluminal angioplasty: Adjunct drug therapy, in Dotter CT, Grüntzig AR, Schoop W, et al (eds): *Percutaneous Transluminal Angioplasty: Technique, Early and Late Results.* Berlin, Springer-Verlag, 1983, pp 84–90.

27. Cunningham DA, Kumar B, Siegal BA, et al: Aspirin inhibition of platelet deposition at angioplasty sites: Demonstration by platelet scintigraphy. *Radiology* 1984; 151:487–490.

28. Brown KT, Schoenberg NY, Moore ED, et al: Percutaneous transluminal angioplasty of infrapopliteal vessels: Preliminary results and technical considerations. *Radiology* 1988; 169:75–78.

29. Gardiner GA Jr, Meyerovitz MF, Stokes KR, et al: Complications of transluminal angioplasty. *Radiology* 1986; 159:201–208.

30. Jensen SR, Voegeli DR, Crummy AB, et al: Iliac artery rupture during transluminal angioplasty: Treatment by embolization and surgical bypass. *AJR* 1985; 145:381–382.

31. Johnston KW, Rae M, Hogg-Johnston SA, et al: Five-year results of a prospective study of percutaneous transluminal angioplasty. *Ann Surg* 1987; 206:403–413.

32. Fraedrich G, Beck A, Bonzel T, et al: Acute surgical intervention for complications of percutaneous transluminal angioplasty. *Eur J Vasc Surg* 1987; 1:197–203.

33. Veterans Administration Cooperative Study Group 141: Comparative evaluation of prosthetic, reversed, and in situ vein bypass grafts in distal popliteal and tibial-peroneal revascularization. *Arch Surg* 1988; 123:434–438.

34. Cole SEA, Baird RN, Horrocks M, et al: The role of balloon angioplasty in the management of lower limb ischemia. *Eur J Vasc Surg* 1987; 1:61–65.

35. Cronin TG, Calandra JD, Sheridan PH, et al: Retrograde puncture of popliteal artery for access to peripheral arteries for laser angioplasty. *Semin Intervent Radiol* 1988; 5:281–282.

36. Martin LG, Casarella WJ, Alspaugh JP, et al: Renal artery angioplasty: Increased technical success and decreased complications in the second 100 patients. *Radiology* 1986; 159:631–634.

37. McDermott JC, Babel SG, Crummy AB, et al: Review of the uses of digital "road map" techniques in interventional radiology. *Ann Radiol* 1989; 32:11–13.

38. Tegtmeyer CJ, Sos TA: Techniques of renal artery angioplasty. *Radiology* 1986; 161:577–586.

39. Zollikofer CL, Salomonowitz E, Castañeda-Zuñiga WR, et al: The relationship between arterial and balloon rupture in experimental angioplasty. *AJR* 1985; 144:777–779.

40. Roubin GS, Douglas JS Jr, King SB, et al: Influence of balloon size on initial success, acute complications, and restenosis after percutaneous transluminal coronary angioplasty: A prospective randomized study. *Circulation* 1988; 78:557–565.

41. Tegtmeyer C: Guide wire angioplasty balloon catheter: Preliminary report. *Radiology* 1988; 169:253–254.

42. Kinney TB, Fan M, Chin AK, et al: Shear force in angioplasty: Its relation to catheter design and function. *AJR* 1985; 144:115–122.

43. Wolf GL, LeVeen RF, Ring EJ: Potential mechanisms of angioplasty. *Cardiovasc Intervent Radiol* 1984; 7:11–17.

44. Tegtmeyer CJ, Kellum CD, Ayers C: Percutaneous transluminal angioplasty of the renal artery. *Radiology* 1984; 153:77–84.

45. van Andel GJ, van Erp WFM, Krepel VM, et al: Percutaneous transluminal dilatation of the iliac artery: Long-term results. *Radiology* 1985; 156:321–323.

46. Krepel VM, van Andel GJ, van Erp WFM, et al: Percutaneous transluminal angioplasty of the femoro-popliteal artery: Initial and long-term results. *Radiology* 1985; 156:325–328.

47. Murray RR Jr, Hewes RC, White RI Jr, et al: Long-segment femoropopliteal stenoses: Is angioplasty a boon or a bust? *Radiology* 1987; 162:473–476.

48. Greminger P, Schneider E, Siegenthaler W, et al: Renovaskuläre hypertonie. *Internist* 1988; 29:246–251.

49. Tisnado J, Vines FS, Barnes RW, et al: Percutaneous transluminal angioplasty following endarterectomy. *Radiology* 1984; 152:361–364.

50. Wilms G, Nevelsteen A, Baert A, et al: Intraoperative angioplasty. *Cardiovasc Intervent Radiol* 1987; 10:8–12.

51. Charlebois N, Saint-Georges G, Hudon G: Percutaneous transluminal angioplasty of the lower abdominal aorta. *AJR* 1986; 146:369–371.

52. Odurny A, Colapinto RF, Sniderman KW, et al: Percutaneous transluminal angioplasty of abdominal aortic stenoses. *Cardiovasc Intervent Radiol* 1989; 12:1–6.

53. Berger T, Sörenson R, Konrad J: Aortic rupture: A complication of transluminal angioplasty. *AJR* 1986; 146:373–374.

54. Tegtmeyer CJ, Kellum CD, Kron IL, et al: Percutaneous transluminal angioplasty in the region of the aortic bifurcation. *Radiology* 1985; 157:661–665.

55. Zeitler E, Richter EI, Roth FJ, et al: Results of percutaneous transluminal angioplasty. *Radiology* 1983; 146:57–60.

56. Rubinstein AJ, Morag B, Peer A, et al: Percutaneous transluminal recanalization of common iliac artery occlusions. *Cardiovasc Intervent Radiol* 1987; 10:16–20.

57. Colapinto RF, Stronell RD, Johnston WK: Transluminal angioplasty of complete iliac obstructions. *AJR* 1986; 146:859–862.

58. Castañeda-Zuñiga WR, Smith A, Kaye K, et al: Transluminal angioplasty for treatment of vasculogenic impotence. *AJR* 1982; 139:371–373.

59. Henriksen LO, Jørgensen B, Holstein PE, et al: Percutaneous transluminal angioplasty of infrarenal arteries in intermittent claudication. *Acta Chir Scand* 1988; 154:573–576.

60. Jørgensen B, Henriksen LO, Karle A, et al: Percutaneous transluminal angioplasty of iliac and femoral arteries in severe lower-limb ischaemia. *Acta Chir Scand* 1988; 154:647–652.

61. Fletcher JP, Fermanis G, Little JM, et al: The role of percutaneous transluminal angioplasty and femoropopliteal bypass in patients with threatened limb. *Vasc Surg* 1988; 22:226–230.

62. Sprayregen S, Veith FJ, Bakal CW: Catheterization and angioplasty of the nonopacified peripheral autogenous vein bypass graft. *Arch Surg* 1988; 123:1009–1012.

63. Kalman PG, Sniderman KW: Salvage of in situ femoropopliteal and femorotibial saphenous vein bypass with interventional radiology. *J Vasc Surg* 1988; 7:429–432.

64. Svigals PJ, McLean GK, Davis JE, et al: Transient hypertension after percutaneous transluminal renal artery angioplasty. *Radiology* 1986; 161:293–294.

65. Klinge J, Mali WPTM, Puijlaert CBAJ, et al: Percutaneous transluminal renal angioplasty: Initial and long-term results. *Radiology* 1989; 171:501–506.

66. Martin LG, Price RB, Casarella WJ, et al: Percutaneous angioplasty in clinical management of renovascular hypertension: Initial and long-term results. *Radiology* 1985; 155:629–633.

67. Miller GA, Ford KK, Braun SD, et al: Percutaneous transluminal angioplasty vs. surgery for renovascular hypertension. *AJR* 1985; 144:447–450.

68. Sos TA, Pickering TG, Sniderman K, et al: Percutaneous transluminal renal angioplasty in renovascular hypertension due to atheroma or fibromuscular dysplasia. *N Engl J Med* 1983; 309:274–279.

69. Archibald GR, Beckman CF, Libertino JA: Focal renal artery stenosis caused by fibromuscular dysplasia: Treatment by percutaneous transluminal angioplasty. *AJR* 1988; 151:593–596.

70. Raynaud A, Bedrossian J, Remy P, et al: Percutaneous transluminal angioplasty of renal transplant arterial stenoses. *AJR* 1986; 146:853–857.

71. Madias NE, Kwon OJ, Millan VG: Percutaneous transluminal renal angioplasty: A potentially effective treatment for preservation of renal function. *Arch Intern Med* 1982; 142:693–697.

72. Courthéoux P, Mani J, Mercier V, et al: L'angioplastie endoluminale percutanée des sténoses des artères rénales sur rein considéré comme unique. *Ann Radiol* 1988; 31:177–180.

73. Vitek JJ: Subclavian artery angioplasty and the origin of the vertebral artery. *Radiology* 1989; 170:407–409.

74. Motarjeme A, Keifer JW, Zuska AJ, et al: Percutaneous transluminal angioplasty for treatment of subclavian steal. *Radiology* 1985; 155:611–613.

75. Rodriguez-Perez JC, Maynar M, Rams A, et al: Percutaneous transluminal angioplasty as best treatment in stenosis of vascular access for hemodialysis. *Nephron* 1989; 51:192–196.

76. Katzen BT: Technique and results of "low-dose" infusion. *Cardiovasc Intervent Radiol* 1988; 11:S41–S47.

77. Zimmerman JJ, Cipriano PR, Hayden WG, et al: Balloon embolectomy catheter used percutaneously. *Radiology* 1986; 158:260–262.

78. Starck EE, McDermott JC, Crummy AB, et al: Percutaneous aspiration thromboembolectomy. *Radiology* 1985; 156:61–66.

79. Kinnison ML, White RI Jr, Bowers WP, et al: Cost incentives for peripheral angioplasty. *AJR* 1985; 145:1241–1244.
80. Diethrich EB, Timbadia E, Bahadir I, et al: Argon laser-assisted peripheral angioplasty. *Vasc Surg* 1988; 22:77–87.
81. Lammer J, Karnel F: Percutaneous transluminal laser angioplasty with contact probes. *Radiology* 1988; 168:733–737.
82. Fleisher HL, Thompson BW, McCowan TC, et al: Human percutaneous laser angioplasty: Patient selection criteria and early results. *Am J Surg* 1987; 154:666–669.
83. Snyder SO Jr, Wheeler JR, Gregory RT, et al: The Kensey catheter: Preliminary results with a transluminal atherectomy tool. *J Vasc Surg* 1988; 8:541–543.
84. Kensey KR, Nash JE, Abrahams C, et al: Recanalization of obstructed arteries with a flexible, rotating tip catheter. *Radiology* 1987; 165:387–389.
85. Zocholl G, Hötker U, Hake U, et al: Gefäßrekanalisation mit einem elektrischen thermokauter-katheter: In-vitro erfahrungen mit einem neuen kathetersystem. *ROFO* 1988; 149:526–528.
86. Strecker EP, Romaniuk P, Schneider B, et al: Perkutan implantierbare, durch ballon aufdehnbare gefäßprothese: Erste klinische ergebnisse. *Dtsch Med Wochenschr* 1988; 113:538–542.
87. Günther RW, Vorwerk D, Bohndorf K, et al: Venous stenoses in dialysis shunts: Treatment with self-expanding metallic stents. *Radiology* 1989; 170:401–405.
88. Putnam JS, Uchida BT, Antonovic R, et al: Superior vena cava syndrome associated with massive thrombosis: Treatment with expandable wire stents. *Radiology* 1988; 167:727–728.

11

Local Thrombolytic Infusion

KEY CONCEPTS

1. Local thrombolytic infusion can produce better clinical results than surgical thrombectomy in many cases of acute embolization or thrombosis.
2. Thrombolysis for bypass occlusion, combined with balloon angioplasty or surgical revision, improves the rate of graft salvage.
3. Results are best for occlusions shorter than 15 to 25 cm, of less than 6 months' duration, and with intact runoff.
4. Urokinase has a more predictable effect, higher rates of lysis, and fewer complications than streptokinase.
5. Whatever the agent used, higher-rate infusions with rapid catheter advancement can decrease infusion time, total dose, and cost.

INDICATIONS

Acute arterial occlusion from embolization or thrombosis can be approached surgically by thrombectomy or bypass graft placement. When possible, thrombectomy is desirable, because the patient's native artery and collateral vessels may be preserved. Thrombectomy, however, is quite dependent on how rapidly the patient reaches the surgeon. Early operation can lead to amputation rates of less than 5%, but delay of as little as 12 hours increases risk of limb loss to 32%.[1] Use of a Fogarty catheter has produced limb salvage in 62% to 96% of patients but may be associated with surgical mortality of 17% and complications in 44%.[1, 2]

Surgery for occluded arterial grafts has an even more dismal prognosis. Thrombectomy and graft revision may have a 6-month limb preservation rate of only 23%.[3]

For these reasons, an effective adjunct or alternative to surgery has been sought. Although systemic administration of thrombolytic (fibrinolytic) drugs has shown benefit in the treatment of severe pulmonary embolism, central venous thrombosis, and acute myocardial infarction, results in the treatment of peripheral arterial occlusion have been disappointing.[4] Moreover, systemic lysis can result in severe bleeding complications.

In 1974, Dotter et al.[5] reported the first local arterial infusion of streptokinase. The rationale for the procedure was to deliver a concentrated dose of enzyme to the occlusion while avoiding the hazards of a systemic lytic state. Initial success was low and complications quite frequent, limiting the general acceptance of local thrombolytic infusion. However, refinements in technique of administration and patient selection, and the availability of the newer agents, urokinase and recombinant tissue-type plasminogen activator (rt-PA), have improved clinical results.

Local streptokinase infusion for embolic arterial occlusion is now associated with a 3% amputation rate and 3.5% hospital mortality.[6] Successful lysis combined with surgical revision in graft occlusion can double graft patency and limb salvage rates in comparison with surgical treatment alone.[3, 7]

Thrombolysis is an alternative therapy for acute arterial occlusions of more than just a few hours' duration in patients without immediate danger of limb loss. It may also benefit those with more chronic occlusion who present a high surgical risk. Although the greatest experience with thrombolytic infusion has been in the arteries of the lower extremity, it has also been used for central venous occlusion, hemodialysis access clots, acute upper extremity ischemia, and in exceptional circumstances for renal and visceral artery occlusions.[8–20] When necessary, surgery can be safely performed shortly after cessation of infusion of any of the available drugs.

PATIENT SELECTION AND PRECAUTIONS

Contraindications to the use of thrombolytics include active hemorrhage, presence of a bleeding diathesis, recent cerebrovascular accident, craniotomy within the previous 2 months, intracranial tumor, tissue necrosis, and life-threatening ischemia. Patients with motor and sensory paresis are at risk for the revascularization syndrome, which may be fatal.[21] Patients with a compartment syndrome are prone to massive myoglobinuria, and should have surgical fasciotomy.[22]

Recent noncranial surgery is only a relative contraindication. Streptokinase has been used less than 3 days after surgery with no wound bleeding.[23] Cardiac or valve thrombi have caused serious complications in isolated cases,[24, 25] but in general they have been associated with a surprisingly low incidence of new emboli during thrombolytic therapy.[4]

Prognostic Factors

Older clots are less susceptible to lysis because of the cross-linking of fibrin chains that occurs with time.[26] Data on the dependence of successful lysis with time are conflicting, but there are strong indications that clots or emboli less than 6 weeks old are twice as likely to lyse with directed infusion as older lesions.[6, 23, 27] Lammer et al.[28] have had a positive response in many chronic occlusions, and suggest that thrombolysis be tried in clots up to 6 months of age.

Length of occlusion also has a bearing on response. Recanalization is twice as likely to succeed if an occlusion is less than 10 cm long.[29] A similar effect was noted by Hess et al.[6] in their series of over 500 patients treated with streptokinase, but better success was noted in lesions up to 25 cm. Absence of any visible runoff on initial arteriography makes successful treatment unlikely, but some patients may still benefit from infusion.[30]

Monitoring and Patient Care

All persons submitting to thrombolytic infusion should be placed in an intensive care unit for the duration of treatment. Intramuscular injections and arterial punctures must be avoided, and venipunctures kept to a minimum. Patients should be closely watched for signs of fluid overload or renal failure, in addition to bleeding. If streptokinase is used, low-grade fever may be expected.

Laboratory studies to be obtained before onset of therapy include hematocrit, thrombin time (TT), and fibrinogen levels. If heparin is given as a simultaneous infusion, activated partial thromboplastin time (PTT) should also be followed. Fibrinogen and TT are checked at 4 hours, and then every 12 hours afterward. A blood sample should be drawn every 4 hours for PTT to allow dosage adjustment of heparin. Changes in fibrinogen and TT indicate a systemic lytic effect. A TT two to three times that of control is considered to be therapeutic, and virtually all patients with local streptokinase infusion will have such prolongation by 48 hours of infusion.[23] Lytic therapy should be adjusted or stopped if serum fibrinogen falls below 100 mg/dL.

AGENTS AND PHARMACOLOGY

One point to be remembered about fibrinolytics agents is that, although they can produce systemic changes in blood coagulation, uninjured vessels don't bleed spontaneously.[26] Thrombolytic drugs attack fibrin plugs, and it is sites of vascular injury that are prone to hemorrhage. All thrombolytic drugs available work by activating the body's endogenous lytic enzyme, plasmin. A systemic lytic state can be reversed by stopping infusion and administering plasma or cryoprecipitate. Epsilon aminocaproic acid (Amicar) is a plasmin inhibitor, and can also reverse the effects of thrombolysis, but it is potentially dangerous and is rarely, if ever, indicated.[31]

Streptokinase

Streptokinase is an indirect plasminogen activator, which also has the effect of depleting plasminogen. Because of this plasminogen depression, systemic heparinization is recommended for several days after successful lysis to prevent reocclusion.[31] In systemic doses streptokinase also depletes fibrinogen and clotting factors V and VIII. Streptokinase has a higher affinity for circulating plasminogen than urokinase or rt-PA and is thus more likely to produce a systemic lytic state.[26] The production of fibrin degradation products (or fibrin split products) is common to all fibrinolytics, and can competitively inhibit fibrin polymerization. The plasma half-life of streptokinase is 23 minutes.

Derived from group C beta-hemolytic streptococci, streptokinase is antigenic and has been associated with anaphylaxis on rare occasions. Antibodies to streptokinase begin rising several days after exposure, and elevation may persist up to 6 months.[26] Virtually everyone has some antibodies to streptococci, and much of the irregular dose-response to systemic streptokinase therapy can be traced to neutralizing antibodies.

Urokinase

Urokinase, unlike streptokinase, is a direct plasminogen activator, which allows more predictability of dose-response.[32] It depletes the same factors that streptokinase does, though it is more clot-specific and is less likely to produce a lytic state. It has a plasma half-life of 16 minutes and, being an endogenous human protein, elicits no immune response.[26]

Lysyl-Plasminogen

This agent has been used by French investigators to potentiate the effects of urokinase.[33] It is produced by the proteolysis of native plasminogen, and the resulting enzyme has a higher affinity for fibrin.

Recombinant Tissue-Type Plasminogen Activator

Another endogenous human protein, rt-PA, is a serine protease produced in quantity by recombinant gene technology.[34] It has a very high affinity for fibrin and low affinity for circulating plasminogen. The plasma half-life of this agent is 5 to 8 minutes.[26]

STREPTOKINASE VS. UROKINASE

In choosing between these two most commonly employed fibrinolytic agents (rt-PA is still in preliminary clinical investigations), one must balance considerations of efficacy, complication rates, previous exposure to streptokinase or high circulating antibodies, and cost. Mean infusion time has been long with successful lysis by streptokinase, up to 120 hours, but urokinase in equivalent low-dose infusions can take nearly as long to produce the desired effect.[32] However, at doses of similar efficacy, streptokinase causes more bleeding and other complications.[27, 32, 35, 36] This reflects the fact that streptokinase will affect fibrinogen levels in most people after just 12 hours of infusion, while urokinase will not.

Even so, streptokinase has been used successfully in many cases, and the higher-dose, rapid catheter advancement techniques popular in Europe decrease treatment time considerably.[6, 28] Urokinase costs about six times as much as streptokinase.[32]

MODES OF INFUSION

In general, the smallest possible catheter should be used to enter the clot. Katzen prefers to use an infusable guidewire, because it is more readily visible with fluoroscopy and is less likely to be coated with thrombus.[15] When possible, placement through an area or reduced flow should be avoided. A sheath or coaxial system can administer a portion of the dose proximally, decreasing the possibility of clot formation on the infusing catheter.

The catheter or guidewire delivering the thrombolytic agent should be embedded in thrombus. Placement in patent vessel above the occlusion may result in loss of infusate into collateral vessels.

Original Low-Dose Protocols

The first doses tried for local therapy were rather arbitrarily set as 5% to 10% of systemic. For streptokinase, with a systemic loading dose of 250,000 units administered over 30 minutes and infusion at 100,000 units/hr, this meant intra-arterial infusion at 5,000 units/hr in a concentration of 100 units/mL.

Systemic urokinase is given at 4,400 units/kg/hr after a bolus of 4,400 units/kg (or 308,000 units for the proverbial 70-kg man). The local dose commonly used for low-dose infusion is 20,000 units/hr.

No loading dose is usually given for local lysis, because it defeats the purpose of avoiding systemic lytic effects. A loading dose is sometimes given for streptokinase infusions in order to overcome circulating antibodies. With both urokinase and streptokinase, lower-dose infusions often take up to 4 or 5 days, and angiography must be repeated every 8 to 12 hours to assess progress and to reposition the catheter. With these protocols not only is hospitalization (especially intensive care monitoring) prolonged, but bleeding complications may be more likely to arise.

Higher Dose Local Streptokinase

In the protocol used by Hess et al.[6] the infusing catheter is placed immediately above the clot, and three separate doses of 1,000 units of streptokinase (dissolved in 2 mL of normal saline) are given at 3-minute intervals. The catheter is then advanced into clot, followed by similar repeated injections and intermittent 1-cm advancement. The final 2 to 3 cm of the occlusion need not be traversed. Any residual nonobstructing thrombus may resolve with the "afterlysis" effect of continued fibrinolysis for some time after infusion ceases. This method may reduce total dose to 20,000 to 30,000 units and keep procedure time under 90 minutes![6] The entire recanalization can be performed without removing the patient from the angiography suite. Local bleeding can be detected and controlled immediately. Such aggressive efforts at fibrinolysis can quickly reveal lesions amenable to balloon angioplasty.[29] Most failures with this protocol are the result of vessel dissection from guidewire/catheter manipulation. Such a complication must be addressed by stopping local infusion.

Many Europeans favor the use of streptokinase with this or similar techniques.[30] Urokinase is reserved for those patients with exposure to streptokinase within the previous several months or who otherwise have high levels of antibody.

Higher Dose Local Urokinase

The method of higher dose infusion of urokinase described by McNamara and Fischer in 1985[21] has been widely applied in the United States. The infusing catheter is placed near the obstructing clot, and a straight guidewire is passed as far as possible into the thrombus. Chances for successful lysis are high if a small channel passing completely through the obstruction can be created. Urokinase (500,000 units in 200 mL of saline) is infused directly into the proximal clot at 4,000 units/min. Response is checked by arteriography every 2 hours, with repeat passage of guidewire and further advancement of the catheter as lysis proceeds. With recanalization and restoration of antegrade flow, infusion is slowed to 1,000 units/min. Infusion is stopped when all clot lyses (an average of 18 hours in their series), and the catheter is removed 1 hour after heparin and urokinase are discontinued.[21] Any necessary anticoagulation can be resumed 12 hours afterward.

If there is distal embolization of clot, the catheter should be advanced to the level undergoing embolization and infusion continued at 4,000 units/min. This maximum rate of infusion should not be exceeded because systemic effects of urokinase rapidly become evident.[37] Some have recommended that the catheter be placed through thrombus at the onset, and 30,000 to 60,000 units of urokinase injected as a bolus during withdrawal, before local infusion into proximal clot is started.[3] Failure to see at least 10% lysis of the occluded segment after 500,000 units is an indication to discontinue lytic therapy.

Heparin helps prevent the formation of clot on the infusing catheter (dropping the incidence of this problem from 29% to 4%), but it must be given in doses high enough to cause threefold to fivefold prolongation of the PTT.[21] McNamara and Fischer have not found an increase in bleeding complications from such anticoagulation, although they caution against treating patients on warfarin or through an aortofemoral graft.[21] Doses of heparin that produce a lesser degree of PTT prolongation are not uniformly effective.[35, 38]

Recombinant Tissue-Type Plasminogen Activator

Recombinant tissue-type plasminogen activator is still in the preliminary stages of clinical investigation, but it may be administered intra-arterially at the rate of 0.05 mg/kg/hr. This dose has been reported to be as effective and as safe as infusion of 0.10 mg/kg/hr.[34]

RESULTS

Published results of local thrombolytic infusion are difficult to compare because of differences in patient selection, agent and technique used, adjunctive surgery or percutaneous angioplasty, and clinical follow-up. Nevertheless, certain trends are evident.

Emboli and in situ thrombi lyse with about equal success, but if the reopened vessel stays patent for 2 weeks, 5-year patency is about 90% for emboli but only 59% for thrombosis associated with a preexisting lesion.[6] McNamara has observed that emboli may need infusion for nearly twice as long as thrombi.[37]

Most series with standard low-dose streptokinase treatment have reported partial or complete lysis in 50% to 56% of cases.[6, 12, 19, 23] Others have had a greater degree of success with the low-dose streptokinase protocol, perhaps related to patient selection.[39, 40] The more rapid infusions with streptokinase have been successful in restoring circulation in 71% to 77% of patients.[29, 30, 41] The higher-dose urokinase methods of infusion have produced lysis in 80% to 90% of cases, and lysis more often has been complete than with streptokinase.[15, 21, 33, 35, 42]

Long-Term Results

Two-year patency of reopened vessels has been reported as 64%, but the balloon dilatation treatment of suitable underlying lesions may improve long-term patency to 82%, a figure comparable to that obtained with successful surgical thrombectomy.[29, 30] Iliac arteries are more likely to stay open than more distal vessels, and the best predictor of reocclusion is the presence of a residual flow-limiting lesion.[43]

Arterial Bypass Grafts

Autologous vein grafts are as likely to respond to fibrinolytics as native arteries, but bypasses of synthetic material often have a poor result.[19, 39] It is unclear whether this difference is from endogenous factors within arteries and veins or because patients with synthetic grafts tend to have more severe peripheral vascular disease. Aortofemoral graft limbs are more successfully treated than distal grafts.[43] The presence of a "stump" at the proximal end of an occluded graft may have positive prognostic significance.[15] As in native arterial occlusions, ultimate results of thrombolysis are improved substantially if an underlying lesion responsible for graft occlusion is found and treated.[3, 44]

Dialysis Access

Standard low-dose thrombolytic therapy for clotted dialysis fistulas or grafts has been disappointing.[12, 45] However, Davis et al.[13] have developed a more intensive technique using crossed catheters "lacing" the thrombus with urokinase. They have managed to lyse thrombus in 90% of the grafts treated, and 62% of their patients have avoided surgical revision for at least 6 months.

Venous Occlusions

Superior vena caval, subclavian, and axillary vein occlusions have been treated with varying success. Some investigators have reported good results for both primary and secondary thromboses in subclavian veins, but all published series have been quite small.[8–10, 12, 23, 46]

Other Vessels

Isolated cases of superior mesenteric artery embolus have been successfully treated by local infusion of streptokinase.[18, 19] Thrombolytic therapy may be instituted only in the absence of bowel necrosis, otherwise the patient should be taken immediately to surgery.

Renal artery occlusions have lysed up to 10 days after thrombosis, and 67% of patients treated have had perfusion reestablished.[15, 16] Streptokinase has produced benefit in patients with acute hand ischemia in a small series.[14] Central venous infusions have been used to successfully treat patients with occlusion of the inferior vena cava and Budd-Chiari syndrome or transplant renal vein occlusion.[11, 17]

COMPLICATIONS

Potential complications of local thrombolytic infusion include hemorrhage, synthetic graft permeation, distal embolization, new thrombus arising on the infusion catheter, limb loss, sepsis, and death. Fortunately, most serious complications are related to hemorrhage and respond to transfusion or surgical evacuation of hematoma.

Bleeding has been more common with streptokinase in prolonged, low-dose infusions. Major hemorrhage has been described in 7% to 32% of patients, with most such series reporting in the 12% to 17% range.[3, 12, 19, 22, 23, 36] Protocols using more rapid infusion of streptokinase have dropped the incidence of severe bleeding to 2% to 4%.[6, 28] In comparison, transfusion or other intervention for hemorrhage has been necessary in 3% to 6% of patients treated with higher dose local urokinase.[3, 15, 21, 27, 33, 36]

Distal macro- or microembolization is not rare, but in most cases resolves with further thrombolysis. Rates of clinical embolization cited have been fairly uniform, from 11% to 14%.[6, 19, 21, 28]

Isolated reported deaths have been due to retroperitoneal hemorrhage or myocardial infarction.[22, 27] Up to 7% hospital mortality may be seen, but many deaths are attributable to other concomittant severe disease.[38] It is rare that limb loss is directly due to choice of thrombolysis for therapy. Most of those undergoing amputation may have been expected to have the same outcome with surgical thrombectomy.

NEW DIRECTIONS

Despite the benefits of local thrombolytic infusion, it is clear that treatment must be made safer, more effective, less labor-intensive, and less expensive in order to gain wide acceptance in the treatment of vascular occlusion. Introduction of rt-PA into practice will provide a fresh alternative. Initial results with peripheral arterial infusion have shown lysis in over 90% and major bleeding in only 5%.[7]

Another area of investigation is the delivery of thrombolytics into clot. Kandarpa et al. have devised a catheter system for pulsatile jet delivery with permeation of thrombus by the active agent.[47] This innovation disrupts clot more extensively than mere passage of a guidewire, and lysis can be accelerated fourfold in laboratory experiments. Application of intraluminal stents may also provide benefit for recanalized vessels, particularly in the central venous system. If these and related developments live up to expectations, local thrombolysis may play a much larger clinical role in the future.

REFERENCES

1. Johnson JA, Cogbill TH, Strutt PJ: Late results after femoral artery embolectomy. *Surgery* 1988; 103:289–293.

2. Genton E, Clagett GP, Salzman EW: Antithrombotic therapy in peripheral vascular disease. *Chest* 1986; 89(suppl):75–81.

3. Koltun WA, Gardiner GA Jr, Harrington DP, et al: Thrombolysis in the treatment of peripheral arterial vascular occlusions. *Arch Surg* 1987; 122:901–905.

4. Marder VJ, Sherry S: Thrombolytic therapy: Current status (part 2). *N Engl J Med* 1988; 318:1585–1595.

5. Dotter CT, Rösch J, Seaman AJ: Selective clot lysis with low-dose streptokinase. *Radiology* 1974; 111:31–37.

6. Hess H, Mietaschk A, Brückl R: Peripheral arterial occlusions: A 6-year experience with local low-dose thrombolytic therapy. *Radiology* 1987; 163:753–758.

7. Graor RA, Risius B, Young JR, et al: Thrombolysis of peripheral arterial bypass grafts: Surgical thrombectomy compared with thrombolysis. *J Vasc Surg* 1988; 7:347–355.

8. Huey H, Morris C, Nichols DM, et al: Low-dose streptokinase thrombolysis of axillary-subclavian vein thrombosis. *Cardiovasc Intervent Radiol* 1987; 10:92–95.

9. Becker GJ, Holden RW, Rabe FE, et al: Thrombolytic therapy for subclavian and axillary vein thrombosis. *Radiology* 1983; 149:419–423.

10. Druy EM, Trout HH, Giordano JM, et al: Lytic therapy in the treatment of axillary and subclavian vein thrombosis. *J Vasc Surg* 1985; 2:821–827.

11. Greenwood LH, Yrizarry JM, Hallett JW Jr, et al: Urokinase treatment of Budd-Chiari syndrome. *AJR* 1983; 141:1057–1059.

12. Becker GJ, Rabe FE, Richmond BD, et al: Low-dose fibrinolytic therapy: Results and new concepts. *Radiology* 1983; 148:663–670.

13. Davis GB, Dowd CF, Bookstein JJ, et al: Thrombosed dialysis grafts: Efficacy of intrathrombotic deposition of concentrated urokinase, clot maceration, and angioplasty. *AJR* 1987; 149:177–181.

14. Tisnado J, Bartol DT, Cho S-R, et al: Low-dose fibrinolytic therapy in hand ischemia. *Radiology* 1984; 150:375–382.

15. Katzen BT: Technique and results of "low-dose" infusion. *Cardiovasc Intervent Radiol* 1988; 11:S41–S47.

16. Morag B, Rubinstein Z, Schneiderman J: Renal artery occlusion: Intra-arterial thrombolytic therapy. *J Intervent Radiol* 1988; 3:77–80.

17. Robinson JM, Cockrell CH, Tisnado J, et al: Selective low-dose streptokinase infusion in the treatment of acute transplant renal vein thrombosis. *Cardiovasc Intervent Radiol* 1986; 9:86–89.

18. Vujic I, Stanley J, Gobien RP: Treatment of acute embolus of the superior mesenteric artery by topical infusion of streptokinase. *Cardiovasc Intervent Radiol* 1984; 7:94–96.

19. Kakkasseril JS, Cranley JJ, Arbaugh JJ, et al: Efficacy of low-dose streptokinase in acute arterial occlusion and graft thrombosis. *Arch Surg* 1985; 120:427–429.

20. Yankes JR, Uglietta JP, Grant J, et al: Percutaneous transhepatic recanalization and thrombolysis of the superior mesenteric vein. *AJR* 1988; 151:289–290.

21. McNamara TO, Fischer JR: Thrombolysis of peripheral arterial and graft occlusions: Improved results using high-dose urokinase. *AJR* 1985; 144:769–775.

22. Lang EK. Streptokinase therapy: Complications of intra-arterial use. *Radiology* 1985; 154:75–77.

23. Risius B, Zelch MG, Graor RA, et al: Catheter-directed low dose streptokinase infusion: A preliminary experience. *Radiology* 1984; 150:349–355.

24. Brewer ML, Kinnison ML, Perler BA, et al: Blue toe syndrome: Treatment with anticoagulants and delayed percutaneous transluminal angioplasty. *Radiology* 1988; 166:31–36.

25. Paulson EK, Miller FJ: Embolization of cardiac mural thrombus: Complication of intraarterial fibrinolysis. *Radiology* 1988; 168:95–96.

26. Marder VJ, Sherry S: Thrombolytic therapy: Current status (part 1). *N Engl J Med* 1988; 318:1512–1520.

27. Grabenwöger F, Dock W, Appel W, et al: Fibrinolysetherapie thromboembolischer gefäßverschlüsse: Primäre erfolgsrate und langzeitergebnisse. *ROFO* 1988; 148:615–618.

28. Lammer J, Pilger E, Justich E, et al: Fibrinolysis in chronic arteriosclerotic occlusions: Intrathrombotic injections of streptokinase. *Radiology* 1985; 157:45–50.

29. Wilms GE, Verhaeghe RH, Pouillon MM, et al: Local thrombolysis in femoropopliteal occlusion: Early and late results. *Cardiovasc Intervent Radiol* 1987; 10:272–275.

30. Do-dai-Do, Mahler F, Triller J, et al: Combination of short- and long-term catheter thrombolysis for peripheral arterial occlusion. *Eur J Radiol* 1987; 7:235–238.

31. Bookstein JJ, Moser KM, Hougie C: Coagulative interventions during angiography. *Cardiovasc Intervent Radiol* 1982; 5:46–56.

32. Belkin M, Belkin B, Bucknam CA, et al: Intra-arterial fibrinolytic therapy: Efficacy of streptokinase vs. urokinase. *Arch Surg* 1986; 121:769–773.

33. Pernes JM, Vitoux JF, Brenoit P, et al: Acute peripheral arterial and graft occlusion: Treatment with selective infusion of urokinase and lysy plasminogen. *Radiology* 1986; 158:481–485.

34. Risius B, Graor RA, Geisinger MA, et al: Thrombolytic therapy with recombinant human tissue-type plasminogen activator: A comparison of two doses. *Radiology* 1987; 164:465–468.

35. Traughber PD, Cook PS, Micklos TJ, et al: Intraarterial fibrinolytic therapy for popliteal and tibial artery obstruction: Comparison of streptokinase and urokinase. *AJR* 1987; 149:453–456.

36. Gardiner GA Jr, Koltun W, Kandarpa K, et al: Thrombolysis of occluded femoropopliteal grafts. *AJR* 1986; 147:621–626.

37. McNamara T: Technique and results of "higher-dose" infusion. *Cardiovasc Intervent Radiol* 1988; 11:S48–S57.
38. Fiessinger J-N, Vitoux J-F, Pernes J-M, et al: Complications of intraarterial urokinase-*lys*-plasminogen infusion therapy in arterial ischemia of lower limbs. *AJR* 1986; 146:157–159.
39. LeBolt SA, Tisnado J, Cho S-R: Treatment of peripheral arterial obstruction with streptokinase: Results in arterial vs. graft occlusions. *AJR* 1988; 151:589–592.
40. Katzen BT, VanBreda A: Low dose streptokinase in the treatment of arterial occlusions. *AJR* 1981; 136:1171–1178.
41. Delcour C, Bellens B, Vandenbosch G, et al: Long-term follow-up of intraarterial infusion of streptokinase in acute lower limb ischemia. *Vasc Surg* 1987; 21:339–343.
42. Barzi F, Lupattelli L, Corneli P, et al: Local fibrinolysis with urokinase in acute and chronic arterial obstruction of the lower limb: Our experience. *J Intervent Radiol* 1988; 3:81–83.
43. McNamara TO, Bomberger RA: Factors affecting initial and 6 month patency rates after intraarterial thrombolysis with high dise urokinase. *Am J Surg* 1986; 152:709–712.
44. Gardiner GA Jr: Thrombolysis of occluded arterial bypass grafts. *Cardiovasc Intervent Radiol* 1988; 11:S58–S59.
45. Young AT, Hunter DW, Castañeda-Zuñiga WR, et al: Thrombosed synthetic hemodialysis access fistulas: Failure of fibrinolytic therapy. *Radiology* 1985; 154:639–642.
46. Fankuchen EI, Neff RA, Collins RA, et al: Urokinase infusion for axillary-subclavian vein thrombosis. *Cardiovasc Intervent Radiol* 1984; 7:90–93.
47. Kandarpa K, Drinker PA, Singer SJ, et al: Forceful pulsatile local infusion of enzyme accelerates thrombolysis: In vivo evaluation of a new delivery system. *Radiology* 1988; 168:739–744.

12 | Embolotherapy

KEY CONCEPTS

1. Embolization can be used for control of bleeding, closing of arteriovenous malformations or fistulas, tumor palliation, organ ablation, and treatment of varicocele.
2. Choice of agent depends on the size of vessel to be occluded, desirability of infarction, permanency of occlusion, and ease of catheter placement.
3. Knowledge of vascular anatomy, variations, and collateral vessels is essential for the prevention of complications.
4. The postembolization syndrome of fever, pain, and leukocytosis may last for several days, and soft tissue gas is often present.
5. Antibiotics are absolutely necessary for splenic embolization.

INDICATIONS

The most widely accepted indication for percutaneous transcatheter embolization is treatment of hemorrhage from trauma or unresectable neoplasm. In cases of pelvic injury involving extensive fractures, surgery is largely ineffective for controlling bleeding, and embolization can be a lifesaving procedure.[1-3] Other situations for which embolotherapy offers distinct advantages over surgery include uncontrollable gastric bleeding, arterial hemorrhage due to pancreatitis, and unremitting hemobilia.[4-8]

In bowel with limited collateral circulation, such as the colon and small bowel, embolization can control massive bleeding as a temporizing measure. Operative mortality is greatly decreased if a patient can undergo transfusion and be stabilized prior to surgery.[9] Embolotherapy eliminates the need for exploration in some cases, but it must be anticipated that bowel ischemia may result.[10, 11] Although bleeding from gastric and esophageal varices has been aggressively treated by percutaneous transhepatic occlusive interventions, results have been almost uniformly disappointing.[12–15] Esophageal sclerotherapy is presently the preferred method of controlling acute variceal hemorrhage. Transhepatic embolization is indicated only for the rare patient with life-threatening bleeding who is unresponsive to endoscopic intervention and who is otherwise a candidate for portosystemic shunt placement. Emergency shunt procedures have a high operative risk; an urgent operation on a hemodynamically stable patient is safer and can prolong life.[16]

Catheter embolization has been effective in controlling massive hemoptysis resulting from tuberculosis, sarcoidosis or other inflammatory disease.[17] It is also a primary treatment for pulmonary arteriovenous malformations (AVMs), which are dangerous even when asymptomatic.[18] Traumatic AV fistulas can be occluded, and AVMs can be palliated by embolotherapy alone or in conjunction with surgery.[19–22]

Embolization for neoplasm can alleviate the symptoms of those with metastatic islet cell or other secreting endocrine tumors, prolong life in patients with unresectable primary or metastatic hepatic malignancies, and relieve intractable pain.[23–27] It can also be employed preoperatively to diminish blood loss in very vascular renal, osseous, or other tumors undergoing resection.

Other indications for which embolotherapy has been used successfully include hypersplenism, recurrent hyperparathyroidism from mediastinal adenomas, malignant hypertension or nephrotic syndrome in end-stage renal disease, and varicocele.[28–31] When surgery has been difficult or impossible, percutaneous techniques have been employed to occlude aneurysms and pseudoaneurysms.[32–34]

Over the years, interventional radiologists have learned the hazards and limitations of embolotherapy. At the same time, however, experience gained has stimulated the continued development of new embolic agents, delivery systems, and applications.

BEFORE PROCEEDING

Therapeutic Goals and Choice of Materials

Before therapeutic embolization is attempted for an individual patient there should be a clear definition of objectives. The interventional radiologist must determine the level and permanency of vascular occlusion needed and the materials most appropriate to the situation at hand.

For traumatic hemorrhage the aim of embolization is not to cause tissue infarction, but rather to diminish perfusion to the point where endogenous hemostatic mechanisms stop the bleeding. Too distal an occlusion, such as that produced by powders, microspheres, or polymerizing fluids, can produce necrosis and major complications. On the other hand, large particles, when used alone, may occlude a vessel feeding the site of extravasation centrally, only to permit the hemorrhage to continue through collateral vessels (a problem common to surgical ligation). As a matter of principle, the agent used should be deposited as selectively and as close to the site of bleeding as possible. The embolizing particles should be appropriate to the size of vessel injured. A nonpermanent material such as gelatin sponge (Gelfoam) is preferable. Ideally, the vessel should be occluded long enough to provide healing, but it should eventually recanalize and provide optimal tissue perfusion for the long term. For example, a man treated for pelvic trauma by bilateral hypogastric artery embolization may suffer erectile dysfunction as a result.[2] Such risks can be minimized (but not eliminated) by knowledge of the materials available and careful attention to technique.

For unresectable malignant neoplasms, complete tissue ablation is usually the goal, although rarely attained in practice. Depending on the organ involved, embolization with small, permanent particles can be used to prevent tumor viability from being maintained by collateral vessels. In hepatic tumors it is unwise to place coils or other large permanent particles in the proper hepatic artery. Repeat embolotherapy may be rendered much more difficult, if not impossible, as a result. Absolute ethanol is extremely toxic, but its careful application in cases of renal cell carcinoma can occasionally produce complete tumor necrosis and clinical remission.[35]

High-flow AV fistulas are best occluded with large permanent devices such as detachable balloons, metallic "spiders," or coils.[8, 19, 36] In contrast, AVMs in an extremity typically have multiple feeding arteries. Occlusion of large vessels leads to recurrence, often through myriad collateral vessels. Lasting palliation requires obliteration of the "nidus" at the core of most AVMs, and small permanent particles or tissue adhesives produce the best results.[21, 22] Judicious application of alcohol may also be effective,[37] and it may become the treatment of choice in symptomatic cavernous hemangiomas of the extremities. A more detailed discussion of specific pathologic and anatomic considerations for various indications can be found later in this chapter.

Postembolization Syndrome

When tumor or other tissue necrosis takes place, both the physician and patient should be aware of its expected consequences. The postembolization syndrome (PES) consists of pain, fever, leukocytosis, nausea, and vomiting.[24, 25, 38] Symptoms arise shortly after embolotherapy and usually resolve within 3 days, but may persist up to 1 week. Leukocytosis may be marked. Only close clinical observation and use of blood cultures can distinguish PES from a complicating infection. Generally, pain can be controlled with intravenous or oral analgesics.

A potentially confusing finding after embolization is the presence of gas in the infarcted soft tissues; sometimes even pneumoperitoneum can be seen. The origin of this gas has been debated. Although some microbubbles are injected during every procedure, particularly when porous materials (gelatin sponge or polyvinyl alcohol) are used, most of the gas evidently arises from the necrotic tissue itself.[35, 39, 40] Precisely because of devascularization of the affected tissues, gas resorption may take several weeks.

Staged Procedures

Complications and patient discomfort may be reduced by performing therapeutic embolization in stages. In many cases, diagnostic angiography should be performed separately in order to reduce the duration of a given procedure and the amount of contrast medium administered at one time. For multiple lesions, particularly pulmonary AVMs or hepatic tumors, or for large pelvic or extremity AVMs, staged embolizations are preferable.[18, 24–26, 37]

Potential Complications

Complications of embolization include infection and undesired ischemia/infarction, in addition to the host of problems that can arise from angiography alone, such as hemorrhage at the puncture site, acute renal failure, and anaphylactic shock.[41, 42] The incidence of complications is dependent on many factors. Experience of the operators, general condition of the patient, lesion treated, choice of materials, and anatomic variants all contribute to outcome. As in any procedure, the patient should be fully informed of potential adverse effects before consent is obtained.

An absolutely essential prerequisite for all occlusive interventions is high-resolution, high-quality diagnostic angiography. The vessels feeding the lesion to be treated must be identified. Large AV communications must be recognized. Anatomy and possible variants must be well understood, and potential problems identified, before embolization is undertaken. If a kidney is to undergo embolization with ethanol, careful attention should be paid to the location of adrenal and gonadal branches of the renal artery.[30] In bronchial artery embolization for hemoptysis, anterior spinal branches must be identified to prevent the disasterous complication of paraplegia.[17] Complications specific to particular lesions, organs, and embolic materials are addressed as these items are discussed.

Use of Antibiotics

Antibiotics have not been given routinely prior to embolotherapy for most indications, but they are indispensable for certain situations. Partial splenic ablation, in particular, demands administration of broad-spectrum antibiotics and strict attention to aseptic technique to prevent splenic abscess formation.[28, 42] Renal abscess can arise if an embolized kidney is associated with stones or an untreated urinary tract infection.[38] Recent reports indicate that unrecognized bacteremia occurs in many angiographic interventions, and sepsis is a leading cause of death directly attributable to therapeutic embolization.[42, 43] For this reason, patients treated for hepatic neoplasms should routinely receive pre-procedural parenteral antibiotics.[42]

MATERIALS AND METHODS

Delivery Systems

In virtually any embolic intervention it is advisable to employ a vascular sheath. Particles and tissue adhesives may jam or occlude a catheter, and a sheath allows its easy replacement. Administering catheters should not contain sideholes. Greater control is afforded by injection through an endhole catheter, and the possibility of unrecognized particle lodging within a sidehole with subsequent unintended embolization can be avoided. When its selective placement is possible, an occlusion balloon catheter adds a margin of safety.[30, 44] Not only is reflux prevented, but the degree of devascularization can also be considerably enhanced.

Coaxial catheter systems have been developed for superselective angiography and vessel occlusion. One that offers distinct advantages over previous designs is the Tracker (Target Therapeutics, Los Angeles,

California),[21, 45] a semirigid 3-F polyethylene catheter tapered to a softer 2.7-F tip. It is easily passed through standard selective catheters, and has great flexibility while retaining a good degree of torque control. A radiopaque marker at the end of the Tracker makes it readily visible; without a marker, small catheters can be very difficult to see on even the highest resolution fluoroscopic monitors. The Tracker is capable of delivering small particles of gelatin sponge or polyvinyl alcohol, tissue adhesives, and fluids.

Flow-directed catheters, such as the Kerber calibrated-leak balloon catheter are also useful for superselective placement.[46] However, only fluids can be injected through this device. Removable-core, injectable guidewires are capable of infusing powders as well as fluids. No doubt there will be further technical developments for improved selective and superselective embolotherapy.

In certain cases, direct percutaneous fine-needle puncture of a lesion is the best or only nonoperative approach possible. Ethanol, bucrylate, and thrombin have been safely injected directly into aneurysms, vascular malformations, and tumors.[32, 37, 47, 48]

The mode of delivery chosen depends on the nature of the lesion being treated, its location, and vascular anatomy. Operator experience and preferences are additional factors. Whatever method is used, placement of the access device must be stable, and the device must accept the occluding agent. If there is any question about catheter–embolizing agent compatibility, the combination should be checked in vitro.

Particulate Agents

Particles are the most commonly employed agents in embolotherapy, and the most versatile. The earliest materials used were endogenous: autologous clot, lyophilized dura mater, and fascia lata. Autologous clot is immediately available at the time of embolization, but its effect may be quite transient, even if it is reinforced by the application of exogenous thrombin. Dura mater and fascia produce more lasting occlusion, but neither is as readily available as many other materials.

Gelatin Sponge (Gelfoam)

A gelatin sponge provided in sheets or powder (Gelfoam) is the most frequently used temporary occlusive material. Sheets can be cut to any desired size, with 2-mm cubes or $2 \times 2 \times 6$-mm "torpedos" easily injectable through standard diagnostic catheters. Particles are soaked in contrast medium diluted with saline and are drawn up in small syringes. Although the particles may jam within a catheter, forceful injection or passage of a guidewire can often clear the lumen. If the catheter tip is not advanced well into the desired vessel or if flow is stagnant, an occlusion balloon should be used to prevent reflux. Gelfoam is quite thrombogenic, and occlusion can be expected to last at least 1 or 2 weeks. Although Gelfoam powder is available, it presents a much greater risk of causing unintended ischemia. Skin necrosis, neural injury, and gallbladder infarction have been reported with the use of powder.[49, 50]

Gelfoam is inherently radiolucent, a problem common to many agents. Mixture with water-soluble contrast material permits monitoring of Gelfoam injection. Tantalum and Lipiodol (iodized oil contrast agent) have also been used to provide radiopacity, which at best is only transient. Once a vessel is occluded, great care must be taken to avoid forceful catheter flushing or selective injection of contrast material into the vessel, because particle reflux can result. If repeat embolotherapy is not a consideration, placement of a central coil can produce permanent occlusion and may prevent delayed retrograde embolization of Gelfoam.[51]

Polyvinyl Alcohol (Ivalon)

Ivalon, like Gelfoam, is supplied in sheets or as small particles. Unlike Gelfoam, it produces a permanent occlusion. Plugs can be cut from sheets, which are constituted of compressed material. Such particles will expand on exposure to blood or other fluids. One mode of administration involves the advance preparation of a plug of Ivalon compressed and dried, mounted on the tip of a guidewire.[52] At the time of embolization the stability of catheter position can be checked by passage of a dummy wire before the Ivalon wire is introduced. After the plug expands it can be stripped from the wire by the introducing catheter.

The small particles are available in a variety of sizes from 0.25 mm to 1.00 mm in diameter. Care should be taken to mix the particles well within 5-mL syringes; otherwise, they have a tendency to aggregate and jam within the catheter. To keep the particles in suspension, two syringes may be joined by a segment of clear connecting tubing, with to-and-fro injection between syringes. Administration into the embolization catheter can follow immediately through a three-way stopcock connection. Ivalon, too, is radiolucent and must be mixed with contrast medium. Because of problems with homogeneity of size, small particles of Ivalon should not be used if any substantial degree of AV shunting is suspected.

Other Particles

Many of the problems attendant with the handling and delivery of gelatin sponge and polyvinyl alcohol are common to other particulate agents. Oxidized cellulose (Oxycel) and microfibrillar collagen (Avitene) are similar in effect to Gelfoam, and the occlusions produced last days to weeks. A newer agent under investigation, glutaraldehyde cross-linked collagen (GAX), produces a more long-lasting occlusion and less of an inflammatory tissue response than most other particulates.[53] It is suspended in contrast medium, can be injected through 2-F catheters or open-ended guidewires, and produces small vessel occlusion. Silicone microspheres occlude small vessels permanently and have been used in the treatment of AVMs. Microspheres, Avitene, and GAX, by their distal site of action, are more likely to produce infarction and necrosis, and neural damage is a possible complication.[41]

Radioactive Emboli

Recent interest in tumor embolization has focused on use of radioactive agents. Yttrium 90 has been incorporated into glass microspheres, and Lipiodol (a fluid) has been labeled with iodine 131 to administer local radiation to hepatic tumors.[54, 55] These therapies are yet in early stages of investigation, and their effectiveness remains unproved. However, Lang and Sullivan have shown that unresectable renal cell carcinomas can undergo embolization with iodine-125 seeds, with palliation as good as that provided by any other currently available treatment.[56]

Metallic Coils and Spiders

Since their introduction in the mid-1970s, Gianturco coils have been a mainstay of embolotherapy for occlusion of large vessels. Originally, wool strands were attached at the tip to enhance thrombosis, but Dacron was later substituted because of the marked granulomatous reactions produced by wool.[41] Dispersal of the fibers along the length of the wire coils has made jamming within catheter less of a problem. In high-flow situations or when a patient's coagulation is impaired, coils do not necessarily thrombose, and combination with gelatin sponge or thrombin injection may be needed.

Coils are now produced in various sizes, and many can be introduced through standard tapered catheters by standard guidewires. In fact, micro-coils and brushes, designed for introduction through the Tracker catheter are now available. Close attention must be paid to the manufacturer's instructions for embolization. As noted earlier, if there is any question as to the compatibility of a catheter-coil-guidewire introducer combination, the system should be tested before its use in a patient. Because coils are introduced by guidewires or pushers, catheter position stability is more critical than it is for many other embolic agents. The introducing wire should be passed through the catheter into the vessel being occluded as a test before the coil is inserted.

Coils should be carefully matched in size to the vessel being treated. A coil that is too large will be deposited in an elongated configuration and may project into a feeding vessel. It can potentially erode through vessel wall, resulting in pseudoaneurysm. If a coil is too small it may fail to lodge, causing unintended embolization by reflux into other vessels. In experienced hands, embolotherapy with coils is quite safe. Chuang et al. noted only eight major complications in over 1,200 patients treated.[57] Even when their coils strayed, they were able to retrieve several by means of intravascular snares.

In cases of large AV fistula, coils alone may be too small to treat the lesion. A metallic "spider" has been designed to serve as a baffle to trap coils or other large particles.[36] Spiders are available in 10-mm and 15-mm diameters, and barbed feet serve to anchor them in the vessel wall. Introduction with a special threaded guidewire permits precise positioning before release of the spider.

Detachable Balloons

Detachable balloons are most useful in the management of large AV communications, such as traumatic AV fistulas or pulmonary AVMs, and varicocele.[8, 18, 58] Their principal advantage is that the operator can check the position and adequacy of occlusion before balloon detachment, and partial balloon inflation can flow-direct the balloon to the site of AV communication. Various balloon systems using coaxial catheter introduction have been described. Silicone balloons should be inflated with an isotonic contrast agent, because silicone is semipermeable. Premature deflation may lead to unintended embolization.[41]

Polymers and Tissue Adhesives

Advantages of polymers include the ability to be injected through small catheters or open-ended guidewires, production of occlusion in vessels of various size, and predictable control of bleeding in patients whose coagulation is impaired. Disadvantages are the rather cumbersome preparations necessary for administration of these agents, and the investigational nature of the polymers developed thus far.

Bucrylate

Bucrylate (isobutyl 2-cyanoacrylate) has been the most widely investigated tissue adhesive in the United States. Its capability of penetrating into small vessels has made it quite useful for obliterating AVMs.[21] However, because of reported carcinogenic effects in animals, bucrylate has been withdrawn from the market.[59] Related compounds developed in Europe, such as *n*-butyl cyanoacrylate (Histacryl), are presently undergoing clinical studies in the United States. A feature common to the compounds of this class is rapid polymerization and adhesion on exposure to ionic fluids. For this reason, delivery systems must be flushed with nonionic dextrose solutions.

Bucrylate must be opacified with agents such as tantalum powder, and the speed of polymerization can be controlled by careful mixture with various amounts of iophendylate (Pantopaque). Polymerization time after injection into a vessel is approximately half that of a given mixture as measured in vitro, and this must be taken into account before embolization.[21] Tissue adhesives must be administered through coaxial catheters (or open-ended wires), and the catheter should be withdrawn immediately after the compound is injected. A complication unique to these agents is the possibility of gluing a catheter in place! Experience in the administration of tissue adhesives is best obtained in the laboratory before one proceeds to clinical application.

Nonadhesive Polymers

Low-viscosity silicone rubber and polyurethane are two other polymerizing fluids that have been studied. As with bucrylate, care must be taken to control the polymerization time so that only the vessels desired undergo embolization. Penetration of capillaries by silicone rubber can result in ischemic skin changes or neural injury.[60] These compounds have not caused the tissue inflammatory responses noted with bucrylate. Lack of tissue adhesion is seen in retraction of the polymers from vessel wall with time.[60] The nonadhesive nature of silicone rubber may also have contributed to the report of a case of pulmonary embolization during treatment of an AVM in an extremity.[61]

Sclerosing and Locally Toxic Fluids

These agents have the property of producing vascular occlusion or tissue infarction by virtue of their direct effects at high concentration. In dilution the effects are substantially decreased. As fluids, they are not as dependent on large catheters or special delivery apparatus as are particulate emboli.

Ethanol

Absolute ethanol produces tissue necrosis directly, with vascular occlusion consequent to stasis.[62] This makes ethanol a very attractive agent for the treatment of unresectable renal tumors.[30] It has also been applied with success in cases of AVM, obliterating the nidus of the lesion.[37] Its use is limited by its very toxicity, for serious complications have arisen from penetration of pure ethanol into collateral vessel beds or reflux into aorta or other vessels.[63–65] Ethanol lacks radiopacity, and a small amount of nonionic contrast agent may be mixed without negating the toxic tissue effects. However, use of an occlusion balloon affords greater control of ethanol administration and is highly recommended. Ethanol should never be injected into mesenteric arteries and must be used with particular care in the extremities.

Other Sclerosants

Sodium tetradecyl sulfate (Sotradecol) and hypertonic glucose have long been injected into varicose veins of the lower extremities, and they have also been applied directly to gastroesophageal varices.[12] In Europe a fatty acid of cod liver oil (Varicocid) has been used successfully to occlude internal spermatic veins associated with varicocele.[31] Ionic contrast material in a concentration of 76% has been effective in ablating the parathyroids and kidney.[29, 66] Again, use of an occlusion balloon is advisable, not only to prevent reflux of the sclerosant but also to prevent dilution by inflowing blood.

Other Methods of Vascular Occlusion

Iodinated contrast medium heated to 100°C occludes vessels by its thermal effects.[67] A particular advantage is that one can fluoroscopically monitor the tissues treated. Electrocoagulation probes also cause thrombosis by heating.[68] Thrombin injection may enhance the effect of other injected agents, or it can precipitate coagulation on its own.[20, 32, 48] Thrombin works best in the presence of stasis; when flow is rapid, it is quickly diluted and inactivated.

SPECIFIC ANATOMIC CONSIDERATIONS

Bronchial and Pulmonary Interventions

Massive Hemoptysis

Untreated massive hemoptysis has high mortality, and embolization of bronchial arteries is an effective treatment that can obviate major surgical intervention. Tuberculosis and other inflammatory diseases are the most common causes of massive hemoptysis. Bronchial arteries are variable in number and distribution (see Chapter 7, Thoracic Aortography and Bronchial Angiography), but there are usually no more than two to each lung, and most arise from the descending thoracic aorta at the level of T5–6. A normal bronchial artery is no larger than 3 mm in diameter. Enlargement can be taken as presumptive evidence of hemorrhagic source if the origin of bleeding has not been identified bronchoscopically. Active extravasation on angiography is rarely recognizable. Peripheral inflammatory lesions can also receive blood supply from intercostal arteries and other chest wall vessels. Innominate, subclavian, and internal thoracic artery injections may be needed for complete evaluation. Massive bleeding from a pulmonary arterial source is distinctly unusual.

Embolization has stopped pulmonary hemorrhage in 75% to 90% of patients, but there is a 20% rate of rebleeding within 6 months.[17] Patients with aspergillomas are most likely to suffer recurrent hemoptysis.

The most feared complication of bronchial artery embolization is spinal cord injury. For this reason, careful angiography prior to embolization should be used to identify a possible anterior spinal branch. A right intercostobronchial trunk is most likely to give rise to such a vessel.[17] Each patient should undergo a neurologic examination before embolization, with possible monitoring of somatosensory evoked potentials during the procedure. Uflacker et al. treated nine patients with spinal branches from the bleeding vessel.[69] They increased the size of the gelatin sponge pledgets injected to 3 mm by 10 mm, producing a more proximal occlusion. Although no neurologic damage occurred, such a procedure should be seen as quite risky.

Bronchial arteries have rich anastomotic connections to pulmonary arteries ranging in size from 72 to 325 μm. Use of small particles or fluids such as alcohol can result in bronchial infarction and death.[64]

Pulmonary Arteriovenous Malformations

Pulmonary AVMs can not only produce problems with blood oxygen desaturation and high-output cardiac failure, but are also likely to cause cerebral emboli and brain abscess. For the latter reasons, even asymptomatic lesions should be treated by embolotherapy. The great majority of these patients have Rendu-Osler-Weber syndrome (hereditary hemorrhagic telangiectasia). White et al. have had a high degree of success in occluding 276 lesions in 76 patients with detachable balloons.[18] By closing all feeding vessels larger than 3 mm, they have seen no recurrence of symptoms in up to 5 years follow-up.

Most lesions are found in the lower lobes and are fed by a single artery. If the vessel is larger than 9 mm, a nest of large coils must be constructed before balloon deposition. The balloon should be placed as peripherally as possible, and the effectiveness of occlusion must be checked by digital subtraction angiography or fluoroscopy before release of the patient.

Renal Embolization

Embolization has been used in kidneys to palliate tumors as well as to decrease operative blood loss at nephrectomy for carcinoma. In cases of end-stage renal disease, ablation can help control hypertension or protein loss from nephrotic syndrome. For these indications complete infarction is desirable. For bleeding or AV fistulas, more direct embolization with larger particles is generally needed.

Wallace et al. suggested that there was prolongation of life in those patients with renal cell carcinoma who underwent embolization before nephrectomy, compared with those having nephrectomy alone.[38] However, this study was uncontrolled, and there has been no subsequent confirmation of any significant effect in such patients. In most institutions, routine preoperative embolization of renal tumors has fallen out of favor. Because of the danger of inadvertent intraoperative embolization of the opposite kidney,[70] any coils placed preoperatively should be several centimeters from the renal artery origin. The surgeon should be aware of the number and position of coils, and the specimen must be examined immediately to verify their removal.

Ethanol is quite useful for producing renal infarction, but it has been implicated in the complications of colonic and testicular infarction.[63, 71] Alcohol can also cause massive release of catecholamine if it enters the adrenal circulation.[72] If ethanol is employed, an occlusion balloon must be placed to prevent reflux. It should remain inflated for several minutes after ethanol injection, and a check with contrast medium should be performed before the balloon is deflated. Ethanol may cause considerable pain when injected into the renal vascular bed, but pain may be prevented to some extent by prior injection of lidocaine through the catheter. Ethanol has been associated with less severe postembolization symptoms than other agents.[35]

Although use of an occlusion balloon can prevent reflux, penetration of collateral channels may be responsible for some cases of colonic infarction. The renal artery very commonly has multiple retroperitoneal anastomoses. The most common of these are transcapsular and adrenal, but aortic, iliac, gonadal, and inferior mesenteric artery connections have been observed.[73] Therefore, caution is still justified with the use of ethanol and other toxic agents.

Hepatic Embolotherapy

For embolization, a most important consideration is the dual hepatic blood supply. Tumors receive their blood almost entirely from hepatic artery, while normal liver parenchyma derives most of its blood from portal vein. As long as the portal vein is open and flow is hepatopedal, embolization can be performed without great danger. However, arterial embolization in the face of portal occlusion or hypotension can be fatal.[51, 74] Those with jaundice or massive replacement of liver by tumor should not be treated because of the risk of hepatic failure. Staged embolization is advisable for many patients with widespread neoplasm.

Embolotherapy has been shown to prolong life in patients with unresectable hepatocellular carcinoma, and it is more effective than chemotherapy alone.[25, 74] Recently it has been found that Lipiodol injected intra-arterially tends to accumulate in primary hepatic tumors and can be used to deliver chemotherapy or local radiation.[55] For small hepatocellular carcinomas not amenable to resection, a promising new treatment is direct injection of ethanol through a fine needle placed percutaneously.[47]

Secreting islet cell tumors and carcinoid tumors unresponsive to medical management can be controlled by embolization in most cases.[23, 24] Although the tumors tend to be indolent in their course, embolization may prolong life in these patients as well.

If the catheter cannot be advanced selectively into proper hepatic artery or its branches, use of an occlusion balloon can reverse flow in the gastroduodenal artery and prevent its embolization. Direct manual compression over the gastroduodenal can also stop antegrade flow, allowing embolization to proceed from the common hepatic artery.[75] The cystic artery is often occluded in the course of hepatic embolization. Even so, clinical cholecystitis is rarely severe, unless a powder is used.[50]

Splenic Embolization

Embolization of the splenic artery has been used to treat hypersplenism and its attendant hematologic abnormalities, while preserving some splenic tissue and function. Painful splenic enlargement can also be resolved by embolization, although complete response may take several weeks.[76] Splenic artery aneurysms can be occluded by careful selective placement of coils or large particles, as long as collateral flow is maintained to the spleen itself.[33, 34]

The overriding danger in splenic embolization is the susceptibility of the organ to abscess if parenchyma is infarcted.[42, 51] Spigos et al. developed a highly aseptic procedure, including full body povidone-iodine baths for the patient prior to embolotherapy, broad-spectrum antibiotic coverage, and injection of Gelfoam particles soaked in an antibiotic solution.[28] As long as strict attention to sterility is observed and no more than two thirds of the spleen is infarcted, risk of infection is low. Potent analgesia or epidural anesthesia may be needed to control pain for several days after treatment. Powder should not be injected into the splenic artery, for it can evoke the complication of pancreatitis.

Mesenteric Procedures

Most gastric hemorrhage is self-limited, and only rarely does such bleeding not respond to conservative measures. In the past, selective vasopressin infusion of the left gastric artery has proved useful for controlling otherwise refractory bleeding from gastritis or other small vessel lesions.[5] Catheter embolization was applied only to those patients who continued to lose blood despite infusion. However, over 70% of such patients could be managed by immediate embolization with fewer complications and fewer episodes of rebleeding.[4, 5] Therefore, embolization with gelatin sponge can be considered a primary interventional treatment for continuing gastric hemorrhage.

However, if an upper gastrointestinal bleeding source is fed by other vessels, such as the gastroduodenal artery, control by embolization is much less likely. Occlusion of the gastroduodenal artery causes immediate collateral perfusion from the superior mesenteric artery by way of the pancreaticoduodenal arcades. Thus, fewer than half of cases of duodenal ulcer bleeding respond to embolotherapy, although the use of tissue adhesives may improve results.[15] As noted earlier, transhepatic occlusion of bleeding gastroesophageal varices is at best a temporizing procedure and has been abandoned in most institutions.

Arterial hemorrhage from erosion by a pancreatic pseudocyst is less common than peptic disease. The splenic or gastroduodenal arteries are most often affected. In this situation embolization with large particles or coils has proved quite effective.[7]

Embolization of small bowel or colonic vessels has been performed with success, although the need for such treatment is exceptional. One must reckon with a 13% to 22% risk of ischemic complications.[9–11] No more than one major mesenteric vessel should undergo embolization, particles should not be placed beyond the marginal arcades into the vasa recta, and the effect of each administered particle must be examined serially by injection of contrast material. Embolization should not be continued if extravasation has ceased. Small bowel vessels do not respond readily to selective arterial vasopressin infusion, but infusion should be attempted in those with colonic bleeding.[15]

Before embolization of the left gastric artery, note should be made of any compromise to collateral circulation, such as severe atherosclerosis or previous surgery. The danger of bowel infarction is elevated in those with limited collaterals.[2, 4] For the same reason, vasopressin infusion should not be reinstituted after embolization therapy, even if bleeding has not stopped.

Pelvic and Retroperitoneal Hemorrhage

Pelvic trauma requires angiographic evaluation for hemorrhage in fewer than 10% of cases, but pelvic bleeding can be deadly.[3] Mortality for patients with pelvic fractures who present in shock has been reported as 42%, compared with 3% in similar patients who are hemodynamically stable.[1] Indications for angiography and embolization include transfusion requirement of more than 4 units over 24 hours, more than 6 units over 48 hours, open pelvic fractures, or expanding hematoma found at laparotomy.[1] In severe blunt trauma such as fall from a height, major bleeding can arise from lumbar arteries, especially in the presence of vertebral fractures.[77]

Evaluation includes abdominal aortography followed by selective injections within the internal iliac arteries, as well as lumbar and other arteries that appear to be possible bleeding sources. As in any suspected case of hemorrhage, filming should be carried out to at least 30 seconds to allow extravasation to be recognized. Arterial stasis or vertebral body "stain" are indirect signs of bleeding.[77] Bleeding can often be controlled with gelatin sponge, but central placement of coils decreases the chance of recurrence. If a bleeding site is found and occluded, clinical stabilization can be expected in 85% to 95% of patients, but a large number may die of associated injuries.[1, 3]

Pelvic embolization is also useful when there is bleeding from bladder, rectal, or uterine neoplasms. In addition, palliation of pain caused by tumor invasion of sacral nerves can be achieved. Because of rich collateral communications, bilateral internal iliac embolization is often necessary in many patients with traumatic or neoplastic bleeding. Powders or deeply penetrating fluids are best avoided because of the risk of neural injury.[49]

Arteriovenous Malformations in Extremities

Treatment of AVMs in an extremity is best reserved for symptomatic lesions, for many AVMs may assume a more aggressive behavior after intervention. Amputation is a last resort, but a real risk when there are large lesions. Vascular occlusion should be directed to sites associated with pain, bleeding, or ulceration. The small-vessel core or nidus must be obliterated for lasting results. Small particles of polyvinyl alcohol or other permanent agents should be directed as selectively as possible, whether through 3-F catheters or direct needle puncture.[21, 22] Good results have also been obtained with bucrylate and the judicious staged application of ethanol.[21, 37] Cavernous hemangiomas can be most thoroughly evaluated by venography. They may be treated by direct needle puncture of the venous spaces and sclerosis with alcohol.

In any AVM one must be aware of major AV communications in order to avoid passing small particles to the pulmonary circulation. Major risks of small vessel occlusion in the extremities include tissue ischemia and nerve palsies.[22]

Varicoceles

Varicocele may be treated because of local symptoms or because of its possible effect on fertility. It has been termed the most common correctible cause of male infertility. Traditional treatment has been surgical ligation of the internal spermatic vein, but recurrence has been fairly high after surgery. The left side is much more commonly abnormal.

Selective internal spermatic venography (by way of the left renal vein) allows detailed delineation of vessel incompetence and anatomy. Venography can show duplications or anomalous communications, which are present in most patients with varicocele.[31] The vein can then be occluded at a location that will not allow reflux through duplications or retroperitoneal communicators. Sclerosing agents, coils, and balloons have been safely applied with long-term clinical and venographic success in up to 97% of cases.[31, 41, 58] The greatest challenge with treatment of varicocele is successful selective placement of the administering catheter.

REFERENCES

1. Mucha P Jr, Welch TJ: Hemorrhage in major pelvic fractures. *Surg Clin North Am* 1988; 68:757–773.
2. Jander HP, Russinovitch NAE: Transcatheter Gelfoam embolization in abdominal, retroperitoneal and pelvic hemorrhage. *Radiology* 1980; 136:337–344.
3. Matalon TSA, Athanasoulis CA, Margolies MN, et al: Hemorrhage with pelvic fractures: Efficacy of transcatheter embolization. *AJR* 1979;133:859–864.
4. Rösch J, Keller FS, Kozak B, et al: Gelfoam powder embolization of the left gastric artery in treatment of massive small-vessel gastric bleeding. *Radiology* 1984; 151:365–370.
5. Eckstein MR, Kelemouridis V, Athanasoulis CA, et al: Gastric bleeding: Therapy with intraarterial vasopressin and transcatheter embolization. *Radiology* 1984; 152:643–646.
6. Kuroda C, Kawamoto S, Hori S, et al: Pancreatic pseudocyst hemorrhage controlled by transcather embolization. *Cardiovasc Intervent Radiol* 1983; 6:167–169.

7. Huizinga WKJ, Kalideen JM, Bryer JV, et al: Control of major haemorrhage associated with pancreatic pseudocysts by transcatheter arterial embolization. *Br J Surg* 1984; 71:133–136.

8. Mitchell SE, Shuman LS, Kaufman SL, et al: Biliary catheter drainage complicated by hemobilia: Treatment by balloon embolotherapy. *Radiology* 1985; 157:645–652.

9. Uflacker R: Transcatheter embolization for treatment of acute lower gastrointestinal bleeding. *Acta Radiol* 1987; 28:425–430, Fasc 4.

10. Palmaz JC, Walter JF, Cho KJ: Therapeutic embolization of the small-bowel arteries. *Radiology* 1984; 152:377–382.

11. Rosenkrantz H, Bookstein JJ, Rosen RJ, et al: Postembolic colonic infarction. *Radiology* 1982; 142:47–51.

12. Benner KG, Keefe EB, Keller FS, et al: Clinical outcome after percutaneous transhepatic obliteration of esophageal varices. *Gastroenterology* 1983; 85:146–153.

13. Yune HY, O'Connor KW, Klatte EC, et al: Ethanol thrombotherapy of esophageal varices: Further experience. *AJR* 1985; 144:1049–1053.

14. Spigos DG, Tauber JW, Tan WS, et al: Umbilical venous cannulation: A new approach for embolization of esophageal varices. *Radiology* 1983; 146:53–56.

15. Feldman L, Greenfield AJ, Waltman AC, et al: Transcatheter vessel occlusion: Angiographic results versus clinical success. *Radiology* 1983; 147:1–5.

16. Levine BA, Gaskill HV, Sirinek KR: Portasystemic shunting remains the procedure of choice for control of variceal hemorrhage. *Arch Surg* 1985; 120:296–300.

17. Stoll JF, Bettman MA: Bronchial artery embolization to control hemoptysis: A review. *Cardiovasc Intervent Radiol* 1988; 11:263–269.

18. White RI Jr, Lynch-Nyhan A, Terry P, et al: Pulmonary arteriovenous malformations: Techniques and long-term outcome of embolotherapy. *Radiology* 1988; 169:663–669.

19. Clark RA, Gallant TE, Alexander ES: Angiographic management of traumatic arteriovenous fistulas: Clinical results. *Radiology* 1983; 147:9–13.

20. Laffey KJ, Bixon R, Martin EC: Thrombin as an adjunct to embolisation in high flow arteriovenous fistulae. *J Intervent Radiol* 1988; 3:27–30.

21. Widlus DM, Murray RR, White RI Jr, et al: Congenital arteriovenous malformations: Tailored embolotherapy. *Radiology* 1988; 169:511–516.

22. Gomes AS, Busuttil RW, Baker JD, et al: Congenital arteriovenous malformations: The role of transcatheter arterial embolization. *Arch Surg* 1983; 118:817–825.

23. Mitty HA, Warner RRP, Newman LH, et al: Control of carcinoid syndrome with hepatic artery embolization. *Radiology* 1985; 155:623–626.

24. Ajani JA, Carrasco CH, Charnsangavej C, et al: Islet cell tumors metastatic to liver: Effective palliation by sequential hepatic artery embolization. *Ann Intern Med* 1988; 108:340–344.

25. Lin D-Y, Liaw Y-F, Lee T-Y, et al: Hepatic arterial embolization in patients with unresectable hepatocellular carcinoma: A randomized controlled trial. *Gastroenterology* 1988; 94:453–456.

26. Chuang VP, Wallace S: Hepatic artery embolization in the treatment of hepatic neoplasms. *Radiology* 1981; 140:51–58.

27. O'Keeffe FN, Carrasco CH, Charnsangavej C, et al: Arterial embolization of adrenal tumors: Results in nine cases. *AJR* 1988; 151:819–822.

28. Spigos DG, Jonasson O, Mozes M, et al: Partial splenic embolization in the treatment of hypersplenism. *AJR* 1979; 132:777–782.

29. Miller DL, Doppman JL, Chang R, et al: Angiographic ablation of parathyroid adenomas: Lessons from a 10-year experience. *Radiology* 1987; 165:601–607.

30. Keller FS, Coyle M, Rösch J, et al: Percutaneous renal ablation in patients with end-stage renal disease: Alternative to surgical nephrectomy. *Radiology* 1986; 159:447–451.

31. Seyferth W, Jecht E, Zeitler E: Percutaneous sclerotherapy of varicocele. *Radiology* 1981; 139:335–340.

32. Cope C, Zeit R: Coagulation of aneurysms by direct percutaneous thrombin injection. *AJR* 1986; 147:383–387.

33. Uflacker R: Transcatheter embolisation of arterial aneurysms. *Br J Radiol* 1986; 59:317–324.

34. Baker KS, Tisnado J, Cho S-R, et al: Splanchnic artery aneurysms and pseudoaneurysms: Transcatheter embolization. *Radiology* 1987; 163:135–139.

35. Ekelund L, Ek A, Forsberg L, et al: Occlusion of renal arterial tumor supply with absolute ethanol: Experience with 20 cases. *Acta Radiol [Diagn]* 1984; 25:195–201, Fasc 3.

36. Lund G, Cragg AH, Rysavy JA, et al: Detachable stainless-steel spider: A new device for vessel occlusion. *Radiology* 1983; 148:567–568.

37. Yakes WF, Pevsner P, Reed M, et al: Serial embolizations of an extremity arteriovenous malformation with alcohol via direct percutaneous puncture. *AJR* 1986; 146:1038–1040.

38. Wallace S, Chuang VP, Swanson D, et al: Embolization of renal carcinoma. *Radiology* 1981; 138:563–570.

39. Carroll BA, Walter JF: Gas in embolized tumors: An alternative hypothesis for its origin. *Radiology* 1983; 147:441–444.

40. Rankin RN: Gas formation after renal tumor embolization without abscess: A benign occurrence. *Radiology* 1979; 130:317–320.

41. Wojtowycz M, Miller FJ: Complications of transcatheter embolization. *Semin Intervent Radiol* 1984; 1:179–188.

42. Hemingway AP, Allison DJ: Complications of embolization: Analysis of 410 procedures. *Radiology* 1988; 166:669–672.

43. Meyer P, Reizine D, Aymard A, et al: Septic complications in interventional angiography: Evaluation of risk and preventive measures—preliminary studies. *J Intervent Radiol* 1988; 3:73–75.

44. Greenfield A, Athanasoulis CA, Waltman AC, et al: Transcatheter embolization: Prevention of embolic reflux using balloon catheters. *AJR* 1978; 131:651–655.

45. Matsumoto AH, Suhocki PV, Barth KH: Technical note: Superselective Gelfoam embolotherapy using a highly visible small caliber catheter. *Cardiovasc Intervent Radiol* 1988; 11:303–306.

46. Kerber C: Balloon catheter with a calibrated leak. *Radiology* 1976; 120:547–550.

47. Livraghi T, Salmi A, Bolondi L, et al: Small hepatocellular carcinoma: Percutaneous alcohol injection—results in 23 patients. *Radiology* 1988; 168:313–317.

48. Rogoff PA, Stock JR: Percutaneous transabdominal embolization of an iliac artery aneurysm. *AJR* 1985; 145:1258–1260.

49. Hare WSC, Holland CJ: Paresis following internal iliac artery embolization. *Radiology* 1983; 146:47–51.

50. Kuroda C, Iwasaki M, Tanaka T, et al: Gallbladder infarction following hepatic transcatheter arterial embolization. *Radiology* 1983; 149:85–89.

51. Trojanowski JQ, Harrist JT, Athanasoulis CA, et al: Hepatic and splenic infarctions: Complications of therapeutic transcatheter embolization. *Am J Surg* 1980; 139:272–277.

52. Tadavarthy SM, Castañeda-Zuñiga W, Zollikofer C, et al: Angiodysplasia of the right colon treated by embolization with Ivalon (polyvinyl alcohol). *Cardiovasc Intervent Radiol* 1981; 4:39–42.

53. Strother CM, Laravuso R, Rappe A, et al: Glutaraldehyde cross-linked collagen (GAX): A new material for therapeutic embolization. *AJNR* 1987; 8:509–515.

54. Herba MJ, Illescas FF, Thirlwell MP, et al: Hepatic malignancies: Improved treatment with intraarterial Y-90. *Radiology* 1988; 169:311–314.

55. Bretagne J-F, Raoul J-L, Bourguet P, et al: Hepatic artery injection of I-131-labeled Lipiodol: Part II. Preliminary results of therapeutic use in patients with hepatocellular carcinoma and liver metastases. *Radiology* 1988; 168:547–550.

56. Lang EK, Sullivan J: Management of primary and metastatic renal cell carcinoma by transcatheter embolization with iodine 125. *Cancer* 1988; 62:274–282.

57. Chuang VP, Wallace S, Gianturco C, et al: Complications of coil embolization: Prevention and management. *AJR* 1981; 137:809–813.

58. Kaufman SL, Kadir S, Barth KH, et al: Mechanisms of recurrent varicocele after balloon occlusion or surgical ligation of the internal spermatic vein. *Radiology* 1983; 147:435–440.

59. Suby-Long T, Bos GD, Rösch J: Biopsy proven eradication of an aneurysmal bone cyst treated by superselective embolization: A case report. *Cardiovasc Intervent Radiol* 1988; 11:292–295.

60. Miller FJ Jr, Rankin RS, Gliedman JB, et al: Experimental internal iliac artery embolization: Evaluation of low viscosity silicone rubber, isobutyl 2-cyanoacrylate, and carbon microspheres. *Radiology* 1978; 129:51–58.

61. Capan LM, Lardizabal S, Sinha K, et al: Acute pulmonary embolism during therapeutic arterial embolization with silicone fluids. *Anesthesiology* 1983; 58:569–571.

62. Buchta K, Sands J, Rosenkrantz H, et al: Early mechanism of action of arterially infused alcohol *U.S.P.* in renal devitalization. *Radiology* 1982; 145:45–48.

63. Siniluoto TMJ, Hellström PA, Päivänsalo MJ, et al: Testicular infarction following ethanol embolization of a renal neoplasm. *Cardiovasc Intervent Radiol* 1988; 11:162–164.

64. Ivanick MJ, Thorwarth W, Donohue J, et al: Infarction of the left main-stem bronchus: A complication of bronchial artery embolization. *AJR* 1983; 141:535–537.

65. Mulligan BD, Espinosa GA: Bowel infarction: Complication of ethanol ablation of a renal tumor. *Cardiovasc Intervent Radiol* 1983; 6:55–57.

66. Siragusa RJ, Merandi S, Hanner JS, et al: Renal ablation with iodinated contrast medium: Initial clinical experience. *Semin Intervent Radiol* 1988; 5:146–148.

67. Cragg AH, Rosel P, Rysavy JA, et al: Renal ablation using hot contrast medium: An experimental study. *Radiology* 1983; 148:683–686.

68. Brunelle F, Kunstlinger F, Quillard J: Endovascular electrocoagulation with a bipolar electrode and alternating current: A follow-up study in dogs. *Radiology* 1983; 148:413–415.

69. Uflacker R, Kaemmerer A, Neves C, et al: Management of massive hemoptysis by bronchial artery embolization. *Radiology* 1983; 146:627–634.

70. Wirthlin LS, Gross WS, James TP, et al: Renal artery occlusion from migration of stainless steel coils. *JAMA* 1980; 243:2064–2065.

71. Cox GG, Lee KR, Price HI, et al: Colonic infarction following ethanol embolization of renal-cell carcinoma. *Radiology* 1982; 145:343.

72. Fink IJ, Girton M, Doppman JL: Absolute ethanol injection of the adrenal artery: Hypertensive reaction. *Radiology* 1985; 154:357–358.

73. Wilkins RA, Sandin B, Price A, et al: Extrarenal arterial connections of the normal renal artery. *Cardiovasc Intervent Radiol* 1986; 9:119–122.

74. Yamada R, Sato M, Kawabata M, et al: Hepatic artery embolization in 120 patients with unresectable hepatoma. *Radiology* 1983; 148:397–401.

75. Kubota H, Nimura Y, Hayakawa N, et al: Hepatic transcatheter arterial embolization with gastroduodenal artery blocking by finger compression. *Radiology* 1989; 170:562–563.
76. Grassi CJ, Boxt LM, Bettman MA: Partial splenic embolization for painful splenomegaly. *Cardiovasc Intervent Radiol* 1987; 10:291–294.
77. Sclafani SJA, Florence LO, Phillips TF, et al: Lumbar arterial injury: Radiologic diagnosis and management. *Radiology* 1987; 165:709–714.

13

Lower Extremity Venography

KEY CONCEPTS

1. Venography remains the "gold standard" for making the diagnosis of lower extremity deep venous thrombosis.
2. Presence of an intraluminal filling defect is necessary for confirmation of acute deep vein thrombosis. Failure to opacify deep veins and sharp cutoffs are less specific signs.
3. Proper technique is essential to prevent artifacts and false positive studies.
4. Low-osmolality contrast media and heparinized saline flush minimize the risk of postvenographic phlebitis.
5. Descending venography is used to assess valvular competence in patients with symptoms of venous stasis.

INDICATIONS

The most common reason for performing leg venography is suspected deep venous thrombosis (DVT). Clinical examination is notoriously unreliable in establishing the diagnosis of DVT, and venography remains the standard test. Roughly only half of those referred to venography will have thrombophlebitis. Conditions capable of mimicking DVT are lymphedema, congestive heart failure, Baker cyst, pelvic tumor, and muscular injury. Other indications for venography include evaluation of deep venous insufficiency, varicose veins (the deep veins must be shown to be patent before the saphenous or other major superficial veins are ligated or stripped), and determination of the suitability of the greater saphenous vein as a potential arterial graft.

ANATOMY

The deep veins of the leg and pelvis parallel the course of the arterial supply (Figs 13–1 and 13–2). The anterior tibial, posterior tibial, and peroneal veins are paired. Popliteal and superficial femoral veins may be duplicated, partially or completely, but in most cases they are single vessels. Muscular veins from the gastrocnemius and soleus are also part of the deep venous system.

A variable number of valves are present in deep veins inferior to the inguinal ligament. These valves permit flow of blood to the heart while preventing distal transmission of the great hydrostatic pressure an uninterrupted column of fluid can produce.

The greater and lesser saphenous veins are part of the superficial venous network. Blood normally flows from the superfical veins into the deep venous system by way of short perforating veins that contain valves for the prevention of backflow.

TECHNIQUE FOR ASCENDING PHLEBOGRAPHY

Many techniques and variations have been devised for lower extremity venography, most of them designed to provide good fill of the deep veins in order to detect clots.[1-8] For DVT studies, venipuncture of the foot is preferable, and many radiologists recommend that the needle tip be directed distally (toward the toes). "Downhill" puncture may produce better opacification of the deep venous system, and allows study of pedal veins.

Semiupright Venography

Semiupright technique is widely used in those patients who are able to stand. With this method of examination contrast material is less likely to layer and incompletely fill a vein, thus reducing chances for misinterpretation. Because contrast material does not clear out as rapidly as when the patient is supine, continuous infusion is unnecessary with the semiupright technique, and the examiner has more time to perform various maneuvers to best demonstrate any abnormalities present.

The patient is placed on a fluoroscopic table tilted 45° from the horizontal and bears weight on the foot not being examined. The injected extremity should be completely relaxed; any muscle contraction can impede filling of the deep veins. Because the normal direction of

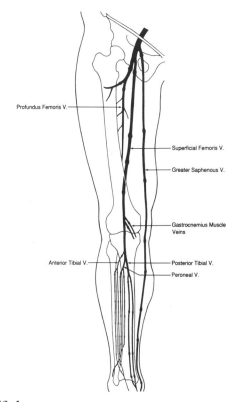

FIG 13–1.
Anterior view of major veins of the lower extremity.

blood flow in the leg is from the superficial into the deep veins, tourniquets are not absolutely necessary in semiupright venography; in fact, they can obstruct filling of some veins, particularly the anterior tibial, soleal, and gastrocnemius. Nevertheless, tourniquets are routinely applied in many institutions, one often placed above the ankle and another above the knee.

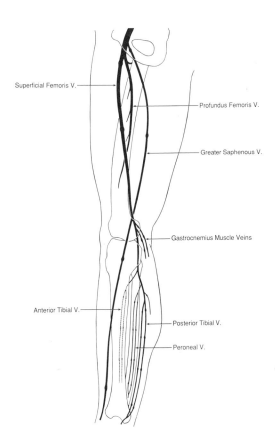

FIG 13–2.
Lateral view of major veins of the lower extremity.

Injection of contrast material may be performed by hand, slow power injection, or drip infusion. Manual injection provides the most immediate control. Observation of flow of the contrast material by fluoroscopy allows optimal timing of film exposure and permits the best projections to be selected. At least two projections of the calf and knee are obtained, including a lateral view. Plantar veins should be examined if there are any symptoms specific to the foot, but most venographers do not routinely study the foot. A single anteroposterior view of the thigh and pelvic vessels is ordinarily sufficient. When used, tourniquets should be released for better filling of the femoral veins. Immediately before filming the iliac veins and distal inferior vena cava, the table is lowered to a horizontal position and the patient's leg is elevated. Large film format, such as 14 × 17 inches with two or three exposures on one film makes interpretation of the study easier. After filming is complete, heparinized saline (5,000 to 10,000 units in 100 to 250 mL of normal saline) is infused through the needle, and the leg maintained in elevation to reduce the risk of postvenography thrombophlebitis. Contrast medium of 43% concentration provides adequate opacification for diagnosis, inducing less discomfort than a 60% concentration and probably decreasing the possibility of thrombosis after venography. Nonionic contrast agents share these advantages over 60% ionic media. The total volume injected per leg is usually on the order of 100 mL.

Methods ancillary to semiupright venography for DVT include calf compression and the Valsalva maneuver for more complete filling of thigh and pelvic vessels, especially the internal iliac vein and profunda femoris. The theoretical possibility of inducing pulmonary embolization by these procedures exists, but Thomas has never encountered such an episode despite performing compression and Valsalva in over 600 studies.[2] If extensive varicose veins are present, the leg can be wrapped in an elastic bandage, permitting better deep venous filling without injection of an excessive volume of contrast material. An elastic bandage can also aid in demonstration of the soleal sinusoids after these have been initially emptied through compression. If the bandage is removed during injection, the sinusoids fill retrograde from the posterior tibial veins.[9] Digital subtraction venography may be helpful in some cases, but multiple projections cannot be obtained during a single injection.

Supine Venography

Venography with the patient supine is preferable for the examination of debilitated or severely ill patients. Most of the technical details are much the same as for semiupright studies, but tourniquets should be used, and continuous infusion of contrast is recommended.[7]

Recently, Smith et al. have described the use of common femoral vein compression to decrease the amount of contrast material needed for supine leg venography.[6] They have also applied the technique at the conclusion of lower extremity runoff arteriography in order to evaluate the saphenous vein as a potential in situ graft.

If DVT is found in one extremity, the other leg need not be studied unless there are compelling reasons. However, routine bilateral simultaneous venography has its advocates.[8]

IF THERE IS NO FOOT VEIN TO BE FOUND

Prolonged elevation of the leg, compressive bandages, or blood pressure cuffs over the foot have been used to displace edema fluid and to uncover foot veins. If these steps don't work, the standard approach in such a situation has been surgical cut-down venotomy over the dorsum of the foot. Another alternative from the early decades of venography is intraosseous injection, but the need for general anesthesia or heavy sedation, as well as the risk of fat embolization, make this an unattractive option.

Puncture of a distal superficial vein or varicosity, with the needle tip directed distally and a tight tourniquet applied proximally, can reliably produce diagnostic deep venous studies up through the iliac veins.[3, 4] This technique can be tried before resorting to a cut-down procedure. Meticulous care must be taken to flush away any residual contrast medium afterward, for varicose veins are very susceptible to thrombosis.

PHLEBOGRAPHIC FINDINGS IN DEEP VEIN THROMBOSIS

The venographic *sine qua non* of clot demonstration is evidence of a filling defect in the contrast material column. Less specific is the finding of an abrupt termination of fill at a constant site. Failure of deep venous filling despite attention to proper technique, particularly in association with a network of collateral channels, is also indicative of thrombosis.

The appearance of the thrombus depends on its age (Fig 13–3). Small clots may lyse within a matter of days. Fresh "free-floating" thrombus indicates formation within the previous week. After the first

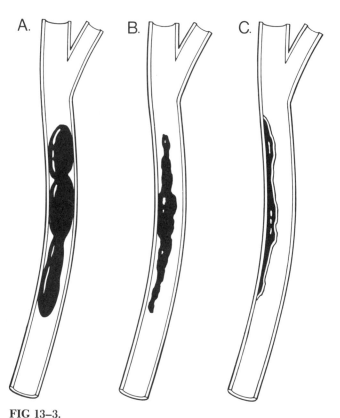

FIG 13–3.
Change in appearance of a thrombus with time. **A,** acute thrombus forms cast of vein. **B,** retraction of subacute clot. **C,** chronic residual mural irregularity.

week, the clot begins to organize—with retraction and mural adherence.[10] This change represents cellular invasion from the vessel wall and capillary proliferation.[11] If adherence is circumferential, the vein becomes occluded. Over the course of weeks to months the vessel may recanalize; the resulting channel often shows irregular walls, loss of valves, and longitudinal "membranes."

The accuracy of venography is difficult to determine, for lack of a more reliable control in vivo. Lund et al. studied 199 lower limbs by postmortem venography followed by meticulous dissection of veins.[12] Thrombi were found in 65 limbs, with the sensitivity and specificity of venography in this context 97% and 95%, respectively.

SOME CLINICALLY RELEVANT INFORMATION

Deep venous thrombosis has a high incidence in some surgical patients, such as those undergoing total hip replacement, but even patients on medical wards (10% to 14%) may develop DVT during hospitalization.[13] Those with congestive heart failure or pneumonia are at substantially higher risk.

Radionuclide fibrinogen studies indicate that the majority of lower extremity thrombi arise in the calf. The work of Nicolaides et al. suggests that clot often forms in the muscular veins of the soleus.[9] In fact, in 18% of their positive venograms, thrombus was confined to the soleus. Although foot veins have been implicated as instrumental in the genesis of DVT, Glanz and Gordon failed to detect any plantar vein thrombosis in isolation in 200 venograms, and routine foot filming does not appear indicated.[14] Untreated DVT is likely to propagate proximally, and although half of pulmonary emboli are said to form in the calf, those arising from more proximal thrombi are much more important clinically. Most patients with clot limited to the calf will have low-probability ventilation-perfusion scans, but up to 35% of those with thrombus extending above calf have evidence of segmental or larger pulmonary emboli, even in the absence of symptoms.[15] The proximal extent of thrombosis determines both the need for therapy and its duration.

Another clinical point of importance is that occluded femoral and iliac veins are less likely to recanalize and are more likely to result in severe debilitation than occluded calf veins.[10, 11] In serious iliofemoral thrombosis, surgery may be considered for clot removal. However, extensive adherence of thrombus can make such a procedure quite difficult, and reocclusion is apt to result from the intimal trauma involved in thrombectomy.[10] Thrombolysis is a therapeutic alternative for iliofemoral occlusion.

Most patients with documented pulmonary emboli will show clot in the lower extremity at venography, but a venogram negative for clot cannot reasonably exclude the possibility of embolus from the diagnosis.[16]

VENOUS INCOMPETENCE AND DESCENDING VENOGRAPHY

The postphlebitic syndrome can be an extremely debilitating condition characterized by edema, pain, varicosities, skin changes, and ulceration resulting from venous stasis and high transmitted venous pressures. These symptoms are the result of valvular thickening, shortening, and incompetence after DVT. Similar symptoms can also be seen as a primary condition in which major valves have stretched and dilated.[17] In fact, primary valvular incompetence appears to be more common than postphlebitic venous stasis.[18] Surgical treatment involves resuspending proximal valves or anastomosing an incompetent vein to another with proximal competent valves in an end-to-side manner (e.g., incompetent superficial femoral vein to intact saphenous vein). If ascending phlebography fails to demonstrate an abnormality responsible for the patient's symptoms, descending venography is indicated.

Descending venography may be performed through a catheter placed from the arm or from the opposite femoral vein,[19, 20] but it can also be done through a needle or cannula placed directly into the common femoral vein of the involved side. A small amount of contrast medium (15 mL) is injected, and the reflux is observed fluoroscopically and graded. The examination proceeds either with the patient upright on a tilt table, or supine and performing the Valsalva maneuver. Grading is from 0 to 5, grade 0 representing no reflux and grade 5 representing reflux to the level of the ankle. Even asymptomatic legs can show reflux to the knee, depending on the technique employed, so surgery is reserved for those with more extensive valvular incompetence.[21]

VARICOSE VEINS

Varicose veins may be secondary to chronic deep venous disease or they may be primary in origin. Primary varicose veins are associated with saphenous vein incompetence and are treated by proximal saphenous vein ligation or stripping.[22] If ascending venography does not show deep venous disease and saphenous incompetence is not clearly present on clinical examination, descending saphenous venography may be performed. The procedure is identical to descending phlebography described earlier, but grading of insufficiency is unnecessary. The very presence of any valvular incompetence determines the need for surgery.

In individuals with recurrent varicose veins, phlebographic identification of incompetent perforating veins is advisable. This may be done by conventional ascending venography, but direct puncture of a varicose group (varicography) provides a clearer and more complete demonstration.[23] A radiopaque ruler should be placed next to the leg to relate the site of disease to bony landmarks. Spot films are obtained during filling, and it is essential that they be exposed early in the examination before overlapping vessels confuse interpretation. It may be necessary to make several venipunctures of isolated varicose groups for a complete study, taking care to flush out contrast material from the previous injection by infusion of heparinized saline before further study. Heparin infusion is of utmost importance for the prevention of superficial thrombophlebitis after varicography.

EVALUATING THE SAPHENOUS AS DONOR VESSEL FOR GRAFT

When the saphenous vein is examined preoperatively, it is useful to do what is explicitly avoided for phlebography in suspected DVT: namely, the distal saphenous vein is punctured directly, the needle is pointed cephalad, no tourniquets are used with the patient supine, and the patient is instructed to engage the quadriceps and calf muscles in sustained isometric contraction against resistance.[5] The result is better opacification of the greater saphenous at the expense of deep and muscular venous filling. The addition of a Valsalva maneuver is also helpful.

TECHNICAL PROBLEMS AND PITFALLS

Errors in diagnosis are commonly the result of incomplete filling of vessels. Defects in the contrast material column are often from the inflow of unopacified blood from a tributary vessel. With fluoroscopy and multiple projections, it can be seen that flow defects, unlike thrombi, are inconstant. Fluoroscopy is essential to establish that contrast medium has reached the vessels being filmed and that opacification is adequate. As mentioned, improper needle placement in a superficial vein, weight-bearing or muscle contraction by the patient, or use of tourniquets can produce artifacts of poor filling that must be recognized.

COMPLICATIONS AND THEIR PREVENTION

As in any study using iodinated radiographic contrast medium, hypersensitivity reactions occasionally occur during venography. Most are mild, but the examiner must be prepared to treat life-threatening situations. Because of the high volumes injected for most studies, especially if bilateral phlebography is necessary, careful attention must be paid to renal function and hydration. Note must be made of diabetes mellitus, multiple myeloma, hyperuricemia, and advanced age, as these all increase the risk of acute renal failure.

Extravasation of contrast medium may occur, and small (less than 10 mL) extravasations do not generally pose a problem. However, tissue necrosis and ulceration have been reported, and risk is higher in those patients with arterial insufficiency.[24] Plastic cannulas are less subject to extravasation. Saline should be infused through a well-secured needle or cannula prior to injection of contrast material, and the injection site must be regularly checked during its infusion. If extravasation does occur, the leg should be elevated, the site massaged, and a warm compress applied to speed absorption.

The complication most specific to leg venography is postprocedural thrombophlebitis. Its incidence is related to the concentration and osmolality of the contrast agent and the duration of its contact with the venous endothelium. Positive radionuclide-labeled fibrinogen studies have been reported after venography in up to 60% of individuals with initially normal venograms, but addition of 10,000 units of heparin to 100 mL of normal saline infusion immediately following phlebography reduces this to 3%.[25] A reduced-osmolality contrast agent is also less likely to produce this complication.[26] No matter what technique or contrast medium is used, it is advisable to keep examination time as short as possible and to elevate the patient's leg immediately after venography.

ALTERNATIVE METHODS FOR DIAGNOSING DEEP VEIN THROMBOSIS

Impedance plethysmography is a noninvasive technique that measures changes in electrical impedance in the leg resulting from changes in blood/fluid volume. It is very sensitive in the detection of proximal (femoral and above) venous occlusion, but subocclusive thrombi or thrombi in the presence of large collaterals or duplicated channels can be missed.[27]

A complementary technique is iodine-125 fibrinogen scanning for active thrombus formation. Fibrinogen scanning requires 2 days to complete and is not a good diagnostic tool for determining proximal clot, but it is well suited for evaluating the popliteal and calf veins. Hull and associates have found the combination of fibrinogen scanning and impedance plethysmography to be cost effective, particularly when obtained on an outpatient basis, and virtually as accurate as venography.[28]

Doppler examination (without ultrasonic imaging) is useful when abnormal, but it lacks sensitivity.[29] Duplex sonography of lower extremity veins, combining real time ultrasonic imaging and Doppler, improves accuracy—with a reported sensitivity of 89% or better for DVT.[30, 31] Still, duplex sonography demands considerable operator experience and is a poor screening test for clots limited to the calf, and the findings in patients with chronic venous disease can be very difficult to interpret.

REFERENCES

1. Rabinov K, Paulin S: Roentgen diagnosis of venous thrombosis in the leg. *Arch Surg* 1972; 104:134–144.
2. Thomas ML: Phlebography. *Arch Surg* 1972; 104:145–151.
3. Tisnado J, Tsai FY, Beachley MC: An alternate technique for lower extremity venography. *Radiology* 1979; 133:787–788.
4. Gordon DH, Glanz S, Stillman R, et al: Descending varicose venography. *Radiology* 1982; 145:832–834.
5. Sapala JA, Szilagyi DE: A simple aid in greater saphenous phlebography. *Surg Gynecol Obstet* 1975; 140:265–267.
6. Smith TP, Cardella JF, Darcy MD, et al: Lower-extremity venography: Value of femoral-vein compression. *AJR* 1986; 147:1025–1026.

7. Kramer FL, Teitelbaum G, Merli GJ: Panvenography and pulmonary angiography in the diagnosis of deep venous thrombosis and pulmonary thromboembolism. *Radiol Clin North Am* 1986; 24:397–418.

8. Rampton JB, Armstrong JD Jr: Bilateral venography of the lower extremities. *Radiology* 1977; 123:802–804.

9. Nicolaides AN, Kakkar VV, Field ES, et al: The origin of deep vein thrombosis: A venographic study. *Br J Radiol* 1971; 44:653–663.

10. Thomas ML, McAllister V: The radiological progression of deep venous thrombus. *Radiology* 1971; 99:37–40.

11. Lipchik EO, DeWeese JA, Rogoff SM: Serial long-term phlebography after documented lower leg thrombosis. *Radiology* 1976; 120:563–566.

12. Lund F, Diener L, Ericsson JLE: Postmortem intraosseous phlebography as an aid in studies of venous thromboembolism. *Angiology* 1969; 20:155–176.

13. Kierkegaard A, Norgren L, Olsson C-G, et al: Incidence of deep venous thrombosis in bedridden non-surgical patients. *Acta Med Scand* 1987; 222:409–414.

14. Glanz S, Gordon DH: Utility of foot venography as part of the routine lower-extremity venogram: A prospective study. *Cardiovasc Intervent Radiol* 1986; 9:15–16.

15. Dorfman GS, Cronan JJ, Tupper TB, et al: Occult pulmonary embolism: A common occurrence in deep venous thrombosis. *AJR* 1987; 148:263–266.

16. Hull RD, Hirsh J, Carter CJ, et al: Pulmonary angiography, ventilation lung scanning, and venography for clinically suspected pulmonary embolism with abnormal perfusion lung scan. *Ann Intern Med* 1983; 98:891–899.

17. Kistner RL: Primary venous valve incompetence of the leg. *Am J Surg* 1980; 140:218–224.

18. Train JS, Schanzer H, Peirce EC, et al: Radiologic evaluation of the chronic venous stasis syndrome. *JAMA* 1987; 258:941–944.

19. Herman RJ, Neiman HL, Yao JST, et al: Descending venography: A method of evaluating lower extremity venous valvular function. *Radiology* 1980; 137:63–69.

20. Taheri SA, Sheehan F, Elias S: Descending venography. *Angiology* 1983; 34:299–305.

21. Thomas ML, Keeling FP, Ackroyd JS: Descending phlebography: A comparison of three methods and an assessment of the normal range of deep vein reflux. *J Cardiovasc Surg* 1986; 22:27–30.

22. Thomas ML, Bowles JN: Descending phlebography in the assessment of long saphenous vein incompetence. *AJR* 1985; 145:1255–1257.

23. Thomas ML, Bowles JN: Incompetent perforating veins: Comparison of varicography and ascending phlebography. *Radiology* 1985; 154:619–623.

24. Spigos DG, Thane TT, Capek V: Skin necrosis following extravasation during peripheral phlebography. *Radiology* 1977; 123:605–606.

25. Minar E, Ehringer H, Sommer G, et al: Prevention of postvenographic thrombosis by heparin flush: Fibrinogen uptake measurements. *AJR* 1984; 143:629–632.

26. Murphy WA, Destouet JM, Gilula LA, et al: Hexabrix as a contrast agent for ascending leg phlebography. *AJR* 1985; 144:1279–1281.

27. Ramchandani P, Soulen RL, Fedullo LM, et al: Deep vein thrombosis: Significant limitations of non-invasive tests. *Radiology* 1985; 156:47–49.

28. Hull R, Hirsch J, Sackett DL, et al: Cost effectiveness of clinical diagnosis, venography, and noninvasive testing in patients with symptomatic deep-vein thrombosis. *N Engl J Med* 1981; 304:1561–1567.

29. Langsfeld M, Hershey FB, Thorpe L, et al: Duplex B-mode imaging for the diagnosis of deep venous thrombosis. *Arch Surg* 1987; 122:587–591.

30. Cronan JJ, Dorfman GS, Scola FJ, et al: Deep venous thrombosis: US assessment using vein compression. *Radiology* 1987; 162:191–194.

31. Vogel P, Laing FC, Jeffrey RB Jr, et al: Deep venous thrombosis of the lower extremity: US evaluation. *Radiology* 1987; 163:747–751.

14 | Pulmonary Angiography

KEY CONCEPTS

1. Ventilation-perfusion scans can direct pulmonary arteriography to the region of greatest suspicion for pulmonary embolism.
2. Beware of patients with severe pulmonary arterial hypertension or left bundle branch block.
3. Selective and subselective arteriography is clearly superior to main pulmonary artery or atrial injections.
4. For unequivocal diagnosis of acute pulmonary embolus, intraluminal clot or an abrupt vascular cut-off must be defined.
5. A negative high-quality pulmonary arteriogram excludes clinically significant pulmonary embolism.

INDICATIONS

Pulmonary arteriography remains the standard for making the diagnosis of acute pulmonary embolism, which is by far the most common indication for the procedure. Occasionally, pulmonary angiography is used to plan or follow treatment for a variety of other conditions, including chronic pulmonary embolism, developmental abnormalities, arterial hypoplasia or stenosis, pulmonary sequestration, aneurysms, arteriovenous malformations, vasculitides, and vascular occlusion by tumor or inflammatory disease. Rarely, catheterization may be used in massive pulmonary embolism for suction embolectomy or other emergency transcatheter intervention.

WHAT TO KNOW ABOUT THE PATIENT BEFORE STARTING

For an invasive study often performed in very ill patients, pulmonary angiography is quite safe. However, appropriate performance of the study with minimum risk to the patient requires knowledge of the patient's clinical presentation—including onset of symptoms, presence of known underlying pulmonary or cardiac disease (especially left bundle branch block), renal insufficiency, diabetes mellitus, multiple myeloma or other risk factors for acute renal failure from exposure to radiographic contrast media, and, naturally, any history of hypersensitivity to iodinated contrast material. If left bundle branch conduction block is present there is real danger of inducing total heart block during catheterization, and a temporary cardiac pacemaker should be placed before the catheter is introduced into the pulmonary artery.

Chest radiographs and radionuclide examinations must be reviewed, with particular attention to location of any focal disease or perfusion abnormalities. A radionuclide ventilation-perfusion scan should be obtained when possible in any patient suspected of pulmonary embolus, because the findings may either obviate angiography or direct the study to the area most likely to provide the diagnosis. Heparin should be discontinued at least 1 hour before angiography is performed, but its administration may be resumed immediately afterward.

TECHNIQUE

Catheterization can be performed from a brachial or jugular approach, but the route most commonly used is the common femoral vein. In the latter location, the femoral artery should be palpated and local anesthetic infiltrated just medial to the pulse. Either a double-wall puncture with Seldinger needle or single-wall puncture with a hollow, thin-wall, 18-gauge needle can be used to gain entry. With the single-wall technique, inadvertent passage through the artery is less likely. The needle is placed under gentle suction as it is advanced (hollow needle) or withdrawn (Seldinger). Free return of venous blood indicates that the tip is within the vessel. The guidewire should not be advanced above the common iliac vein before catheter insertion, in case of a major clot in the inferior vena cava. Patency of the iliac vein and inferior vena cava are checked by hand injections of contrast material under fluoroscopic guidance as the catheter is advanced into the right atrium.

The catheter placed should be of large bore (7 F or greater), with multiple sideholes and a pigtail tip. Large catheter caliber allows the rapid injection necessary to obtain adequate vascular opacification, and it makes catheter recoil less violent. Straight endhole catheters have been known to cause myocardial injury and pericardial extravasation through recoil and jet effect.

Grollman-type catheters are fairly stiff, have good torque control, and have a sidearm that may be directed through the tricuspid valve. Once the right ventricle is entered, such a catheter is rotated 180° to direct the tip toward the pulmonic valve. The catheter is then advanced into the pulmonary circulation. An alternative method employs a soft pigtail catheter (such as a 7-F Kifa) and a tip-deflecting guidewire. The guidewire tip is placed to the level of the pigtail, but not beyond the catheter tip. In the atrium the deflecting wire can be used to direct the catheter through the tricuspid valve by fixing the curve of the wire and feeding the catheter forward off the wire (Fig 14–1). Once into the right ventricle, the catheter and wire are rotated 180° and deflection is repeated toward the pulmonic valve. Similar manipulations can then place the catheter into the right or left pulmonary artery as desired. Occasionally the catheter may coil in an enlarged atrium as attempts are made to advance it; in such an event the catheter can be straightened by introducing the stiff end of a guidewire up to the pigtail. The catheter may then be slipped off the wire and advanced without buckling.

During all manipulations in the right side of the heart, an observer must constantly monitor the patient's electrocardiogram and call out any induced ectopic beats. If a run of ventricular tachycardia is encountered, it can usually be reversed by immediately withdrawing the catheter and instructing the patient to <u>cough</u>. Rarely, pharmacologic infusion (lidocaine or other antiarrythmic agent) or electroversion may be necessary. The examiner must always be prepared to treat a possible cardiac or pulmonary arrest.

The catheter tip is best placed initially within the lower lobe pulmonary arterial trunk, unless the major abnormality seen on prior radionuclide studies involves the upper lobe. Pulmonary arterial pressure readings are obtained, and a hand injection of contrast material is used to gauge blood flow. For full-lung opacification in a relatively healthy individual, between 30 and 45 mL of 76% contrast medium is injected at a rate of 20 to 35 mL/sec, depending on cardiac output. In patients with severe underlying cardiac or pulmonary disease, the amount and rate of injection should be decreased accordingly. The equipment used must allow a rapid exposure sequence of 14 × 14-

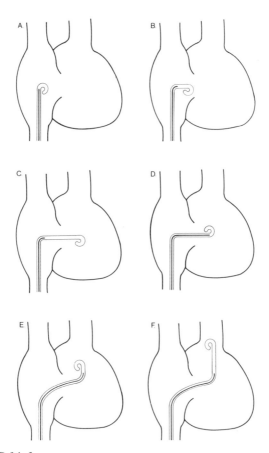

FIG 14–1.
Pulmonary artery catheterization by use of a deflecting wire. **A,** pigtail catheter and wire in right atrium. **B,** activation of deflecting wire. **C,** catheter advanced through tricuspid valve as wire is held stationary with active deflection. **D,** wire advanced and catheter rotated 180°. **E,** activation of deflecting wire. **F,** catheter fed off of deflecting wire through pulmonic valve.

inch films. A typical study may include three films per second for 3 seconds, followed by one film per second for the next 6 seconds. Such a sequence normally provides opacification of the pulmonary veins, left side of the heart, and aorta (revealing any unsuspected vascular abnormality, such as aortic dissection). On subsequent injections in a given lung, the last three films may be omitted.

At least two views of each lung should be obtained. Simultaneous biplane filming can decrease length of the study and the amount of contrast material injected. The ipsilateral posterior oblique view tends to produce the least amount of vascular overlap. If any injection demonstrates embolus, the examination can usually be stopped. In difficult cases, the pigtail catheter may be replaced with a straight or occlusion balloon catheter over an exchange guidewire, and subselective studies performed with magnification filming. Balloon occlusion studies may be more sensitive than conventional angiograms.[1] At some institutions such injections with magnification are performed routinely in the region of greatest perfusion scan abnormality, if selective injections fail to reveal embolus.[2] Single-projection angiograms with main pulmonary artery or right atrial injection are inadequate to exclude pulmonary embolus.[3]

There are various alternatives to these techniques. The use of cineangiography has some advantages, including the recognition of to-and-fro motion of emboli and the ability to resolve problems with overlap by using respiratory motion,[4] but these are counterbalanced by the smaller field size, lower spatial resolution, and higher radiation dose. Digital subtraction angiography has been employed with success by some,[5–8] but its role has not been clearly established. It is probably most appropriate in high-risk patients for whom the concentration and amount of contrast material per injection can be decreased, particularly if radionuclide study suggests larger central emboli. The digital filming device must be capable of a rapid frame rate to allow selection of an appropriate mask image. One technique to be avoided is the use of "bedside" pulmonary angiography through indwelling central venous catheters. In one comparison of conventional pulmonary angiography and injection of wedge catheters in 21 patients, the latter had a sensitivity of 19%, specificity of 60%, and accuracy of 29% for detection of pulmonary embolus.[9]

ACUTE PULMONARY EMBOLISM

The estimated incidence of pulmonary embolism in the United States is about 600,000, and perhaps 90% survive the initial insult.[10] In the

1950s only 12% of major emboli were diagnosed antemortem, and although diagnosis has improved, a more recent large autopsy series indicated that only 30% are currently being recognized.[11] The elderly, and those with concomittant pneumonia or congestive heart failure, are less likely to be given the correct diagnosis. About a third of those with pulmonary embolus will have recurrent emboli, and mortality in untreated pulmonary embolus has been reported from 18% to 38%.[2] With proper treatment, mortality can be reduced to as low as 3%.[12] Clinical signs and symptoms, laboratory results, and chest radiographs have been notoriously unreliable in securing the diagnosis.

A confident diagnosis of pulmonary embolus is required, because therapy has its own risks. Standard treatment comprises intravenous administration of heparin, followed in 1 week by oral Coumadin (warfarin), the latter being continued for 3 to 6 months. Heparin may be the leading cause of drug reaction in hospitalized patients, and anticoagulation in treatment of pulmonary embolus results in major hemorrhage in 5% to 10% of cases.[13]

The development of radionuclide lung scanning has provided clinicians a useful, low-morbidity tool to screen for embolic disease, but its accuracy has been quite disparate in various reports.[2, 10, 14, 15] Nevertheless, those sources giving inferior results have generally relied upon perfusion-only scanning.[2, 14] In a prospective comparison of radionuclide scanning, pulmonary angiography, and venography in 139 patients, Hull et al.[15] found 30 of 35 patients with at least one segmental or larger ventilation-perfusion mismatch to have embolus on angiographic study. Refined criteria for scan interpretation have produced a sensitivity of 97%, a specificity of 94%, and an accuracy of 96% for readings of high or low probability in another series of patients.[10]

Lower extremity venography in patients suspected of suffering pulmonary embolus is helpful in the institution of treatment only if fresh thrombus is found above the knee. A large percentage of those with acute thrombus extending proximally will have embolus, even if no pulmonary symptoms are evident.[12] Conversely, absence of thrombus on leg venography does not exclude pulmonary embolus. Up to 18% of those with documented pulmonary embolus will have no evidence of leg clot, and noninvasive examinations (less sensitive for deep venous thrombosis than venography) may be normal in more than half![16] Potential other sources for embolization of thrombus include the pelvic, renal, and subclavian veins. Confounding the issue still further is the observation that 15% of patients found to have thrombotic-embolic disease will have a pulmonary arteriogram that is negative for disease, but a positive leg venogram.[15]

Pulmonary angiography remains the standard by which all other studies are judged. To make the diagnosis of pulmonary embolus angiographically, intraluminal filling defects or abrupt vascular cutoffs must be identified. Indirect signs, such as diminished capillary stain or delayed opacification, are very nonspecific. Common conditions in those found by angiography not to have pulmonary embolus include congestive heart failure, angina, myocardial infarct, pneumonia, atelectasis, pleurisy, bronchospasm, and bronchiectasis.[2, 17] Angiographically, atelectasis is shown by crowded vessels and an intense capillary stain; pneumonia produces markedly slowed blood flow with intrinsically normal vessels and no marked blush.[17] Pulmonary hypertension can produce occasional arterial occlusion from intimal and medial hypertrophy, but such occlusions arise in vessels less than 0.5 mm in diameter. Bronchiectasis is characterized by slow flow or nonopacification because of bronchial artery communications. Emphysematous blebs displace vessels, and there is decreased arborization. Patients with carcinoma may have flow slowed to the entire involved lung, independent of arterial encasement or obliteration.[17]

Injections into the main pulmonary artery may demonstrate emboli as small as 2 mm in diameter, but superselective and magnification techniques can enable detection of clots down to 0.5 to 1.0 mm.[2, 18] Interobserver agreement is very good for segmental or larger emboli, but interpretation of angiograms is more difficult for subsegmental filling defects.[19]

In a study of 180 patients followed at least 6 months after negative high-quality pulmonary arteriograms, none had signs or symptoms of recurrent embolization.[2] These results support the contention that pulmonary angiography is highly sensitive for detection of clinically relevant pulmonary embolus, and a patient should not be placed on anticoagulants in the face of a disease-free angiogram (assuming there is no other indication for anticoagulation).

In obtaining a pulmonary arteriogram, one must be aware of when the patient's symptoms arose. Fred et al. have reported that emboli can lyse completely in 1 to 3 weeks.[20] Another study with 15 patients undergoing serial angiography showed little change in findings in the 1st week, and only rare disappearance of clots before two weeks.[21] In any event, it is rarely necessary to perform angiography for diagnosis in the early morning hours unless there is a contraindication to anticoagulation with heparin.

Pulmonary infarction does not occur with large central occluding emboli but rather in distal occlusions, and is probably related to increased segmental bronchial artery blood flow.[22] True pulmonary infarction is less common than embolus with pulmonary hemorrhage. The latter tends to resolve quickly, while the former almost never occurs in the absence of heart disease. It is felt that elevated pulmonary venous pressure is required to produce a true infarction.

In massive pulmonary embolism, catheter methods may be the most expeditious means of saving a patient's life. Catheter embolectomy using a special suction device has been successfully used in individuals in shock or cardiopulmonary arrest.[23] Recent modifications in fascial dilator and sheath systems allow introduction of such suction catheters without a surgical venous cut-down.[24] Alternative ways of quickly improving pulmonary perfusion include breaking up central thrombi with the catheter, guidewire, or angioplasty balloon catheter. Selective thrombolysis is not very effective in such dire situations.

CHRONIC PULMONARY EMBOLISM

Chronic pulmonary embolism presents a different clinical and angiographic picture. Symptoms of progressive exertional dyspnea, right-sided heart failure, and cyanosis arise insidiously. Anticoagulants are not an effective treatment, and surgical embolectomy is recommended.[25] Preoperative pulmonary and bronchial angiograms may be obtained to demonstrate anatomy and the likelihood of postoperative improvement. Chronic pulmonary embolism is manifested by arterial webs, stenoses, or irregular occlusions.[17]

PULMONARY ANGIOGRAPHY IN NONEMBOLIC DISEASE

Arteriography can be used to confirm the presence of pulmonary arteriovenous malformations, which can be effectively treated by transcatheter embolization. Untreated lesions can produce cyanosis, hemoptysis, and systemic emboli (often septic). Mycotic or other pulmonary arterial aneurysms may require angiographic evaluation.[26]

Rarely, Takayasu arteritis can present with occlusion of major pulmonary arterial branches.[27, 28] Primary or metastatic intraluminal tumors can simulate pulmonary embolism.[28] Congenital webs, enlarged nodes, or fibrosing mediastinitis can produce pulmonary venous obstruction and secondary pulmonary arterial hypertension. In such cases segmental injections and measurement of pulmonary wedge pressures may be vital for making the correct diagnosis.[29]

COMPLICATIONS

Reactions to contrast material are similar in incidence and severity to those elicited by excretory urography or other contrast infusion studies. In a review of 1,350 procedures at Duke University hospitals, 11 episodes of hypersensitivity to contrast material were noted, including 4 severe reactions.[30] In the same review there were 11 significant arrythmias and 5 cardiac arrests; all patients were successfully resuscitated.

At Duke, three deaths have been directly attributable to pulmonary angiography (mortality of 0.2%), all in patients with pulmonary arterial systolic pressures greater than 70 mm Hg and with right ventricular end-diastolic pressures 20 mm Hg or more. These individuals suffered irreversible right-heart failure after angiography. Risk of mortality in such patients is 2% to 3% and does not appear related to the site and amount of injection.[30, 31] One similar death has been reported after a single subselective injection of 10 mL of contrast medium by hand.[14] Nevertheless, prudence dictates that high-risk patients be studied by subselective injections in which small amounts of contrast material are used. Nonionic or dilute contrast agents may be safer. Special care should be exercised in patients that have been given amiodarone for cardiac arrythmias.[32]

The risk of performing pulmonary angiography in any given patient must be weighed against that of inappropriate treatment. As noted earlier, long-term anticoagulation can cause life-threatening hemorrhage, and the decision to assume anticoagulant therapy should not be made lightly.

REFERENCES

1. Ferris EJ, Smith PL, Lim WN, et al: Radionuclide-guided balloon occlusion pulmonary cineangiography: An adjunct to pulmonary arteriography. *Am Heart J* 1984; 108:539–542.
2. Novelline RA, Baltarowich OH, Athanasoulis CA, et al: The clinical course of patients with suspected pulmonary embolism and a negative pulmonary arteriogram. *Radiology* 1978; 126:561–567.
3. Price L: False negative angiogram in pulmonary embolism. *Chest* 1985; 88:139–141.
4. Meyerovitz MF, Levin DC, Harrington, DP, et al: Evaluation of optimized biplane pulmonary cineangiography. *Invest Radiol* 1985; 20:945–949.

5. Pond GD, Ovitt TW, Capp MP: Comparison of conventional pulmonary angiography with intravenous digital subtraction angiography for pulmonary embolic disease. *Radiology* 1983; 147:345–350.

6. Hirji M, Gamsu G, Webb WR, et al: EKG-gated digital subtraction angiography in the detection of pulmonary emboli. *Radiology* 1984; 152:19–22.

7. Harder T, Lackner K, Vatter J: Digitale subtraktionsangiographie (DSA) der lunge. *ROFO* 1984; 140:425–430.

8. Bargon G, Arlart IP: Die intravenöse digitale subtraktionsangiographie (DSA) zur darstellung der pulmonalgefäße. *ROFO* 1984; 140:431–435.

9. LePage JR, Gracia RM: The value of bedside wedge pulmonary angiography in the detection of pulmonary emboli: A predictive and prospective evaluation. *Radiology* 1982; 144:67–73.

10. Spies WG, Burstein SP, Dillehay GL, et al: Ventilation-perfusion scintigraphy in suspected pulmonary embolism: Correlation with pulmonary angiography and refinement of criteria for interpretation. *Radiology* 1986; 159:383–390.

11. Goldhaber SZ, Hennekens CH, Evans DA, et al: Factors associated with correct antemortem diagnosis of major pulmonary embolism. *Am J Med* 1982; 73:822–826.

12. Dorfman GS, Cronan JJ, Tupper TB, et al: Occult pulmonary embolism: A common occurrence in deep venous thrombosis. *AJR* 1987; 148:263–266.

13. Glenny RW: Pulmonary embolism: Complications of therapy. *South Med J* 1987; 80:1266–1276.

14. Marsh JD, Glynn M, Torman HA: Pulmonary angiography: Application in a new spectrum of patients. *Am J Med* 1983; 75:763–770.

15. Hull RD, Hirsh J, Carter CJ, et al: Pulmonary angiography, ventilation lung scanning, and venography for clinically suspected pulmonary embolism with abnormal perfusion lung scan. *Ann Intern Med* 1983; 98:891–899.

16. Scigala EM, McDonnell AM, Hadcock WE, et al: Prevalence of deep venous thrombosis in patients with proven pulmonary embolism. *Bruit* 1984; 8:222–224.

17. Bookstein JJ, Silver TM: The angiographic differential diagnosis of acute pulmonary embolism. *Radiology* 1974; 110:25–33.

18. Bookstein JJ, Feigin DS, Seo KW, et al: Diagnosis of pulmonary embolism: Experimental evaluation of the accuracy of scintigraphically guided pulmonary arteriography. *Radiology* 1980; 136:15–23.

19. Quinn MF, Lundell CJ, Klotz TA, et al: Reliability of selective pulmonary arteriography in the diagnosis of pulmonary embolism. *AJR* 1987; 149:469–471.

20. Fred HL, Axelrad MA, Lewis JM, et al: Rapid resolution of pulmonary thromboemboli in man. *JAMA* 1966; 196:1137–1139.

21. Dalen JE, Banas JS Jr, Brooks HL, et al: Resolution rate of acute pulmonary embolism in man. *N Engl J Med* 1969; 280:1194–1199.

22. Dalen JE, Haffajee CI, Alpert JS, et al: Pulmonary embolism, pulmonary hemorrhage, and pulmonary infarction. *N Engl J Med* 1977; 296:1431–1435.

23. Moore JH Jr, Koolpe HA, Carabasi RA, et al: Transvenous catheter pulmonary embolectomy. *Arch Surg* 1985; 120:1372–1375.

24. Cope C: Venous cannula for emergency catheter pulmonary embolectomy. *Radiology* 1986; 161:553.

25. Mills SR, Jackson DC, Sullivan DC, et al: Angiographic evaluation of chronic pulmonary embolism. *Radiology* 1980; 136:301–308.

26. SanDretto MA, Scanlon GT: Multiple mycotic pulmonary artery aneurysms secondary to intravenous drug abuse. *AJR* 1984; 142:89–90.

27. Hayashi K, Nagasaki M, Matsunaga N, et al: Initial pulmonary artery involvement in Takayasu arteritis. *Radiology* 1986; 159:401–403.

28. Cassling RJ, Lois JF, Gomes AS: Unusual pulmonary angiographic findings in suspected pulmonary embolism. *AJR* 1985; 145:995–999.

29. Bowen JS, Bookstein JJ, Johnson AD, et al: Wedge and subselective pulmonary angiography in pulmonary hypertension secondary to venous obstruction. *Radiology* 1985; 155:599–603.

30. Mills SR, Jackson DC, Older RA, et al: The incidence, etiologies, and avoidance of complications of pulmonary angiography in a large series. *Radiology* 1980; 136:295–299.

31. Perlmutt LM, Braun SD, Newman GE, et al: Pulmonary arteriography in the high-risk patient. *Radiology* 1987; 162:187–189.
32. Wood DL, Osborn MJ, Rooke J, et al: Amiodarone pulmonary toxicity: Report of two cases associated with rapidly progressive fatal adult respiratory distress syndrome after pulmonary angiography. *Mayo Clin Proc* 1985; 60:601–603.

15 | Vena Cava Filters

KEY CONCEPTS

1. Inferior vena cava (IVC) filters are used to prevent pulmonary emboli (PE) in those who have recurrent PE on anticoagulants or in whom anticoagulation is contraindicated.
2. Caval size, patency, and the presence of any venous anomalies must be documented prior to filter insertion.
3. Infrarenal placement, as close as possible to the lowest renal vein, is optimal.
4. Use of a guidewire and meticulous attention to technique prevent the most serious complication: accidental filter discharge or embolization into the heart.
5. Recurrent PE and IVC occlusion are rare after proper placement of a Kimray-Greenfield filter.

INDICATIONS

Standard therapy for pulmonary embolism (PE) or lower extremity deep venous thrombosis (DVT) is intravenous heparin followed by longer term oral Coumadin (warfarin). Such a regimen may reduce the estimated 30% mortality of untreated PE to 8%.[1] Patients in dire straits from massive embolization or with large iliofemoral thromboses may be treated more aggressively by systemic thrombolytic drugs. However, there are many patients for whom anticoagulant and thrombolytic therapy are contraindicated or have failed to prevent recurrent PE.

Anticoagulation is contraindicated in the presence of active or recent hemorrhage, especially intracranial bleeding. Peptic ulcer disease and alcoholism predispose patients on Coumadin to major bleeding problems. Although some might dispute the risks posed to those with primary or metastatic intracranial tumors, most physicians would avoid anticoagulation drugs in such patients.[2] Recent or planned surgery may be complicated by the need for anticoagulation, and some patients, such as those undergoing hip arthroplasty in the face of a previous history of DVT, are particularly vulnerable to postoperative thromboembolic disease. Others whose problems are difficult to manage by anticoagulation alone include those with malignant tumors and paraplegics.[3]

Furthermore, up to 28% of those on drug therapy for PE may have recurrent emboli.[1] Often this can be attributed to poor control and inadequate anticoagulation, but there are cases in which the best medical management is ineffective.

It is for these indications that surgical and nonsurgical interventions have been devised to interrupt the path between peripheral thrombi and the pulmonary circulation.

BACKGROUND

The first surgical approaches to thromboembolic prophylaxis included bilateral femoral vein ligation below the junction with saphenous vein, but recurrent emboli were still observed. Inferior vena cava (IVC) ligation similarly resulted in clinical failure in 4% to 50% of cases while causing venous stasis problems in a large number of patients.[4] These procedures did not address the problem of collateral drainage. Collateral veins (pelvic and lumbar) will enlarge markedly with time and can thus provide an alternative pathway for large embolizing clots. For this same reason balloon occlusion of the vena cava is not an ideal measure.

Surgical techniques for caval fenestration by sutures or specially designed clips were developed in order to maintain IVC flow through a set of smaller parallel channels. Recurrent PE was reduced to the order of 4%, and caval patency maintained in 76% to 80% of patients.[4] However, long-term problems with venous stasis were not overcome by caval fenestration. Moreover, ligation or clip placement requires major surgery, with its attendant morbidity in patients who are often seriously ill. Reported mortality of IVC plication patients has been 12%.[5]

CAVAL FILTERS

The problems of recurrent emboli, caval patency–venous stasis, and operative risk have been addressed by the design of a number of caval filtering devices (Fig 15–1). Several features desired of any venous filter include ease and stability of placement, efficacy of clot trapping, long-term patency, and lack of associated complications (caval perforation, migration, infection, or thrombosis of the access vein). An ideal filter would also be removable percutaneously, permitting use for temporary indications.

Mobin-Uddin Filter

The first filter to receive widespread application was the Mobin-Uddin "umbrella," which was introduced in 1967.[5] It was the first to allow introduction through a peripheral venous cut-down, which posed much less risk to the patient than an abdominal operation. The initial filter was 23 mm in diameter, but this was quickly increased to 28 mm when migration was found to be a major problem. Although the instances of migration were decreased to less than 1%, IVC thrombosis occurred in 33% to 85% of patients.[6]

Greenfield Filter

The Greenfield (or Kimray-Greenfield) filter represented a major improvement over the Mobin-Uddin filter. The Greenfield filter is comprised of six wires joined at the hub and radiating in a cone of 35° with the hook-tipped wires 6 mm apart at the base. The hub points in the direction of blood flow, and trapped emboli are directed toward the center of the vessel. This has the effect of maintaining flow through the IVC and subjecting the trapped clot to endogenous thrombolysis. The more open design of the filter has resulted in a 1-year caval patency rate of 96%, while decreasing the incidence of recurrent PE to 4%.[7]

Nevertheless, the efficacy of the Greenfield filter is a function of proper alignment with the vena cava. A tilt of 10° to 20° with respect to the IVC axis may make the filter completely ineffective for all but the largest emboli.[8] Significant tilting has been observed in 12% to 16% of cases,[6, 9] and has been directly implicated in cases of recurrent PE.[10]

Methods for percutaneous insertion of both the Mobin-Uddin and Greenfield filters have been devised, obviating the need for surgical cut-down and greatly speeding placement.[5, 11] Still, the large carrier

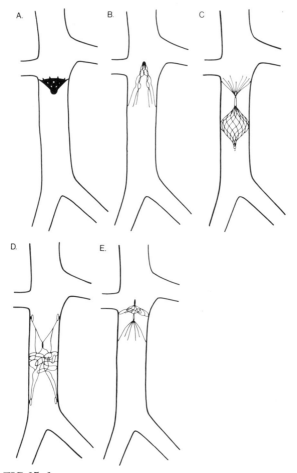

FIG 15–1.
Various inferior vena caval filters in place inferior to the renal veins.
A, Mobin-Uddin. **B,** Greenfield. **C,** Günther. **D,** bird's nest. **E,** Simon nitinol.

capsules must be placed through 24-F (8-mm) venotomies, essentially limiting access to the right jugular or common femoral veins. Other technical problems can be encountered because of capsule size, and the local trauma involved can produce bleeding or thrombophlebitis. Because of these problems, another generation of vena cava filters has been produced, and the devices are currently undergoing laboratory investigation and clinical trials. Also, a titanium Greenfield filter that can be inserted through a 14-F sheath is under development.[12]

Bird's Nest Filter

The bird's nest filter is descriptively named. After discharge, it forms a tangle of wires resembling a nest. The filter is comprised of four fine stainless steel wires that can be inserted through a 12-F sheath.[13] It is fixed in place by pairs of cranial and caudal hooks and can be released in larger diameter IVCs than the Greenfield or other filters can. Control during filter discharge is maintained by a guidewire-pusher that must be rotated to fully disengage the filter. Reported incidences of IVC thrombosis and recurrent pulmonary embolism are comparable to those occurring after Greenfield filter placement.[13] The bird's nest filter was released for general clinical use by the U.S. Food and Drug Administration (FDA) in 1989.

Amplatz Filter

This filter is comprised of 18 wires formed into a "spider" and is capable of insertion through a 14-F sheath.[14] The Amplatz filter is positioned from above by a threaded guidewire, or from the femoral approach by a snare fastened to a hook. The Amplatz filter has the advantage of possible retrieval through the common femoral vein, as long as retrieval is attempted within 2 weeks of placement. Filters left in position for a longer period become permanently fixed within the IVC.

Günther Basket Filter

The Günther filter can be inserted by way of a 10-F catheter and consists of 12 heparin-coated stainless steel wires formed into a basket.[15] This filter can also be removed within 10 days of insertion by introducing a snare from the femoral route.

Simon Nitinol Filter

The Simon filter is composed of nitinol, a nickel-titanium alloy that has a heat-dependent memory.[16] The configuration is somewhat similar to that of the Greenfield filter, but with the addition of an overlapping open-wire cap to the hub. The cap permits the filter to align itself with the IVC immediately upon discharge. As long as the filter is cooled in iced water it is quite pliable and can be straightened in its 8-F carrier. Exposure to body temperature at the time of placement causes the filter to resume the shape it "memorized" at manufacture.

Early evaluation suggests a slightly higher rate of IVC thrombosis with this filter (7%) than with the Greenfield filter. One-fourth of the patients examined by caval imaging at follow-up had signs of thrombi trapped by the filter.[16]

Vena Tech LGM Filter

This filter is a stamped, six-pronged cone made of stainless steel.[17] It is prepackaged in a loading syringe and inserted through a 12-F sheath. One-year caval patency has been 92%, and recurrent PE has been noted in 2%.[17] The Vena Tech filter has been recently released for clinical use by the FDA.

TECHNIQUE OF PERCUTANEOUS GREENFIELD FILTER PLACEMENT

Before placing an IVC filter, some would advocate documenting the lower extremities as source of emboli in all cases.[1] However, absence of leg clots on venography does not necessarily exclude a lower extremity source, and some PE undoubtedly arises in pelvic veins. Therefore, the decision to perform lower extremity venography before filter placement should be made on an individual basis. Similarly, some radiologists recommend pulmonary arteriography in every patient.[11] In all events, the risks of DVT and PE, their documentation by imaging studies, and the contraindications to conventional medical therapy must all be taken into consideration.

For the currently available Greenfield filter, either a right jugular approach or common femoral approach is necessary. Some of the newer filters allow insertion through a left jugular or subclavian vein.

Whatever approach is used, an IVC cavogram is obtained initially to document caval patency and size. A very large IVC, or megacava (over 30 mm in size), will not allow a Greenfield filter to anchor, and embolization of the filter itself into the heart becomes a real danger.[18] In such a case, one therapeutic alternative is placement of filters in both common iliac veins. In a similar (but quite rare) vein, a duplicated IVC can be identified and treated with filters in each vessel.

The location of the renal veins must also be documented. Ideal filter position is immediately inferior to the renal veins. If the IVC should occlude at the site of filter placement, renal vein obstruction would not necessarily follow. An anatomic variant that must be kept in mind is the circumaortic renal vein, a venous ring found in up to 11% of renal venograms.[19] The retroaortic segment often drains into IVC at the level of L-3. If a filter were placed between the two segments of the ring, a large unfiltered collateral pathway to the pulmonary circulation would be immediately available to emboli. When necessary, such as with renal vein thrombi or caval occlusion to the level of the kidneys, Greenfield filters can be safely placed in the suprarenal IVC.[6, 7]

The use of a Bell-Thompson ruler, or radiopaque letters or numbers taped to the patient's back, permits the level of renal veins to be determined at a glance. During placement, the Greenfield filter must be positioned and discharged quite rapidly; otherwise, thrombus can form about its legs and prevent complete expansion and fixation. The use of radiopaque markers expedites the procedure considerably.

Femoral Approach

Placement from the right femoral vein is preferred by many radiologists, who find it a more familiar and comfortable approach than the jugular vein. Also, patient comfort, as well as field sterility, may be maintained more easily. Air embolus is much less likely to occur with transfemoral insertion. Disadvantages of the approach include the possibility of iliofemoral thrombus obstructing the vein, difficulty in advancing the Greenfield filter capsule out of the pelvis and into IVC, and greater risk of postinsertion thrombosis. When necessary, the left common femoral vein can be used for access, but greater difficulty is posed by vessel tortuosity and the crossing of iliac artery.

Prior to tract dilatation, one should load the filter in its carrier assembly, flush the sideports, and lock the pusher down to prevent accidental discharge during insertion. If the operator is not familiar with the procedure, or has not placed a filter in some time, it is good practice to rehearse the steps of introduction, insertion, and filter discharge immediately before the venotomy is dilated.

After a 1- to 2-cm incision is made in the skin, the vein is opened either by serial insertion of Teflon fascial dilators, or by an 8-mm balloon. Histologic and clinical studies support use of a balloon as being less traumatic to endothelium.[20, 21] Limiting the passage of dilators or balloon should diminish the possibility of subsequent femoral vein thrombosis. With a long, stiff guidewire positioned in the IVC, the final 24-F dilator or balloon is removed. Simple compression over the site prevents bleeding while a 24-F Coons-Amplatz sheath is loaded. The sheath has a Silastic cuff, which is clamped about the guidewire immediately after the sheath is in position and the introducing dilator has been removed.

At this point the filter within its carrier is passed over the guidewire. A stiff, straight-tipped exchange wire is preferred so that the wire tip need not be withdrawn from the IVC at any time during loading. Wire stiffness minimizes buckling and obstruction to advancement. A J-tip may catch the filter hub during wire withdrawal, possibly inducing filter tilt.

The Silastic cuff is unclamped, and a quick rush of blood clears any thrombus that may have formed within the sheath. The carrier is introduced in a smooth motion. Once the capsule passes beyond the tip of the sheath, blood return is again observed through the sheath. The sheath is then withdrawn from the vein, and pressure is applied to the patient's groin. The sideports of the filter carrier are flushed with saline, and the filter is advanced over wire to the designated spot. If possible, the tip of the Greenfield filter should project between the renal veins, limiting the potential IVC "dead space" in which clot can form proximal to the filter.

The pusher assembly is then unlocked. Rather than pushing the filter from the capsule, it is better to withdraw the capsule while keeping both pusher and filter stationary. This maneuver, as well as keeping the wire through the filter, lessens the chance of filter tilt.[7] The wire is carefully withdrawn while the pusher is kept in contact with the filter. No attempt should be made to push or reposition the filter after discharge because of the risk of caval perforation.[6]

Finally, the capsule is removed and compression applied for 10 to 15 minutes. Surgical bandages (Steri-Strips) or sutures can be used to close the skin once hemostasis is achieved. Abdominal radiographs are obtained to document filter position.

Jugular Placement

Use of the right jugular vein for access avoids the possibility of trying to catheterize an occluded femoral vein. Symptomatic jugular vein occlusion is less likely to occur after percutaneous filter placement than postinsertion femoral vein thrombosis. However, extra care must be taken to prevent air embolism by placing the patient in a Trendelenberg position or by elevating the legs, and by having the patient perform a Valsalva maneuver during insertion of the sheath and carrier assembly.

The eustachian valve (valve of the IVC) at the junction of right atrium and IVC can occasionally cause difficulties. A stiff exchange wire should be used, and its tip securely positioned in an iliac vein to ensure filter discharge in IVC. Mistaken placement of a filter into hepatic, renal, or even gonadal veins has occurred.[14] If a stiff guidewire and careful fluoroscopic observation are not employed, the wire and filter can buckle into the right ventricle. Fluoroscopy is mandatory; the great majority of misplacements into the heart have resulted from lack of radiologic guidance.[22]

The Greenfield filter carrier for transjugular placement is different than the one used for a femoral approach. One must be sure that the filter is loaded properly and that the pusher is locked securely before the capsule is inserted into the IVC. In other respects, the steps in filter placement are identical to those described above for the femoral approach.

FOLLOW-UP, COMPLICATIONS, AND OTHER CONSIDERATIONS

The patient should be observed for several days to detect signs of postinsertion thrombosis, noted in 3% to 12% of transfemoral Greenfield filters insertions,[21, 23–25] or the more serious complication of IVC occlusion. Venography has shown that postinsertion thrombosis may actually affect up to 41% of placements, although it may be minimally symptomatic in many patients.[26] On the other hand, some patients with symptoms of thrombosis will actually have patent veins, and not all cases of venous stasis should be ascribed to caval filter insertion.[21] In any event, anticoagulation may be resumed within 24 hours, but use of thrombolytics should be avoided after filter placement.[27] An additional risk of femoral placement is the development of arteriovenous fistula, which may present as lower extremity ischemia.

If a Greenfield filter is severely tilted (or if thrombus is found to arise from the filter or more proximal IVC) another filter may be placed above it, in the suprarenal cava. Cases of inappropriate filter discharge can sometimes be managed conservatively, but filters within the heart should be removed. There has been a report of percutaneous retrieval by means of a wire snare and specially modified sheath,[28] but most patients with intracardiac filters will need open-heart surgery.

Placement of a Greenfield filter in the superior vena cava has been reported in conjunction with documented pulmonary emboli from an upper extremity source.[29]

Distal migration of Greenfield filters (up to 2 cm) has been commonly observed with time, as has been perforation of the IVC wall by filter struts.[30] Distal migration and caval perforation have also been noted with Günther filters.[15] Although caval wall perforation is usually innocuous, it may lead to adhesions, bowel obstruction, or peritonitis, particularly in quadriplegic patients engaging in vigorous maneuvers during pulmonary physical therapy.[30]

In vitro characteristics of various filters have been compared by Katsamouris et al., using clots of 2-mm to 7-mm diameter.[8] In their study the Mobin-Uddin and Greenfield filters were ineffective when tilted, and caused the most turbulent downstream flow. The other filters all captured clots larger than 2 mm diameter, whether or not filter configuration was optimal. The Mobin-Uddin and Amplatz filters produced the largest pressure gradients after trapping a few clots, while the bird's nest filter accommodated the most thrombi before causing a pressure rise.

Despite the implications of these data, the Greenfield filter has been quite effective clinically and reliably captures thrombi 6 to 8 mm in diameter in vivo, even when tilted.[31] The utility and safety of the newer filters remain to be proved in wider clinical practice. Accumulation of clinical experience and longer term follow-up should allow a future rational choice of optimal filter design.

REFERENCES

1. Glenny RW: Pulmonary embolism: Complications of therapy. *South Med J* 1987; 80:1266–1276.
2. Olin JW, Young JR, Grator RA, et al: Treatment of deep venous thrombosis and pulmonary emboli in patients with primary and metastatic brain tumors: Anticoagulants or inferior vena cava filter? *Arch Intern Med* 1987; 147:2177–2179.
3. Golueke PJ, Garrett WV, Thompson JE, et al: Interruption of the vena cava by means of the Greenfield filter: Expanding the indications. *Surgery* 1988; 103:111–117.
4. Coleman CC: Overview of interruption of the inferior vena cava. *Semin Intervent Radiol* 1986; 3:175–187.
5. Coleman CC, Castañeda-Zuñiga WR, Amplatz K: Mobin-Uddin vena caval filters. *Semin Intervent Radiol* 1986; 3:193–195.
6. Kanter B, Moser KM: The Greenfield vena cava filter. *Chest* 1988; 93:170–175.
7. Greenfield LJ, Michna BA: Twelve-year clinical experience with the Greenfield vena caval filter. *Surgery* 1988; 104:706–712.

8. Katsamouris AA, Waltman AC, Deichatsios MA, et al: Inferior vena cava filters: In vitro comparison of clot trapping and flow dynamics. *Radiology* 1988; 166:361–366.

9. Greenfield LJ, Peyton R, Crute S, et al: Greenfield vena cava filter experience: Late results in 156 patients. *Arch Surg* 1981; 116:1451–1456.

10. Schanzer H, Knight R: Recurrent pulmonary embolism from thrombi in Greenfield filter: Case report. *Vasc Surg* 1988; 22:110–113.

11. Denny DF, Cronan JJ, Dorfman GS, et al: Percutaneous Kimray-Greenfield filter placement by femoral vein puncture. *AJR* 1985; 145:827–829.

12. Greenfield LJ, Cho KJ, Pais O, et al: Preliminary clinical experience with the titanium Greenfield vena caval filter. *Arch Surg* 1989; 124:657–659.

13. Roehm JOF Jr, Johnsrude IS, Barth MH, et al: The bird's nest inferior vena cava filter: Progress report. *Radiology* 1988; 168:745–749.

14. Darcy MD, Hunter DW, Lund GB, et al: Amplatz retrievable vena caval filter. *Semin Intervent Radiol* 1986; 3:214–219.

15. Fobbe F, Dietzel M, Korth R, et al: Günther vena caval filter: Results of long-term follow-up. *AJR* 1988; 151:1031–1034.

16. Simon M, Athanasoulis CA, Kim D, et al: Simon nitinol inferior vena cava filter: Initial clinical experience. *Radiology* 1989; 172:99–103.

17. Ricco JB, Crochet D, Sebilotte P, et al: Percutaneous transvenous caval interruption with the "LGM" filter: Early results of a multicenter trial. *Ann Vasc Surg* 1988; 3:242–247.

18. Molina EJ: Interruption of the inferior vena cava for prevention of pulmonary embolism. *Semin Intervent Radiol* 1986; 3:188–192.

19. Beckmann CF, Abrams HL: Circumaortic venous ring: Incidence and significance. *AJR* 1979; 132:561–565.

20. Dorfman GS, Esparza AR, Cronan JJ: Percutaneous large bore venotomy and tract creation: Comparison of sequential dilator and angioplasty balloon methods in a porcine model. *Invest Radiol* 1988; 23:441–446.

21. Dorfman GS, Cronan JJ, Paolella LP, et al: Iatrogenic changes at the venotomy site after percutaneous placement of the Greenfield filter. *Radiology* 1989; 173:159–162.

22. Villard J, Detry L, Clermont A, et al: Huit filtres de Greenfield dans les cavités cardiaques droites: Traitment chirurgical. *Ann Radiol* 1987; 30:102–104.

23. Denny DF Jr, Dorfman GS, Cronan JJ, et al: Greenfield filter: Percutaneous placement in 50 patients. *AJR* 1988; 150:427–429.

24. Welch TJ, Stanson AW, Sheedy PF, et al: Percutaneous placement of the Greenfield vena caval filter. *Mayo Clin Proc* 1988; 63:343–347.

25. Mewissen MW, Erickson SJ, Foley WD, et al: Thrombosis at venous insertion sites after inferior vena caval filter placement. *Radiology* 1989; 173:155–157.

26. Kantor A, Glanz S, Gordon DH, et al: Percutaneous insertion of the Kimray-Greenfield filter: Incidence of femoral vein thrombosis. *AJR* 1987; 149:1065–1066.

27. Novelline RA: Practical points on transvenous insertion of inferior vena cava filters. *Cardiovasc Intervent Radiol* 1980; 3:319–324.

28. Tsai FY, Myers TV, Ashraf A, et al: Aberrant placement of a Kimray-Greenfield filter in the right atrium: Percutaneous retrieval. *Radiology* 1988; 167:423–424.

29. Pais SO, De Orchis DF, Mirvis SE: Superior vena caval placement of a Kimray-Greenfield filter. *Radiology* 1987; 165:385–386.

30. Balshi JD, Cantelmo NL, Menzoian JO: Complications of caval interruption by Greenfield filter in quadraplegics. *J Vasc Surg* 1989; 9:558–562.

31. Thompson BH, Cragg AH, Smith TP, et al: Thrombus-trapping efficiency of the Greenfield filter in vivo. *Radiology* 1989; 172:979–981.

16

Lymphography

KEY CONCEPTS

1. Despite widespread use of computed tomography in the staging of tumors, lymphography remains useful, particularly in Hodgkin lymphoma and testicular tumors.
2. For best accuracy, both lymphatic flow and nodal patterns must be analyzed together.
3. Tumors can produce obstruction, marginal filling defects, enlargement, and diffuse "foamy" patterns.
4. Central defects are often the result of scar or fatty replacement.
5. Diffuse enlargement with homogeneous uptake of contrast material may represent benign "reactive" hyperplasia.

Lymphography has been in clinical use for over 40 years, and for much of that time, it was the only way in which to evaluate nonpalpable lymph nodes short of surgery. In more recent years it has been superceded to a large extent by computed tomography (CT) and ultrasonography. Magnetic resonance imaging and lymphoscintigraphy are under investigation, but are unlikely to provide major advances in the near future.

Lymphography has been virtually abandoned at many institutions because it is an invasive procedure demanding considerable time and technical skill.[1] Even so, it maintains a role in oncologic centers and teaching hospitals, for lymphography remains the sole imaging study that can detect neoplastic involvement of nonenlarged nodes.

Bipedal lymphography is best suited for conditions that may affect femoral, inguinal, external or common iliac, and para-aortic nodes. The thoracic duct and some mediastinal nodes can also be imaged. However, internal iliac, high para-aortic, retrocrural, and mesenteric nodes are usually poorly opacified, if at all. Lymphography does not enable detection of pathologic processes in renal or hepatic hilar nodes, and it does not provide information about possible extranodal disease. On the other hand, lymphographic contrast material may persist for 1 or 2 years, allowing tumor response to treatment (or relapse) to be followed easily and inexpensively by periodic abdominal and pelvic radiographs.[2]

INDICATIONS

Hodgkin Lymphoma

In patients with Hodgkin lymphoma presenting for staging, up to 10% will have lymph nodes of normal size containing tumor.[3] Because CT criteria of abnormality are based on size alone, disease in these patients would be staged inaccurately if CT were not followed by lymphography or surgery. Lymphography can be quite sensitive. Castellino et al.[4] reported no false negative examinations in 111 patients undergoing surgical node dissection after a lymphogram interpreted as showing no malignant disease. Overall accuracy for lymphography in Hodgkin lymphoma has been reported as between 82% and 95%, compared with CT accuracy of 75% to 84%.[5] Use of lymphography as an initial staging tool for Hodgkin patients, supplemented with thoracic and abdominal CT, remains justified.

Non-Hodgkin Lymphoma

Staging laparotomy is seldom necessary for non-Hodgkin lymphoma nowadays. Because of the tendency of this disease to involve retrocrural, high para-aortic, and mesenteric nodes, as well as liver, spleen and other organs, CT is the initial diagnostic procedure of choice. Lymphography may be reserved for those with normal or equivocal CT findings, in whom the results would affect therapy.[6]

Pelvic Malignancies

In testicular tumors, lymphography has a lower sensitivity, ranging from 54% to 89%, and a specificity on the order of 88%.[7] While CT does not improve sensitivity to nodal mestastases, it is superior at the level of the thoracolumbar junction.[8] More bulky metastases are generally seen in patients with embryonal carcinoma. Positive studies can be confirmed by needle biopsy. Lymphograms can direct surgical dissection or radiation portals, and their use in evaluating testicular tumors is still favored.[7, 9]

For prostatic carcinoma, both CT scanning and lymphography are quite specific, but sensitivity is lacking, generally less than 60%.[7, 10] Patients with positive nodes can be spared lymphadenectomy, and Flanigan et al. assert that outpatient lymphography with directed needle biopsy can save costs when compared to CT screening.[10]

The benefits of screening are more controversial in cervical carcinoma. With 18% to 25% sensitivity, lymphography may be worse at clinical staging than physical examination alone.[11, 12] Nevertheless, some have reported acceptable overall accuracy (91%) and continue to use lymphography on a routine basis.[13] The value of the study in various other pelvic tumors, such as vulvar carcinoma and bladder carcinoma, remains in debate.[14, 15]

TECHNIQUE

Because some degree of venous embolization of the oily contrast medium is encountered in all patients, lymphography should be avoided in those with severe pulmonary disease or intracardiac shunts. Older individuals with marked tremors may be impossible to study.

Prior to lymphography, patients should be questioned about any hypersensitivity reactions on previous exposure to iodinated contrast materials or vital dyes. They should be advised that lymphography is a 2-day test, requiring up to 4 or 5 hours on the first day for isolation of lymphatic vessels, cannulation, and injection, and return the following day for filming of the static nodal phase. They should also be warned of the temporary discoloration produced by the vital dye (Evans blue, methylene blue, isosulfan blue, among others used) injected into the feet. Because of the length of time required for contrast injection, patients should void before being placed on the procedure table, and they may bring reading material to occupy the time.

Vital dye is injected intradermally into the webs between the toes. A 25-gauge needle is used to administer about 0.5 mL at each site. The webs between the first three toes, or the most medial and lateral webs may be chosen. Having the patient then walk about for several minutes may improve opacification of the lymphatics in the foot. After 10 to 15 minutes, both feet are sterilely prepared and draped.

The course of the lymphatic channels is made evident by uptake of dye, and a vessel over the dorsum of the foot can be selected. After generous administration of 1% lidocaine, a shallow transverse incision is made, taking care not to cut completely through dermis over the lymphatic. A hemostat can be passed gently under the incompletely incised portion from the margin of the incision. The scalpel can then cut down upon the hemostat, avoiding inadvertent injury to the lymphatic vessel. The vessel is isolated by meticulous dissection with blunt forceps. Spreading a hemostat repeatedly under the vessel may help clean fat from its surface.

A sterile hairclip may be used to isolate and stabilize the vessel for cannulation. Ligatures can also be placed, but they are not absolutely necessary. The lymphatic may be distended by "milking" the dorsum of the foot, compressing skin distally to proximally with a blunt instrument. A 30-gauge needle with connecting tubing filled with iodized oil contrast agent (Lipiodol or Ethiodol) is advanced, bevel up, into the channel. Gentle injection will confirm intraluminal placement by absence of leakage. If a few small air bubbles are in the line, they may help recognition of successful cannulation.[16]

With satisfactory needle placement, the clip is closed, and the connecting tubing is secured by sterile tape. The contrast agent is then injected by means of a 10-mL syringe placed under a 10-lb weight. The patient is monitored every 15 minutes during injection, and an early film over the lower leg is obtained to confirm lymphatic filling and exclude the possibility of lymphatic-venous communication. Injection is complete when 6 to 10 mL of contrast medium per foot have been introduced. A slightly larger amount may be given if cannulation is successful in one side but not the other. There is normally some cross-filling of periaortic lymphatics, an effect that may provide adequate information about abdominal nodes from a unilateral injection.

After injection and needle removal, incisions are closed by interrupted sutures and dressed. Anteroposterior (AP) and oblique films of the pelvis and abdomen are obtained, in addition to lateral view of the abdomen and an AP view of the chest. These allow assessment of lymphatic channels, including displacement or collateral formation. The same studies are repeated on the following day for evaluation of lymph node characteristics. The studies may be repeated at intervals of several months for up to a year or more to detect new or recurrent disease in opacified nodes.

Adverse reactions are rare. Other than hypersensitivity reactions to the iodized oil or vital dye, major risks include local infection at the incision site or massive oil embolism if venous communications are present. Normally, patients may experience mild fever and cough the night after the procedure.

INTERPRETATION

Normal lymph nodes are oval or kidney-shaped and may vary in length from 2 mm to 30 mm.[17] Lymph nodes are composed of follicles (predominately located in the cortex) and sinuses that drain the afferent channels entering at the periphery. Lymph from the afferent channels enters marginal sinuses and flows through a reticular network of medullary sinuses into the hilus of the node. The hilus gives rise to the efferent channels. Oily lymphographic contrast material is normally homogeneously distributed throughout a node, which has a fine reticular appearance.

Pathologic nodes may be enlarged or normal-sized, with focal or diffuse filling defects. In general, nodes with all diameters greater than 2.0 cm are abnormal, although size criteria do vary somewhat between chains.[9] Particular care must be made in interpreting inguinal nodes, which are prone to repeated or chronic inflammation.

Non-neoplastic disease can produce patterns that confound interpretation. Nodes may normally develop central scarring or fat deposition, and experienced lymphographers can usually recognize such cases of fibrolipomatosis. Many false positive examinations are caused by sinus histiocytosis (packing of sinuses by proliferating histiocytes) and follicular hyperplasia. These conditions are grouped under the term "nonspecific reactive change." One may see focal or diffuse defects. Still, in many cases of inflammatory change, nodes maintain a homogeneous but "magnified" structure.[17] A similar nodal pattern can be found in some patients with lymphocytic leukemia. Still, most types of lymphoma can be distinguished from inflammatory nodes by the gradations of disease displayed in different nodal groups involved by lymphoma.[4]

Metastases from carcinoma often produce marginal deposits, which must exceed the size of a couple of lymphoid follicles in order to be recognized. In practical terms, a metastasis must be at least 5 mm in diameter to be detected in the best of circumstances. Central defects larger than 10 mm are suspicious for malignancy.

Lymphomas usually produce enlarged nodes with a "foamy" or "lacy" appearance or conglomerate masses with irregular or disorganized uptake of contrast material. However, nodes involved by Hodgkin lymphoma may not produce a "classic" pattern in up to 29% of cases.[18] They may instead show irregular filling defects more typical for carcinoma. In other patients lymphatic flow may be obstructed to the degree that the affected nodes do not opacify at all.

Because flow patterns provide information about the presence or absence of neoplastic disease, early lymphangiographic studies are critical for interpretation. An obstructive pattern is manifested by collateral vessel filling, delayed flow (channels should not retain contrast material at 24 hours in most healthy people), dermal backflow, or opening of lymphaticovenous anastomoses.[18] Channels may be displaced about enlarged, nonopacified nodes. Small filling defects are much more likely to represent tumor metastases when they are associated with obstructed flow.

If the patient's primary diagnosis is in doubt, surgical or needle biopsy of abnormal nodes is warranted. The differential diagnosis of the patterns described here is a large one, including such varied conditions as inflammation from leaking abdominal aortic aneurysm or spinal osteomyelitis, Whipple's disease, Waldenström's macroglobulinemia, sarcoidosis, syphilis, tuberculosis, and collagen vascular diseases.[19]

CONCLUSION

Despite the use of cross-sectional imaging techniques, lymphography maintains a place in staging of Hodgkin lymphoma and various other neoplasms. Its use is probably best confined to major cancer referral and treatment centers, where the volume of procedures is sufficient to maintain the operator's technical expertise, as well as the radiologist's interpretive skills.

REFERENCES

1. Dixon AK: The current practice of lymphography: A survey in the age of computed tomography. *Clin Radiol* 1985; 36:287–290.
2. Pera A, Capek M, Shirkhoda A: Lymphangiography and CT in the follow-up of patients with lymphoma. *Radiology* 1987; 164:631–633.
3. Dooms GC, Hricak H: Radiologic imaging modalities, including magnetic resonance, for evaluating lymph nodes. *West J Med* 1986; 144:49–57.
4. Castellino RA, Billingham M, Dorfman RF: Lymphographic accuracy in Hodgkin's disease and malignant lymphoma with a note on the "reactive" lymph node as a cause of most false-positive lymphograms. *Invest Radiol* 1974; 9:155–165.

5. Enig B, Bjerregaard Jensen B, Hjøllund Madsen E, et al: Detection of neoplastic lymph nodes in Hodgkin's disease and non-Hodgkin lymphoma. Comparison between tomography and lymphography. *Acta Radiol [Oncol]* 1985; 24:491–495.

6. Strijk SP: Lymphography and abdominal computed tomography in the staging of non-Hodgkin lymphoma. *Acta Radiol [Diagn]* 1987; 28:263–269.

7. von Eschenbach AC, Jing BS, Wallace S: Lymphangiography in genitourinary cancer. *Urol Clin North Am* 1985; 12:715–723.

8. Lien HH, Kolbenstvedt A, Talle K, et al: Comparison of computed tomography, lymphography, and phlebography in 200 consecutive patients with regard to retroperitoneal metastases from testicular tumor. *Radiology* 1983; 146:129–132.

9. Deprez-Curely JP: Lymphography in testicular tumours. *Prog Clin Biol Res* 1985; 203:243–252.

10. Flanigan RC, Mohler, JL, King, CT, et al: Preoperative lymph node evaluation in prostatic cancer patients who are surgical candidates: The role of lymphangiography and computerized tomography scanning with directed fine needle aspiration. *J Urol* 1985; 134:84–87.

11. Vercamer R, Janssens J, Usewils R, et al: Computed tomography and lymphography in the presurgical staging of early carcinoma of the uterine cervix. *Cancer* 1987; 60:1745–1750.

12. Feigen M, Crocker EF, Read J, et al: The value of lymphoscintigraphy, lymphangiography and computer tomography scanning in the preoperative assessment of lymph nodes involved by pelvic malignant conditions. *Surg Gynecol Obstet* 1987;165:107–110.

13. Smales E, Perry CM, MacDonald JS, et al: The value of lymphography in the management of carcinoma of the cervix. *Clin Radiol* 1986; 37:19–22.

14. Strijk SP, Debruyne FMJ, Herman CJ: Lymphography in the management of urologic tumors. *Radiology* 1983; 146:39–45.

15. Weiner SA, Lee JKT, Kao M-S, et al: The role of lymphangiography in vulvar carcinoma. *Am J Obstet Gynecol* 1986; 154:1073–1075.

16. Staton R: Lymphography. *Radiol Technol* 1984; 55:233–238.

17. Wiljasalo M: Lymphographic differential diagnosis of neoplastic diseases: Roentgen anatomy and roentgen pathology of the lymph glands. *Acta Radiol* (Suppl) 1965; 247:12–19.
18. Koehler PR, Salmon RB: Lymphographic patterns in lymphoma, with emphasis on the atypical forms. *Radiology* 1966; 87:623–629.
19. Parker BR, Blank N, Castellino RA: Lymphographic appearance of benign conditions simulating lymphoma. *Radiology* 1974; 111:267–274.

17 | Needle Biopsy

KEY CONCEPTS

1. Percutaneous needle biopsy can be used to diagnose primary or metastatic malignancy, tumor recurrence, or infection.
2. Diagnostic yield is lower for lymphoproliferative disorders and benign disease, but can be increased by the use of larger cutting needles.
3. Incidence of pneumothorax is higher in patients with chronic obstructive lung disease.
4. Bowel can be traversed with relative impunity by needles 20 gauge or smaller.
5. Mortality is extremely low with fine-needle biopsy, the greatest hazards being air embolism, pancreatitis, and pulmonary or hepatic hemorrhage.

Percutaneous needle biopsy has been applied for many years to superficial, palpable masses or in an undirected manner to liver parenchyma. However, the diagnostic potential of needle biopsy was not fully appreciated until the 1970s, when cross-sectional imaging became available. Presently, fine-needle techniques allow the great majority of soft tissue abnormalities to be sampled safely, with good prospects of obtaining diagnostic material.

Major indications include suspicion of primary or metastatic malignancy, confirmation of tumor recurrence or metastasis in a patient with known primary tumor, and diagnosis of infection.

COMMON PRINCIPLES

The success and risk of a needle biopsy procedure depend critically on a number of factors. The patient must be able to cooperate with the physician obtaining the sample; in many cases, the patient's respiration must be suspended during needle placement. The organ or lesion to undergo biopsy must be identifiable on fluoroscopy, ultrasound, (US), or computed tomography (CT) (needles are being designed for use with magnetic resonance imaging). The operator should be skilled at directing the needle accurately, avoiding major vessels, bowel, and other potential intervening hazards where possible. Once obtained, the specimen must be handled correctly: either examined on the spot by a cytopathologist or taken immediately to the surgical pathologist with relevant history and any special instructions. If infection is suspected, samples should be obtained for culture and appropriate staining. Accuracy of diagnosis is quite dependent on the experience and interest of the pathologist examining needle specimens.

There is no strong correlation of diagnostic accuracy or incidence of complications with needle size. However, mortality is a real risk with the use of large cutting needles (14 gauge). Fatality rates have been reported as 0.5% for hepatic, and as high as 3.8% for pancreatic biopsies with use of such needles.[1]

Multiple Samples

Multiple samples increase accuracy. Sampling the periphery of a lesion is quite valuable, particularly for large masses which may have central necrosis. The negative predictive value of a biopsy is much lower than its positive predictive value. As a consequence, it is often necessary to repeat a biopsy in the face of negative results.[2, 3] The low morbidity and cost of needle biopsy make a second or third biopsy procedure preferable to thoracotomy or laparotomy for diagnosis.

Nonmalignant Lesions

Needle biopsy is most diagnostic for malignant neoplasms (see "Site-Specific Considerations," later in this chapter, for accuracy in specific sites). A definitive positive diagnosis of benign lesion can rarely be made in more that 65% of cases in even the best of circumstances.[4, 5] For suspected infection the situation is somewhat better. Sensitivity of needle aspiration has been 76% in 46 cases of pulmonary infection reported by Conces et al.[6] Needle biopsy of lymph nodes has also proved useful in diagnosing mycobacterial infection in patients with acquired immune deficiency syndrome.[7] If suspicion of infection is strong but no material is obtained from a suspected fluid collection through a fine needle, it is often wise to use an 18-gauge needle or to place a 5- to 8-F catheter for aspiration (see Chapter 18, Percutaneous Abscess Drainage). Although needle biopsy has its limitations in making a positive diagnosis of benign conditions, it very rarely results in a falsely positive diagnosis of malignancy.[8]

Approach

If it can be avoided, multiple tissue planes should not be crossed. For example, most adrenal biopsies can be performed by a posterior approach, staying entirely within the retroperitoneum. The diaphragm and pleural space are best avoided for liver biopsies. The number of pleural reflections traversed in a pulmonary procedure may affect the likelihood of pneumothorax. In a parenchymal organ such as liver, it is wise to traverse some normal parenchyma to reach an abnormality. Puncture of a vascular lesion at a free surface can result in peritoneal hemorrhage.

Patient Preparation

Before any biopsy procedure the patient's history, clinical status, and radiographic studies are reviewed to determine if needle biopsy is possible and appropriate. Hematocrit, platelet count, and coagulation studies should be obtained. Those with thrombocytopenia or abnormal coagulation can be prepared with platelet transfusions or fresh frozen plasma. Pulmonary function tests may be helpful in assessing risk of lung biopsy in a given individual.[9] As a precaution, oral intake should be restricted to clear liquids in the 4 to 6 hours preceding biopsy.

The procedure, its purpose, attendant risks, and alternatives are explained to the patient before consent is obtained. It should be made clear that multiple needle passes are customarily needed to obtain adequate material. Most biopsies can be performed on an outpatient basis; however, those undergoing biopsy must understand the possibility of hospital admission for complications, as well as the potential necessity of repeat biopsy if results are negative or unsatisfactory. Outpatients should be prepared to spend several hours in the radiology department or other designated area for observation. They should be accompanied by a family member or friend, who can drive the patient home, observe symptoms, and provide assistance should any problems be encountered during the trip.

It is advisable to place an intravenous line prior to percutaneous needle biopsy. In many cases premedication is not necessary, but very apprehensive patients may be given midazolam (1 to 3 mg intravenously in doses of 0.5- to 1.0-mg increments) for sedation, and fentanyl (50 to 200 μg intravenously in doses of 25- to 50-μg increments) for pain. Oxygen, naloxone, and other resuscitative drugs and equipment should be at hand for any emergency.

Biopsy Guidance

The size, location, and depth of a lesion, as well as its detectability by a given imaging modality, determine the best method for guidance. Many abdominal lesions can be approached equally well with CT or US; instrument availability and operator preference are often determining factors. Fluoroscopy is most commonly used for pulmonary, pleural, and osseous lesions. Lymph nodes opacified by lymphographic contrast material can be sampled percutaneously under fluoroscopic direction. Rarely, a bone lesion may be detected (and marked on the skin for needle biopsy) by radionuclide scanning.[10]

When an approach is chosen, the skin is prepared, anesthetized with lidocaine, and a small incision made. For fine-needle biopsy a short, larger needle, such as an 18-gauge thin-wall vascular puncture needle, may be placed through the incision into subcutaneous tissues to stabilize the trajectory of the fine needle introduced coaxially.

No matter what imaging technique is employed, use of tandem needles can save time. When one needle is either in or near the target, a second can be placed adjacent to it, to mark depth and angle of approach for further sampling when the first needle is removed. With a tandem technique, a fine needle can be placed with whatever number of passes necessary. A larger needle can then be advanced safely alongside to obtain a core for histologic sampling.

Fluoroscopy

When fluoroscopy is used, a C-arm helps enormously, both for detecting a lesion and for targeting. When the tube and image intensifier are lined up with the skin entry site, the needle hub and tip can be superimposed over the lesion. Changing to the lateral projection then confirms when the tip has been advanced to the proper depth for sampling. If a true lateral projection cannot be obtained, rotating the tube through various obliquities can be used to see if the needle tip "stays" with the target during rotation. In many cases the target can be felt as a change in resistance when it is encountered by the advancing needle. Oblique or lateral fluoroscopy during actual sampling can also confirm accurate placement by the observation of lesion motion with needle penetration.

Ultrasound

With US, the skin site is determined and marked, and the angle of approach and depth are noted. The needle can then be placed without direct observation. Alternatively, a biopsy guide can be employed, or the needle placed "freehanded" with simultaneous sonographic monitoring. If placement is to be monitored, the scanning head should be covered with a sterile sheath, sonographic contact being maintained by sterile gel. If a sterile sheath is unavailable or deemed cumbersome, scrupulous cleansing of the scanning head with povidone-iodine and absolute alcohol has been shown to be safe.[11] The cable connecting the head to the rest of the sonographic unit can be draped with towels secured by sterile clips.

The availability of electronic delineation of the tract of a US biopsy guide is most helpful, for it is often quite difficult to image the needle tip. Jiggling the needle, sliding its stylet in and out, or injecting a small amount of saline are ways of increasing its sonographic visibility. Reading et al. have recently described the use of a screw stylet to greatly enhance the echogenicity of biopsy needles.[12] Overall, US has produced good results in guiding biopsy procedures in even very small abdominal lesions.[11] It has also been valuable in approaching pleural lesions.[13]

Computed Tomography

Computed tomography is well suited to approaching small, deep lesions, especially those surrounded by large vessels or bowel. The needle tip can be unequivocally demonstrated, and depiction of surrounding structures is not hindered by the presence of gas or bone. Biopsy is most readily performed when the approach can be made in an axial plane; off-axis biopsy trajectories are much more problematic.

A difficulty with CT is the amount of time expended with needle placement, scanning, and repositioning. Rapid image reconstruction is indispensable for CT biopsy to be practical. Catheters or other radiopaque markers may be placed on the patient during localization scans to help plan an approach. A CT stereotactic device has been developed for percutaneous needle biopsy that can greatly decrease the number of needle passes needed and can accurately guide nonaxial trajectories.[14]

Biopsy Needles

There are a dazzling variety of needles commercially available. Aspiration needles differ chiefly in tip and trocar configuration. Tips may be beveled, pencil- or diamond-shaped, blunt, hooked, spiraled, notched, or trephined (see Figure 17–1). Some are supplied with attached aspirating syringes or have a specially designed self-aspirating mechanism. Larger needles may have an inner notched trocar, over

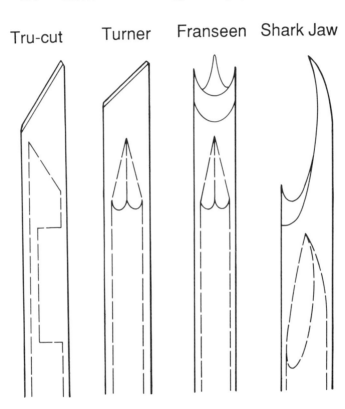

FIG 17–1.
Four of the many types of biopsy needles available (not drawn to scale).
The Tru-cut is a cutting-gap needle, the Franseen is trephined, and
the other two represent beveled cutting needles.

which a cutting cannula is slipped. Some needles are adapted to particular sites, such as bone, lung, and pleura (see "Site-Specific Considerations," later in this chapter.) Coaxial systems have been fabricated to allow multiple samples to be obtained with a single needle placement.[15] A blunt-tip needle has been designed by Hawkins et al. to deflect from, rather than puncture, muscular arteries, thus minimizing the possibility of major bleeding.[16]

In general, larger needles obtain larger specimens, and one needle of a given size has no great advantage over another in terms of material obtained, particularly for cytologic study. Histologic analysis is more dependent on sample size. In this regard, in vitro experiments by Andriole et al. have indicated some benefit related to the degree of bevel in an instrument.[17] Their studies also found that cutting-gap and trephined needles, such as the Lee and Franseen needles, tended to give larger samples for their size, while screw-stylet needles obtained smaller amounts of tissue.

The length and bore of a needle should correspond to the depth of a lesion, clearance considerations (CT gantry, image intensifier), and the anticipated ease of hitting a target. A large superficial mass can be easily entered with a short-large-gauge needle. However, the vascularity of a lesion must also be taken into consideration. For example, an abnormality suspected to be a cavernous hemangioma or metastatic renal cell carcinoma should be approached gingerly, usually with a needle no larger than 20 gauge. Fine (22-gauge or smaller) needles, particularly those with beveled stylets, may not follow a straight course if directed through any great length of soft tissue. The particular equipment chosen for the task at hand depends on the operator's experience and personal preference.

Sampling Technique

For simple aspiration or cutting needles, a syringe can be attached directly to the hub. A Luer-lok makes coupling more secure. The larger the syringe, the greater the suction that can be generated. With suction applied, the needle is advanced into the target, either in a single motion or with a 1- to 2-cm to-and-fro excursion. Rotation may be helpful with cutting needles. When an 18-gauge or smaller needle is used with a 50-mL syringe, no more than 10 mL of suction is needed to obtain maximal samples.[18] Suction should be maintained or slowly released during withdrawal.

After needle removal the syringe can then be detached and used to draw up several milliliters of heparinized saline (nonbacteriostatic, if infection is a consideration). The saline is injected through the needle into a sterile sample cup or tube. The stylet or trocar can be reinserted to expel any remaining material. At the time the needle is being cleared, single drops can be applied to slides and spread immediately for air-drying or alcohol fixation.

If two people are available to perform an aspiration, it is more convenient to attach the syringe to a length of connecting tubing. This allows the operator to use two hands to make the needle pass, while the assistant provides suction. The syringe can be filled with 5 to 10 mL of heparinized saline, so that suction is efficiently generated and the sample can be immediately expelled from the needle by injection.

Needles with syringes premounted, such as the Menghini Surecut, make it possible to conduct a biopsy easily with one hand. However, the added length and weight of such devices may make checking needle position by CT or fluoroscopy more difficult. For other, more complicated, biopsy devices the manufacturer's instructions should be consulted prior to use.

Aspiration vs. Core, or Cytology vs. Histology

In addressing the relative merits of cytologic and histologic study in needle biopsy, it may be helpful to define a few terms. "Fine-needle biopsy" can be defined rather arbitrarily as that performed with needles of 1.0 mm diameter or less.[1] In practical terms, this means needles smaller than 19 gauge (Table 17–1). Cutting needles are those designed to procure a core of tissue, and cores can be obtained with needles as small as 22 gauge. Many authors, however, imply larger cutting devices (14 to 19 gauge) when they refer to cutting needles.[4, 19] Very fine bore needles are used to aspirate aggregates of cells for cytologic examination. Histologic study depends on the preservation of tissue architecture. In many cases material obtained with a needle can be submitted for both cytologic and histologic review.

Cytologic study is highly sensitive for detection of carcinoma, but histologic study is more reliable in the diagnosis of lymphoma, thymoma, or benign lesions.[4] For lymphoma, needle biopsy is of greatest value in the event of recurrence, where histologic slides on file can be compared with the needle specimen.[20] Hodgkin lymphoma is especially difficult to diagnose by percutaneous methods. At our institution, Lieberman et al. found histologic work-up more useful for detection of various types of malignancy (sensitivity of 83% vs. 62% for cytology), particularly in determining tissue of origin, but the addition of cytologic study can improve accuracy.[21]

Sample Handling and Evaluation

One's colleague in the pathology laboratory should be consulted for

TABLE 17–1.
Conversion of Needle Gauge to
Diameter Equivalents

Needle Gauge	Millimeters	Inches
23	0.64	0.025
22	0.71	0.028
21	0.81	0.032
20	0.89	0.035
19	1.07	0.042
18	1.24	0.049
17	1.47	0.058
16	1.65	0.065
15	1.83	0.072
14	2.11	0.083

the preferred handling of specimens at a given institution. It should be noted that immediate placement of a specimen in buffered formalin prevents its use for immunohistochemical studies or for touch preparations. The same holds true if an entire core is submitted for frozen section review.

The attendance of a cytopathologist at the time of biopsy or preparation of frozen sections can determine not only the adequacy of the specimen, but also in many instances the diagnosis. If more tissue is needed, it can be obtained immediately. Repeating a biopsy at another time increases cost and is inconvenient for both patient and physician.

The pathologist should be given all pertinent clinical history, including any previous malignancy and its treatment, so that special stains can be used when needed. For suspected infection, both culturing and staining should be done, for one or the other may be negative for disease in nearly half of cases.[6] Special media may be needed to transport specimens for anaerobic, fungal, viral, and mycobacterial culture, and these must be obtained before needle aspiration.

Potential Complications

Needle biopsy can be complicated by bleeding, local infection, pneumothorax, air embolism, bile leakage, peritonitis, pancreatitis, or tumor seeding of the biopsy tract. The last is often of great concern to clinicians, but is a very rare consequence of aspiration biopsy.[22] All in all, use of fine needles (20 gauge or smaller) is extraordinarily safe, with severe complications reported in 0.04% to 0.05% and a mortality rate of 0.004% to -0.008% in two large reviews including a combined total of over 65,000 abdominal biopsies.[1, 22] For cutting needles, severe complications are twice as common as for fine aspiration needles (but still quite unusual), and a mortality of 0.027% has been reported for over 11,000 abdominal procedures.[22]

A prospective study of nearly 400 procedures by Yankaskas et al. showed that 13% of patients had pain requiring analgesics after abdominal or pelvic biopsy with standard needles.[23] A drop in hematocrit felt to represent postbiopsy bleeding was found in nine patients. The fall in hematocrit was noted from 3 to 72 hours after the procedure, and appeared more likely in those with cancer. Even so, most of the patients described had clinically silent bleeding, and the incidence of minor complications is a function of how closely patients are observed.

SITE-SPECIFIC CONSIDERATIONS

Thoracic Biopsy

For primary lung tumors a primary goal of biopsy is to distinguish small cell from nonsmall cell malignancy, a distinction important for therapy. This determination is usually possible with needle sampling. The sensitivity of needle aspiration for primary or metastatic carcinomas in lung is over 90%.[5, 20] Bronchoscopy is preferred for central endobronchial lesions, but most other thoracic masses can be safely approached by percutaneous needles.

If a lesion is suspected to be an arteriovenous malformation, peripheral pulmonary arterial aneurysm, or is adjacent to large vessels in the hilus, dynamic contrast material-enhanced CT is warranted prior to biopsy. Mediastinal biopsies generally call for CT guidance. As noted earlier, large cutting needles may be needed to enable a diagnosis of thymoma or lymphoma.[4] Small lesions near the heart or great vessels are best approached with fine needles. For pleural biopsy, the Cope needle can be used when no discrete mass is detected, and has been valuable in the diagnosis of mesothelioma, tuberculosis, and lymphoma.[13] This needle should be placed immediately superior to the rib, avoiding the inferior course of the neurovascular bundle.

Pneumothorax

Whether pneumothorax is considered a complication or an anticipated adverse side-effect of pulmonary needle biopsy is semantic. The incidence of pneumothorax after biopsy is correlated with the depth of the lesion and number of needle passes made, but is not related to the size of the needle.[24] A rate of 10% to 35% can be expected when fluoroscopy is used for guidance.[24] The incidence is higher when the procedure is performed with CT (43% to 57%), related perhaps to the longer duration of the procedure or to the smaller size or increased difficulty of lesions referred for CT-guided biopsy.[5, 25] Also at greater risk are those patients with abnormal pulmonary function tests. Fish et al. showed a pneumothorax rate of 46% in those with FEV_1 (1-second forced expiratory volume) less than 70% predicted, compared with an incidence of only 19% for those with normal pulmonary tests.[9] In the latter group, pneumothorax was almost always inconsequential.

Upright inspiratory and expiratory chest radiographs are obtained immediately after the biopsy procedure, and repeat films obtained from 1 to 4 hours later; outpatients should be observed for about 4 hours. When pneumothorax occurs, 90% of instances are evident immediately, and only 2% first appear more than hour later.[26] If a patient develops delayed symptoms, radiographs should be obtained without hesitation.

Chest Tube Placement

The necessity of intervention for pneumothorax has been reported in 3% to 20% of all thoracic biopsies,[5, 9, 20, 25, 26] and the operator must always be prepared to insert a chest tube. Criteria for placement include any pneumothorax associated with dyspnea or distress, or any pleural air collection shown to be increasing in size on serial chest films. Many small apical pneumothoraces do not need aspiration, but the size at which intervention is warranted in the absence of symptoms is a matter of judgment.

A Heimlich chest tube is small, effective, and easy to insert. With the patient upright, the tube and its trocar are introduced anteriorly through the second or third intercostal space at the midclavicular line. The catheter is fed off the trocar when loss of resistance signals entry into the pleural space. The tube then can be attached to the one-way valve supplied, to suction, or a three-way stopcock for immediate aspiration with a large syringe. Care must be taken that all connections are secure, and that any stopcock included in the line cannot be accidentally opened to atmospheric pressure.

Other Pulmonary Complications

Other hazards of lung biopsy include air embolism and pulmonary hemorrhage, both potentially fatal. Biopsy on deep inspiration or with the Valsalva maneuver should be avoided, and the patient should be instructed to suppress coughing as much as possible. All increase the possibility that air may enter pulmonary veins from the lung.[27] Air embolism should be treated by placing the patient in a left lateral decubitus and Trendelenberg position. The patient may need transfer to an institution with a hyperbaric chamber. Hemoptysis is seen to a minor degree in many patients, but massive bleeding is, fortunately, quite unusual and has been reported with 18-gauge or larger needles. The patient should be given oxygen and placed with the involved lung dependent. In severe cases selective intubation of the uninvolved lung may be required for tamponade of the hemorrhage.

Bone

Biopsy of osseous lesions has been traditionally performed with very large trephines or boring needles, such as the Turkel, Ackermann, Craig, and Jamshidi. However, smaller needles have been recently developed that can be used with a motorized drilling attachment.[28] Most bone lesions presenting for needle biopsy are lytic, and if enough destruction is present, standard fine needles can be used without difficulty.

Diagnostic accuracy with large trephined needles has been 81% to 95%, while major bleeding has been encountered in 2%[10]. The number of passes into a vertebral body should be limited to two, and a smaller needle should be used if possible, for the risk of neurologic complication is not negligible (8%).[10] One point to keep in mind: about one third of patients with known cancer and a solitary bone lesion will have a benign lesion.

If blood returns through the needle, about 5mL should be aspirated and allowed to clot. This can be submitted as a tissue specimen, with both sections and smears prepared. Clot examination may be more sensitive for metastatic carcinoma than the study of bone tissue. Hewes et al. found that in those cases of osseous neoplasm from which both tissue and clot had been obtained, only 39 of 54 would have been positive for malignancy had examination of blood been omitted.[29]

Disk space biopsy is indicated for patients with narrowing or destruction in conjunction with pain, fever, leukocytosis, or elevated erythrocyte sedimentation rate. Use of a large trephined needle can be diagnostic in 68%, with infection the most common finding.[30]

Abdomen and Pelvis

Ileus may occasionally be encountered after needle placement through peritoneum. Frank peritonitis has been described as complicating 0.3% of biopsies performed with 18- and 19-gauge cutting needles.[19]

Liver

Hepatic percutaneous needle biopsy has over 90% sensitivity for tumor.[20] The presence of ascites has long been considered a relative contraindication, but it has recently been shown not to pose any additional hazard.[31]

Although most hepatic hemangiomas can be diagnosed by noninvasive means, many cannot be distinguished from possible malignancy without biopsy. As long as needles no larger than 20 gauge are used and care is taken to traverse normal parenchyma in reaching the lesion, percutaneous biopsy of hemangioma is safe.[32] The presence of endothelial elements in an aspirate supports the diagnosis. However, the pathologist must be alerted to the possibility that the lesion is a hemangioma for a positive diagnosis to be made.

Hemorrhage and bile leakage are the greatest potential problems associated with hepatic biopsy. Hepatocellular carcinoma and various metastatic lesions may be very vascular, and bleeding from such tumors is a prime cause of death due to cutting needle biopsy.[22] Bleeding risk is elevated in the presence of cirrhosis and hepatic dysfunction, both common conditions in those referred for diagnosis.

One method of preventing hemorrhage is to plug the biopsy tract. Chuang and Alspaugh have performed embolizations of the hepatic tract of high-risk patients by preloading a 6-F sheath over the cannula of a 14-gauge Tru-Cut needle.[33] Serial injections of gelatin sponge (Gelfoam) during withdrawal prevented significant bleeding in all 22 patients undergoing the procedure. At the University of Wisconsin – Madison, we have modified this technique by placing a 16-gauge short venous catheter into the biopsy cannula after specimen (but not cannula) removal. In this manner, gelatin foam or embolic coils can be introduced without a preloaded sheath in place.

Alternative approaches to hepatic biopsy include the use of a percutaneous transjugular venous needle, which can be guided into hepatic veins, and placement of sheathed needles, brushes, or biopsy forceps through existing percutaneous biliary drainage tracts.[2, 34]

Pancreas

Fine-needle biopsy can provide a positive diagnosis of pancreatic carcinoma in 73% to 85% of cases.[2, 3] This compares favorably with surgical biopsy, which may have a sensitivity of only 76%.[2] False negative studies can be related to the marked desmoplastic character of many pancreatic malignancies. This fibrotic reaction may be responsible for the lower incidence of pancreatitis after needle biopsy of larger lesions. In the series described by Mueller et al., all five patients who developed severe pancreatitis after fine-needle biopsy had lesions smaller than 3 cm in diameter.[35] In fact, those with normal pancreatic tissue may be at greater risk for this severe and occasionally fatal complication. Thus, caution should be exercised in pursuing biopsy in patients with questionable pancreatic abnormalities.

Adrenal

Adrenal biopsy can be very accurate, with one series reporting definitive diagnosis in 13 of 14 patients studied.[36] Pitfalls include the inability to distinguish normal adrenal from adenoma microscopically or to distinguish primary adrenal carcinoma from metastatic renal cell carcinoma. Successful needle biopsy of pheochromocytoma has been performed, but premedication with alpha- and beta-blockers (phenoxybenzamine and propranolol) is advisable.[37]

Other Sites

Needle biopsy has not been widely used for the diagnosis of renal neoplasm, although renal cyst aspiration was in vogue for a number of years. The main function of renal biopsy at present is to diagnose diffuse parenchymal disease or renal transplant rejection. In these cases, large cutting needles are generally needed. Ultrasound is commonly used to direct biopsy toward the lower pole of kidney, avoiding major vessels. If bleeding is encountered, it may be treated by tract embolization.

Percutaneous biopsy has been used extensively for masses or lymph nodes suggesting recurrent gynecologic, genitourinary, or colonic tumor after surgical resection. Masses involving or extending into inferior vena cava can be sampled by long intravenous needles.[34] Presently, rectal sonographic probes are being used to guide needle aspiration of focal prostatic lesions. The potential benefits and place of prostatic biopsy need yet to be defined in the wider population suffering prostatic enlargement.

REFERENCES

1. Livraghi T, Damascelli B, Lombardi C, et al: Risk in fine-needle abdominal biopsy. *J Clin Ultrasound* 1983; 11:77–81.
2. Cohan RH, Illescas FF, Braun SD, et al: Fine needle aspiration biopsy in malignant obstructive jaundice. *Gastrointest Radiol* 1986; 11:145–150.

3. Teplick SK, Haskin PH, Kline TS, et al: Percutaneous pancreaticobiliary biopsies in 173 patients using primarily ultrasound or fluoroscopic guidance. *Cardiovasc Intervent Radiol* 1988; 11:26–28.

4. Goralnik CH, O'Connell DM, El Yousef SJ, et al: CT-guided cutting-needle biopsies of selected chest lesions. *AJR* 1988; 151:903–907.

5. vanSonnenberg E, Casola G, Ho M, et al: Difficult thoracic lesions: CT-guided biopsy experience in 150 cases. *Radiology* 1988; 167:457–461.

6. Conces DJ Jr, Clark SA, Tarver RD, et al: Transthoracic aspiration needle biopsy: Value in the diagnosis of pulmonary infections. *AJR* 1989; 152:31–34.

7. Bottles K, McPhaul LW, Volberding P: Fine needle aspiration biopsy of patients with the acquired immunodeficiency syndrome (AIDS): Experience in an outpatient clinic. *Ann Intern Med* 1988; 108:42–45.

8. Nash JD, Burke TW, Woodward JE, et al: Diagnosis of recurrent gynecologic malignancy with fine-needle aspiration cytology. *Obstet Gynecol* 1988; 71:333–337.

9. Fish GD, Stanley JH, Miller KS, et al: Postbiopsy pneumothorax: Estimating the risk by chest radiography and pulmonary function tests. *AJR* 1988; 150:71–74.

10. Mink J: Percutaneous bone biopsy in the patient with known or suspected osseous metastases. *Radiology* 1986; 161:191–194.

11. Reading CC, Carboneau JW, James EM, et al: Sonographically guided percutaneous biopsy of small (3 cm or less) masses. *AJR* 1988; 151:189–192.

12. Reading CC, Carboneau JW, Felmlee JP, et al: US-guided percutaneous biopsy: Use of a screw biopsy stylet to aid needle detection. *Radiology* 1987; 163:280–281.

13. Mueller PR, Saini S, Simeone JF, et al: Image-guided pleural biopsies: Indications, technique, and results in 23 patients. *Radiology* 1988; 169:1–4.

14. Onik G, Cosman ER, Wells TH Jr, et al: CT-guided aspirations for the body: Comparison of hand guidance with stereotaxis. *Radiology* 1988; 166:389–394.

15. Frederick PR, Miller MH, Bahr AL, et al: Coaxial needles for repeated biopsy sampling. *Radiology* 1989; 170:273–274.

16. Hawkins IF Jr, Akins EW, Mladinich C, et al: Transvisceral access using a blunt needle: Technical note. *Semin Intervent Radiol* 1988; 5:149–151.

17. Andriole JG, Haaga JR, Adams RB, et al: Biopsy needle characteristics assessed in the laboratory. *Radiology* 1983; 148:659–662.

18. Hueftle MG, Haaga JR: Effect of suction on biopsy sample size. *AJR* 1986; 147:1014–1016.

19. Berger H, Permanetter W, Steiner W, et al: Feinnadel- und schneidebiopsietechnik in der perkutanen punktion abdomineller raumforderungen. *Radiologe* 1988; 28:265–268.

20. Koss LG: Aspiration biopsy: A tool in surgical pathology. *Am J Surg Pathol* 1988; 12(suppl 1):43–53.

21. Lieberman RP, Hafez GR, Crummy AB: Histology from aspiration biopsy: Turner needle experience. *AJR* 1982; 138:561–564.

22. Weiss H, Düntsch U, Weiss A: Risiken der feinnadelpunktion: Ergebnisse einer umfrage in der BRD (DEGUM-umfrage). *Ultraschall Med* 1988; 9:121–127.

23. Yankaskas BC, Staab EV, Craven MB, et al: Delayed complications from fine-needle biopsies of solid masses of the abdomen. *Invest Radiol* 1986; 21:325–328.

24. Westcott JL: Percutaneous transthoracic needle biopsy. *Radiology* 1988; 169:593–601.

25. Harter LP, Moss AA, Goldberg HI, et al: CT-guided fine-needle aspirations for diagnosis of benign and malignant disease. *AJR* 1983; 140:363–367.

26. Perlmutt LM, Braun SD, Newman GE, et al: Timing of chest film follow-up after transthoracic needle aspiration. *AJR* 1986; 146:1049–1050.

27. Cianci P, Posin JP, Shimshak RR, et al: Air embolism complicating percutaneous thin needle biopsy of lung. *Chest* 1987; 92:749–751.

28. Hauenstein KH, Wimmer B, Beck A, et al: Knochenbiopsie unklarer knochenläsionen mit einer neuen 1,4 mm messenden biopsiekanüle. *Radiologe* 1988; 28:251–258.

29. Hewes RC, Vigorita VJ, Frieberger RH: Percutaneous bone biopsy: The importance of aspirated osseous blood. *Radiology* 1983; 148:69–72.

30. Armstrong P, Chalmers AH, Green G, et al: Needle aspiration/biopsy of the spine in suspected disc space infection. *Br J Radiol* 1978; 51:333–337.

31. Murphy FB, Barefield KP, Steinberg HV, et al: CT- or sonography-guided biopsy of the liver in the presence of ascites: Frequency of complications. *AJR* 1988; 151:485–486.
32. Cronan JJ, Esparza AR, Dorfman GS, et al: Cavernous hemangioma of the liver: Role of percutaneous biopsy. *Radiology* 1988; 166:135–138.
33. Chuang VP, Alspaugh JP: Sheath needle for liver biopsy in high-risk patients. *Radiology* 1988; 166:261–262.
34. Krieves D, Keller FS, Dotter CT, et al: Percutaneous intravenous biopsy. *Diagn Imaging* 1980; 49:297–302.
35. Mueller PR, Miketic LM, Simeone JF, et al: Severe acute pancreatitis after percutaneous biopsy of the pancreas. *AJR* 1988; 151:493–494.
36. Heaston DK, Handel DB, Ashton PR, et al: Narrow gauge needle aspiration of solid adrenal masses. *AJR* 1982; 138:1143–1148.
37. Koenker RM, Mueller PR, vanSonnenberg E: Interventional radiology of the adrenal glands. *Semin Roentgenol* 1988; 22:314–322.

18 | Percutaneous Abscess Drainage

KEY CONCEPTS

1. Most intra-abdominal and pelvic abscesses can be drained by percutaneous catheters.
2. Abscesses with fistulas can also be treated but need a longer period of drainage.
3. If a patient does not improve within 24 hours of drainage, the reason for inadequate response must be found and corrected.
4. Catheters can be removed when the patient is afebrile, leukocytosis has resolved, and tube output is negligible.
5. A fistula should be suspected if drainage continues to be more than 50 mL/day after the first few days.

INDICATIONS

Any abdominal, pelvic, or other soft tissue abscess that cannot be readily treated by simple incision and drainage may be initially approached by percutaneous catheters. Untreated abdominal abscesses are almost invariably fatal, and mortality even with surgery may be quite high—from 17% in single abdominal abscess to 43% to 80 % in patients with multiple abscesses.[1-3]

CONTRAINDICATIONS AND CAUTIONS

There are no absolute contraindications to percutaneous catheter drainage, but care must be taken to avoid bowel, spleen, major vessels, or the diaphragm during tube placement. Transgression of the diaphragm is likely to lead to pleural empyema. Severe coagulopathy is a relative contraindication to catheter placement. Multiloculated or multiple abscesses will require multiple drains. Foreign bodies must be removed if present, for they serve as foci of infection. Abscesses around infected vascular grafts must be managed surgically. Infected hematomas or other collections containing very viscous or debris-laden fluid are likely not to respond. Suspected ecchinococcal cysts should not be approached because of danger of peritoneal spillage and shock. If no clinical improvement is seen within 24 to 48 hours of tube placement, surgery or other percutaneous measures must be vigorously pursued.

CHOOSING AN ENTRY SITE

Percutaneous drainage has been made possible by the advent of cross-sectional imaging, allowing not only the earlier detection of infected fluid collections, but also fine-needle aspiration for diagnosis, and the planning of percutaneous catheter approach. Ideally, the site of entry should allow the most direct approach (extraperitoneal if possible), while avoiding vessels and viscera.

As noted, the diaphragm should not be traversed. Posteriorly, the pleural reflection is at the level of the 12th rib; in the midaxillary line, pleura extends to the tenth rib.[4, 5] For subphrenic lesions, sharply cephalad angulation may be needed. Alternatively, a subxiphoid approach can be used.

At times, it may be necessary to place catheters transgluteally to treat pelvic abscess.[6] In such cases, the route chosen should be low in the greater sciatic foramen, close to the sacrum above the sacrospinous ligament. In this fashion the sacral plexus and sciatic nerve can be avoided. The transgluteal approach tends to be more painful for the patient, and because of the amount of muscle being traversed, a stiffening cannula may be required for catheter insertion.

Although catheters have been placed through stomach or bowel in order to reach an abscess, this is usually inadvertent and not planned.[7] Recognition of such transgression may take several days, but fortunately, most such cases are managed successfully by catheter alone. Manipulations should be avoided for at least 2 weeks (to allow a fibrous tract to form), and the catheter should be slowly withdrawn over 2 to 3 days. If the stomach is violated, the patient should have nothing by mouth for 24 hours prior to catheter removal.

A transhepatic approach for otherwise inaccessible abscesses of the lesser sac has been used successfully and without complication by Mueller et al.[8] Precautions in such cases have included avoidance of central portions of the liver, restriction of catheter size to 9 F or smaller, and gradual catheter withdrawal over a period of days after clinical resolution of the abscess. Patients with cirrhosis, portal hypertension, bleeding disorder, or biliary obstruction should be treated surgically.

In choosing a site for percutaneous drainage, physical factors and patient comfort should be kept in mind. A drain is best placed in the most dependent portion of the abscess, in order to have gravity assist in its evacuation. However, a direct posterior approach is seldom optimal, because the catheter will not allow the patient to lie supine comfortably, and kinking of the tube can result.

PLACING THE CATHETER

Before draining a fluid collection, it is best to place a diagnostic needle (often 20 or 18 gauge, if a 22-gauge needle yields no aspirate) in order to determine if the fluid is infected. If the fluid is not obviously purulent, Gram staining should be performed immediately, and samples sent for culture, sensitivity, and special stains (acid-fast bacilli, potassium hydroxide preparations, and so forth). Some sterile fluid collections, hematomas in particular, are better left alone; introducing a catheter may complicate the condition by allowing bacteria access.

Patients referred for suspected abscess are almost invariably receiving intravenous antibiotics. However, if none have been given, a broad-spectrum antibiotic covering gram-negative and anaerobic organisms should be started before proceeding. Any anticoagulation should be stopped, and recent prothrombin and partial thromboplastin time values should be available.

There are two basic methods for placing a drainage catheter percutaneously: trocar technique and Seldinger technique. The former is generally reserved for large, relatively superficial lesions with no critical intervening structures. The catheter is threaded over a matched trocar, and after appropriate sterile preparation of the skin, local anesthesia, and small incision, it is advanced until pus can be aspirated through the trocar or until imaging shows that the tip is well into the fluid space. The trocar is then stabilized, while the catheter is slipped over it to curl into the abscess. With the Seldinger technique a needle is first placed within the fluid, a guidewire is coiled into the abscess, and serial dilators are used before the drainage catheter is introduced over the wire.

In general, catheters should be no smaller than 8 F to ensure drainage of infected fluid, and it is clear that 5- to 6-F tubes are inadequate.[3] Larger sump tubes, such as the 12- to 14-F vanSonnenberg catheters, have much to recommend them.

After as much purulent material as possible has been aspirated, a sinogram can be obtained. No more than 20 to 30 mL of contrast material should be injected at the time of initial drainage, to avoid the risk of bacteremia and sepsis. Contrast injection may demonstrate fistulous communication with biliary ducts, pancreatic duct, or bowel. This sinogram also serves as a baseline for future studies. If an abscess is large and contains septations or a large amount of debris, a curved catheter may be rotated gently or a soft-tipped guidewire curled within the cavity to break up any loculations.

The catheter may be sutured directly to the skin. As an alternative, an enterostomy stoma ring can be applied to the entry site, and the catheter secured to the ring by sutures. Connecting tubing should be secured with tape to make transmission of inadvertent tension to the catheter less likely.

CATHETER MANAGEMENT AND END-POINTS

The management of drainage catheters is an area in which there is no great consensus. Many put catheters to dependent drainage or low intermittent suction. Depending on the viscosity of the draining material, the catheter and sump port can each be flushed with 5 mL of saline every day or every nursing shift. Strict records of tube input and output are mandatory. If constant drip infusion through one port or catheter is elected (with suction through another), input/output observation becomes critical. The amount infused should never exceed the volume drained. Antibiotics are chosen according to the sensitivity of the organisms cultured, and clinical response is measured by the return to normal of the patient's temperature and leukocyte count within several days. The use of acetylcysteine (Mucomyst) or antibiotic infusion is controversial, and no consistent benefit has been demonstrated.[3]

The net output of the catheter should quickly drop to less than 50 mL/day, otherwise fistula should be suspected. Serial sinograms are obtained every several days to confirm optimal catheter placement and decrease in cavity size, and to detect any fistula. When tube output approaches zero and the cavity is essentially collapsed about the drain, the catheter may be removed. If the lesion is deep, it is advisable to withdraw the catheter in stages over 2 or 3 days. Successful drainage is usually accomplished within 1 to 2 weeks in uncomplicated cases. Those individuals able to care adequately for their catheters may be discharged and followed as outpatients until tube removal is deemed appropriate.

FISTULAS

In the past, underlying fistula was considered a contraindication to percutaneous drainage, but experience at many centers has shown that an abscess with fistula can be managed just as successfully as an uncomplicated abscess. In the great majority of cases fistula is not suspected and not found until several days after catheter placement. If it is demonstrated, an additional catheter should be placed as close as possible to the communication. Bowel rest or biliary diversion (in the case of biliary fistulas) may be appropriate. Fistulas associated with bowel or biliary obstruction will not resolve until the obstruction is relieved. Laboratory analysis for fluid amylase or bilirubin may confirm pancreatic or biliary communication. Successful management of fistulas with percutaneous catheters usually requires twice as much time as simple abscess drainage, often longer than 1 month.

SPECIAL SITUATIONS

Although considered unconventional therapy, percutaneous catheter drainage has been used to drain splenic abscess in high-risk patients, drain mediastinitis from esophageal rupture, and drain lymphoceles and pancreatic pseudocysts, among other lesions. There has been good success in resolving thoracic empyema, as long as fluid has been present less than 1 month.[9, 10] Sterile but symptomatic lymphatoceles can often be resolved by needle aspiration alone.[11] Diverticular and periappendiceal abscesses respond well to catheter therapy, allowing what might be a two- or three-stage surgical approach to become a single-stage elective operation after resolution of the acute inflammatory process.[12–14]

Crohn disease patients are prone to develop abscess complicated by fistula (usually to the ileum). Catheter drainage may be appropriate for palliation, although fistulas tend to recur despite adequate drainage, treatment with parenteral nutrition, and bowel rest.[15] The diseased segment of bowel must be resected for definitive therapy. Those without bowel fistulas respond well to percutaneous drainage alone.[15, 16] Creation of an enterocutaneous fistula does not appear to be a problem in such patients.

Infected pancreatic pseudocysts or peripancreatic fluid collections can be managed by catheter alone in many cases.[3, 17, 18] Success is most likely when a single fluid pocket is present. Uninfected pseudocysts may be aspirated for diagnostic purposes, but it is best not to leave a catheter in place. Pancreatic phlegmons must be surgically debrided.

SUCCESS RATES

Percutaneous catheter drainage alone has been successful in treating 70% to 91%[2, 6, 19–22] of abdominal and pelvic abscesses in several large series. This compares favorably with surgery, and mortality for percutaneously drained patients is on the order of 1% to 10%.[19] Catheter therapy may resolve fistulas in 77% to 86%.[6, 20, 23, 24] Even those with multiloculated or difficult to drain collections have had "cure" in 64% of cases.[1] Therefore, it is reasonable that percutaneous drainage be attempted in most cases of percutaneously accessible abscesses, and that surgery be reserved for the small number of those who fail to respond.

REASONS FOR FAILURE

Catheter drainage of abscess may fail from a complication of entry, such as hemorrhage; premature removal of the catheter; inadequate size or number of catheters; poorly liquefied necrotic debris; or failure to recognize a fistula. Fistulas from malignancy or radiation therapy are unlikely to respond. Infected hematomas have been notoriously difficult to drain, but infusion with thrombolytic agents shows some promise in improving evacuation.[25] If a patient does not respond positively within 1 day of tube placement, a technical problem should be sought immediately. If no remediable problem is uncovered, surgery should not be delayed.

REFERENCES

1. van Waes PFGM, Feldberg MAM, Mali WPTM et al: Management of loculated abscesses that are difficult to drain: A new approach. *Radiology* 1983; 147:57–63.

2. Serrano A, Dahl EP, Rubin RH, et al: Eclectic drainage of subphrenic abscesses. *Arch Surg* 1984; 119:942–945.

3. Pruett TL, Simmons RL: Status of percutaneous catheter drainage of abscesses. *Surg Clin North Am* 1988; 68:89–105.

4. Nichols DM, Cooperberg, PL, Golding RH, et al: The safe intercostal approach? Pleural complications in abdominal interventional radiology. *AJR* 1984; 141:1013–1018.

5. Neff CC, Mueller PR, Ferrucci JT Jr, et al: Serious complications following transgression of the pleural space in drainage procedures. *Radiology* 1984; 152:335–341.

6. Butch RJ, Mueller PR, Ferrucci JT Jr et al: Drainage of pelvic abscesses through the greater sciatic foramen. *Radiology* 1986; 158:487–491.

7. Mueller PR, Ferrucci JT Jr, Butch RJ, et al: Inadvertent percutaneous catheter gastroenterostomy during abscess drainage: Significance and management. *AJR* 1985; 145:387–391.

8. Mueller PR, Ferrucci JT Jr, Simeone JF, et al: Lesser sac abscesses and fluid collections: Drainage by transhepatic approach. *Radiology* 1985; 155:615–618.

9. Westcott JL: Percutaneous catheter drainage of pleural effusion and empyema. *AJR* 1985; 144:1189–1193.

10. Hunnam GR, Flower CDR: Radiologically-guided percutaneous catheter drainage of empyemas. *Clin Radiol* 1988; 39:121–126.

11. Jensen SR, Voegeli DR, McDermott JC, et al: Percutaneous management of lymphatic fluid collections. *Cardiovasc Intervent Radiol* 1986; 9:202–204.

12. vanSonnenberg E, Wittich GR, Casola G, et al: Periappendiceal abscesses: Percutaneous drainage. *Radiology* 1987; 163:23–26.

13. Mueller PR, Saini S, Wittenburg J, et al: Sigmoid diverticular abscesses: Percutaneous drainage as an adjunct to surgical resection in 24 cases. *Radiology* 1987; 164:321–325.

14. Neff CC, vanSonnenberg E, Casola G, et al: Diverticular abscesses: Percutaneous drainage. *Radiology* 1987; 163:15–18.

15. Lambaise RE, Cronan JJ, Dorfman GS, et al: Percutaneous drainage of abscesses in patients with Crohn disease. *AJR* 1988; 150:1043–1045.

16. Casola G, vanSonnenberg E, Neff CC, et al: Abscesses in Crohn disease: Percutaneous drainage. *Radiology* 1987; 163:19–22.

17. Gerzof SG, Johnson WC, Robbins AH, et al: Percutaneous drainage of infected pancreatic pseudocysts. *Arch Surg* 1984; 119:888–893.

18. Stanley JH, Gobien RP, Shabel SI, et al: Percutaneous drainage of pancreatic and peripancreatic fluid collections. *Cardiovasc Intervent Radiol* 1988; 11:21–25.

19. Olak J, Christou NV, Stein LA, et al: Operative vs. Percutaneous drainage of intra-abdominal abscesses. *Arch Surg* 1986; 121:141–146.

20. Kerlan RK Jr, Jeffrey RB Jr, Pogany AC, et al: Abdominal abscess with low-output fistula: Successful percutaneous drainage. *Radiology* 1985; 155:73–75.

21. vanSonnenberg E, Wittich GR, Casola G, et al: Complicated pancreatic inflammatory disease: Diagnostic and therapeutic role of interventional radiology. *Radiology* 1985; 155:335–340.

22. Gordon DH, Macchia RJ, Glanz S, et al: Percutaneous management of retroperitoneal abscesses. *Urology* 1987; 30:299–306.

23. Mueller PR, Ferrucci JT Jr, Simeone JF, et al: Detection and drainage of bilomas: Special considerations. *AJR* 1983; 140:715–720.

24. Papanicolaou N, Mueller PR, Ferrucci JT Jr, et al: Abscess-fistula association: Radiologic recognition and percutaneous management. *AJR* 1984; 143:811–815.

25. Vogelzang RL, Tobin RS, Burstein S, et al: Transcatheter intracavitary fibrinolysis of infected extravascular hematomas. *AJR* 1987; 148:378–380.

19

Percutaneous Gastrostomy

KEY CONCEPTS

1. Percutaneous gastrostomy is an alternative for long-term feeding or gastric decompression.
2. Patients with gastroesophageal reflux or who are at risk for aspiration pneumonia can be fed safely by a transgastric jejunal tube.
3. Ultrasound and fluoroscopy are used to avoid liver and colon during tube placement.
4. Most patients can be fed the day following gastrostomy.
5. Tubes must be carefully secured both internally and externally for 2 weeks to prevent the possibility of displacement before a fibrous tract has formed.

Percutaneous gastrostomy is a quick, simple, and safe method for obtaining access to the stomach. The prime indication is for long-term feeding, but it may also be used for gastric decompression, or rarely for simplified dilation of upper gastrointestinal stricture when a per oral approach has failed.

ENTERIC NUTRITIONAL SUPPORT

Malnutrition is increasingly recognized as a problem for many hospitalized and chronically ill patients.[1-3] Those who are unable to take nourishment for themselves, but who have an otherwise intact and functioning gastrointestinal tract, may benefit from tube feeding. In most cases, when the need for nutritional support is of limited duration, small-bore nasoenteric feeding tubes are quite adequate. However, if tube feeding is necessary for more than 4 to 6 weeks, gastrostomy provides a well-tolerated alternative.

Nearly all patients fed through a nasal tube are distressed by nasopharyngeal irritation, mouth breathing, and feelings of limited mobility.[4] Reflux about a tube traversing the gastroesophageal junction can become a problem. Nasoenteric tubes are also prone to accidental removal and recurrent occlusion. Many of these difficulties can be avoided by gastrostomy. One factor appreciated by patients discharged with gastric feeding tubes is the ability to appear in public without feeling excessively self-conscious.

Gastrostomy as an operative procedure has been performed for over a century, in many cases with local anesthesia alone; however, its widespread acceptance has been limited by reports of up to 35% mortality and 56% morbidity.[5] More recent improvements in surgical technique and patient preparation have reduced procedure-related mortality to 1.8% or less.[5–7]

A percutaneous, trocar approach to gastrostomy was described in 1967 by Jascalevich,[8] but over a decade passed before its first clinical application.[9] About the same time, endoscopy was enrolled by a number of investigators to guide percutaneous placement.[10, 11] It was soon recognized that the procedure could be simplified by use of fluoroscopy and a small nasogastric tube.[12–14] The placement of long feeding tubes by a percutaneous transgastric approach has permitted the extension of the technique to patients with a history of gastroesophageal reflux and aspiration pneumonia.[15, 16]

PATIENT SELECTION

Those likely to benefit from percutaneous gastrostomy for feeding include individuals with impaired swallowing due to neurologic disease or obstructing oropharyngeal/esophageal neoplasms, burn or trauma patients, cancer patients suffering anorexia and inanition, and patients with pharyngeal or esophageal fistula. The appropriateness of feeding enterostomy in a given case is a judgment to be made by the radiologist, patient, referring physician and patient's family, after a clear delineation of the potential risks and benefits of the procedure. Percutaneous gastrostomy should *not* be performed as a mere short cut for long-term care; debilitated patients fed through tubes need careful monitoring for potential nutritional problems and signs of aspiration.[17]

Gastrostomy for feeding is contraindicated in the presence of gastric outlet or more distal bowel obstruction (gastrostomy for decompression may be indicated in such circumstances). The severely neurologically impaired, as well as those with known gastroesophageal reflux or poor gastric emptying, should not have stomach feedings. Rather, they should have transgastric jejunal tubes, because the more distal the infusion, the less likely the occurrence of reflux and aspiration.[1, 18]

Although percutaneous gastrostomy is possible in those with previous partial gastrectomy, it is more difficult and entails higher risk. The presence of ascites is a contraindication, because of the potential for chronic external leakage of ascitic fluid. As in any percutaneous interventional procedure, anticoagulant medication should be stopped prior to percutaneous gastrostomy, and any coagulation abnormalities must be reversed.

TECHNIQUE

The patient should be given nothing by mouth for at least 8 hours prior to gastrostomy placement. Antibiotics are not routinely given. Care should be taken to ensure that no organ or bowel interposes itself between the anterior abdominal wall and stomach. To that end, real-time ultrasound is used to mark the edge of left lobe of liver on the skin. Transverse colon usually contains enough gas to be readily apparent on fluoroscopy. Dilute barium given the previous day may be helpful in delineating the colon, but in some patients it may be necessary to instill air or radiographic contrast material by way of the rectum. If it is suspected that small bowel lies anterior to stomach (a very rare situation), cross-table lateral radiographs may be obtained. A soft, small-bore nasogastric tube is inserted, if one is not already in place.

A subcostal approach to the left of midline is chosen with the assistance of fluoroscopy. The outer third of the rectus abdominis muscle should be avoided, because of the course of the inferior epigastric artery. Often the site chosen will overlie colon or small bowel loops, but gastric insufflation reliably creates a "window" for tube placement (Fig 19–1). The epigastrium is sterily prepared and draped, local anesthesia is infiltrated, and a small skin incision made.

At this point, any retained gastric fluid is aspirated by way of the nasogastric tube, which is then used to inflate stomach with 600 to 1,000 mL of air. The patient is instructed to hold the distention as well as possible. Glucagon (0.5 to 1.0 mg) may be given intravenously prior to insufflation, in order to prevent air from escaping through pylorus.

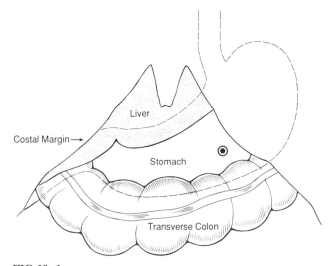

FIG 19–1.
Anatomic relationships relevant to gastrostomy placement. Dot marks an appropriate entry site.

However, some patients experience retching and vomiting with its administration, and glucagon is not absolutely necessary for successful gastrostomy.

With stomach inflated, a 5-F Teflon sheath needle is advanced by a quick, sharp jab to a depth of 4 to 5 cm. Stomach wall is quite compliant, but if distention is adequate no difficulty in entry should be encountered. Limiting the depth of the jab prevents penetration of the posterior gastric wall and possible pancreatic or splenic vessel injury. After removal of the needle trocar, water-soluble contrast material is injected through the sheath. Intragastric position is confirmed by delineation of ruggae and pooling in the fundus. A stiff guidewire is advanced through the sheath to make several coils within stomach. Teflon dilators are used to prepare for placement of a locking loop catheter, such as a Cope nephrostomy catheter.

Catheter introduction is greatly assisted by a central stiffener and a tapered tip to the catheter. Cope nephrostomy catheters ordinarily are

8 F to 12 F in diameter, but can be ordered in sizes to 16 F. Fluoroscopy (with oblique angulation to avoid exposing the operator's hands) is advisable, because tenting of stomach wall can be observed about dilators and catheters as they are inserted. Gastric entry of tubes is appreciated by fluoroscopy, as the tenting "gives way." Also, with fluoroscopy one can recognize any buckling and intraperitoneal looping of the guidewire immediately. If, despite the precautions taken, small bowel or transverse colon has interposed itself, fluoroscopic monitoring may lead to early recognition of the situation before a larger catheter is inserted.

No sedation is usually needed, and most patients have minimal, if any, discomfort from the procedure. If there are complaints of pain, peritoneal irritation from looping of guidewire should be suspected. Percutaneous gastrostomy can usually be completed within 10 to 15 minutes of gastric insufflation. If stomach tenting prevents insertion of dilators or catheter, more air should be injected through the nasogastric tube.

Once the gastric tube is in place, it must be secured both internally (locking loop) and externally by suture to the patient's skin or to a stoma ring applied to the patient's skin. The nasogastric tube may be removed. The patient is kept at bed rest with frequent monitoring of vital signs, and the gastric tube should be maintained on low suction until the next morning. If no problems arise, feedings can be initiated by graduated instillation first of saline, and later of nutrient solutions. Tube changes should be avoided during the first 1 to 2 weeks, in order to allow a fibrous tract to form around the catheter.[19] For the same reason, gastrostomy tubes should not be removed before 2 weeks have elapsed.

VARIATIONS

Percutaneous gastrostomy can be performed by trocar-loaded catheter, avoiding intermediate dilatation of the tract.[20] Large intragastric balloons have been devised to provide a target for needle or trocar,[21] but such special devices are largely unneeded. Gastrostomy catheters can be secured internally by balloons or "mushroom" devices, as well as by locking loops. Sheaths generally are necessary for introduction of Foley catheters or other large soft tubes.

Placement of a transgastric jejunal feeding tube at the time of gastrostomy requires more time, skill, and exposure to radiation.[15, 16] A fundal loop of wire can be used to direct the guiding catheter toward the pylorus. A curved metal cannula can serve the same purpose. Unless the stomach is secured to anterior abdominal wall during these manipulations, there is danger of wire coiling in peritoneum and loss of gastric access. This is particularly true if a fibrous tract has not yet formed. Coaxial manipulations through a large-bore gastrostomy balloon catheter help one avoid this danger.[22] Toggle-like and nylon-T anchoring devices have been used to approximate anterior gastric wall to anterior abdominal wall (gastropexy), facilitating introduction of large catheters, as well as the maneuvers needed for small bowel intubation.[23, 24]

In the presence of complete esophageal obstruction, percutaneous gastrostomy may be performed without a nasogastric tube. Under fluoroscopic or computed tomographic guidance a 22-gauge needle can be advanced percutaneously into the collapsed stomach. When the tip is confirmed to be intragastric by injection of water-soluble contrast material, air is injected through the fine needle. After sufficient gastric distention, tube placement proceeds in the manner described.

COMPLICATIONS

There have been no early procedure-related fatalities described after fluoroscopically guided percutaneous gastrostomy, although they will undoubtedly occur with wider application of the technique. Complications have been reported in 4% to 16% of patients, with about 3% needing surgical treatment.[13, 15, 16]

Transient peritoneal signs and mild fever may be seen in 5% to 10% of patients, but symptoms usually resolve within 1 to 2 more days of tube suction and use of intravenous antibiotics. Peritoneal air is to be expected and is of no concern unless the amount of pneumoperitoneum is increasing.[25] Despite early fears that lack of gastropexy would lead to internal leakage and diffuse peritonitis, such complications are quite rare. External leakage of gastric contents has complicated up to 7% of surgical gastrostomies,[26] and can also occur with percutaneous gastrostomy.[27]

Patients commonly have minor hematomas of the abdominal wall or gastric mucosa.[25] Major hemorrhage requiring transfusion arises in fewer than 1% of patients,[28] an incidence comparable to hemorrhage from surgical gastrostomy.[7, 29]

Subcutaneous emphysema should alert one to improper tube positioning; the catheter should have external fixation for at least 2 weeks in addition to internal retention devices.[27] If a balloon-tip catheter is used, external fixation must be maintained to prevent distal migration. Displaced balloon catheters can cause pyloric or biliary obstruction.

Catheters placed on traction may produce erosion or ulceration of the stomach wall. Those patients with transgastric jejunal tubes should have positioning checked after any episodes of vomiting. If the tube has moved proximally, danger of aspiration is increased. Infectious complications are rare after percutaneous gastrostomy, most commonly minor skin infections that respond to local care alone.

A major cause of morbidity and mortality from use of surgical gastrostomy for feeding is aspiration pneumonia,[17, 30] and one can anticipate the same consequence for percutaneous techniques. If a patient gives any indication of reflux, a transgastric jejunal tube must be placed.

GASTRIC OR SMALL BOWEL FEEDING

Although small bowel alimentation is safer, gastric feedings are preferable for patients not at risk for aspiration. Blenderized food is much less expensive than the liquid formulas needed for jejunal infusion. Moreover, jejunal feedings may cause diarrhea and electrolyte disturbances.

COMPARISON OF GASTROSTOMY METHODS

Percutaneous gastrostomy is as safe as surgical tube placement, and prospective studies have documented that percutaneous endoscopic gastrostomy is considerably less expensive than surgical gastrostomy.[31] Although similar prospective comparisons with fluoroscopically guided gastrostomy are lacking, it can be assumed that the nonendoscopic procedure will result in even further cost savings.

In some endoscopic techniques the gastrostomy tube is inserted perorally and thus the wound is exposed to the anaerobic flora of the mouth.[32] Prophylactic use of antibiotics does not change the incidence of catheter colonization but instead selects for resistant strains. It is noteworthy that a death from necrotizing fasciitis has resulted from endoscopic gastrostomy.[33] Furthermore, endoscopy requires a level of patient sedation that is unnecessary for the fluoroscopic technique. Some types of tubes actually require repeat endoscopy for replacement or removal!

Endoscopic approaches are ineffective in the face of esophageal occlusion. Radiologic gastrostomy can be performed in such patients, it can proceed at the bedside, and use of fluoroscopy can prevent the small bowel or colon perforation seen after some endoscopic procedures.[34]

These considerations suggest that percutaneous gastrostomy under radiologic guidance should be the method of choice for creating a feeding enterostomy.

REFERENCES

1. Torosian MH, Rombeau JL: Feeding by tube enterostomy. *Surg Gynecol Obstet* 1980; 150:918–926.

2. Meguid MM, Eldar S, Wahba A: The delivery of nutritional support: A potpourri of new devices and methods. *Cancer* 1985; 55(suppl):279–289.

3. Heymsfield SB, Bethel RA, Ansley JD, et al: Enteral Hyperalimentation: An alternative to central venous hyperalimentation. *Ann Intern Med* 1979; 90:63–71.

4. Padilla GV, Grant M, Wong H, et al: Subjective distresses of nasogastric tube feeding. *JPEN* 1979; 3:53–57.

5. Kumar SS: Tube gastrostomy: A routine adjunct in major abdominal operations. *Am Surg* 1985; 51:201–203.

6. Ruge J, Vasquez RM: An analysis of the advantages of Stamm and percutaneous endoscopic gastrostomy. *Surg Gynecol Obstet* 1986; 162:13–16.

7. Shellito PC, Malt RA: Tube gastrostomy: Techniques and complications. *Ann Surg* 1985; 201:180–185.

8. Jascalevich ME: Experimental trocar gastrostomy. *Surgery* 1967; 62:452–453.

9. Preshaw RM: A percutaneous method for inserting a feeding gastrostomy tube. *Surg Gynecol Obstet* 1981; 152:659–660.

10. Gauderer MWL, Ponsky JL: A simplified technique for constructing a tube feeding gastrostomy. *Surg Gynecol Obstet* 1981; 152:83–85.

11. Russell TR, Brotman M, Norris F: Percutaneous gastrostomy: A new simplified and cost-effective technique. *Am J Surg* 1984; 149:132–137.

12. Wills JS, Oglesby JT: Percutaneous gastrostomy. *Radiology* 1983; 149:449–453.

13. Wills JS, Oglesby JT: Percutaneous gastrostomy: Further experience. *Radiology* 1985; 154:71–74.

14. Tao HH, Gillies RR: Percutaneous feeding gastrostomy. *AJR* 1983; 141:793–794.
15. Ho C-S, Gray RR, Goldfinger M, et al: Percutaneous gastrostomy for enteral feeding. *Radiology* 1985; 156:349–351.
16. Alzate GD, Coons HG, Elliott J, et al: Percutaneous gastrostomy for jejunal feeding: A new technique. *AJR* 1986; 147:822–825.
17. Campbell-Taylor I, Fisher RH: The clinical case against tube feeding in palliative care of the elderly. *J Am Geriatr Soc* 1987; 35:1100–1104.
18. Gustke RF, Varma RR, Soergel KH: Gastric reflux during perfusion of the proximal small bowel. *Gastroenterology* 1970; 59:890–895.
19. Johnston WD, Lopez MJ, Kraybill WG, et al: Experience with a modified Witzel gastrostomy without gastropexy. *Ann Surg* 1982; 195:692–699.
20. vanSonnenberg E, Wittich GR, Cabrere OA, et al: Percutaneous gastrostomy and gastroenterostomy 2: Clinical experience. *AJR* 1986; 146:581–586.
21. vanSonnenberg E, Cubberley DA, Brown LK, et al: Percutaneous gastrostomy: Use of intragastric balloon support. *Radiology* 1984; 152:531–532.
22. Wojtowycz MM: Coaxial tube system for transgastric jejunal feeding *J Intervent Radiol* 1989; 4:46–48.
23. Brown AS, Mueller PR, Ferrucci JT Jr: Controlled percutaneous gastrostomy: Nylon T-fastener for fixation of the anterior gastric wall. *Radiology* 1986; 158:543–545.
24. Cope C: Suture anchor for visceral drainage. *AJR* 1986: 146:160–162.
25. Wojtowycz M, Arata JA Jr, Micklos TJ, et al: CT findings after uncomplicated percutaneous gastrostomy. *AJR* 1988; 151:307–309.
26. Wilkinson WA, Pickleman J: Feeding gastrostomy: A reappraisal. *Am Surg* 1982; 80:273–275.
27. Wojtowycz M, Arata JA Jr: Case report: Subcutaneous emphysema after percutaneous gastrostomy. *AJR* 1988; 151:311–312.
28. Rose DB, Wolman SL, Ho CS: Gastric hemorrhage complicating percutaneous transgastric jejunostomy. *Radiology* 1986; 161:835–836.
29. Wasiljew BK, Ujiki GT, Beal JM: Feeding gastrostomy: Complications and mortality. *Am J Surg* 1982; 143:194–195.

30. Swartzendruber FD, Laws HL: The superior feeding gastrostomy. *Am Surg* 1982; 80:276–278.
31. Stiegmann G, Goff J, VanWay C, et al: Operative versus endoscopic gastrostomy: Preliminary results of a prospective randomized trial. *Am J Surg* 1988; 155:88–91.
32. Jonas SK, Neimark S, Panwalker AP: Effect of antibiotic prophylaxis in percutaneous endoscopic gastrostomy. *Am J Gastroenterol* 1985: 80:438–441.
33. Greif JM, Ragland JJ, Ochsner MG, et al: Fatal necrotizing fasciitis complicating percutaneous endoscopic gastrostomy. *Gastrointest Endosc* 1986; 32:292–294.
34. Bui HD, Dang CV, Schlater T, et al: A new complication of percutaneous endoscopic gastrostomy. *Am J Gastroenterol* 1988; 83:448–451.

20

Biliary Interventions

KEY CONCEPTS

1. Percutaneous transhepatic cholangiography (PTHC) is an excellent method for examining biliary anatomy, but it has largely been superseded by cross-sectional imaging and endoscopic retrograde studies with contrast material.
2. Fine-needle PTHC is successful in nearly all patients with dilated ducts, and in 70% to 80% of those without ductal dilatation.
3. Success of PTHC is proportional to the number of fine-needle passes made, but complications do not increase with up to a dozen attempts.
4. Broad-spectrum antibiotics should be started at least 1 hour before cholangiography or other biliary interventions.
5. Percutaneous biliary drainage (PBD) is used primarily to palliate unresectable malignant disease, but those patients with a very limited life expectancy or with otherwise asymptomatic jaundice should not be subjected to PBD.
6. A T-tube tract must mature at least 5 to 7 weeks before it can be safely used to remove retained biliary stones.

The era of nonsurgical intervention in the biliary tree began with the description of percutaneous biliary drainage (PBD) by Molnar and Stockum in 1974.[1] Transhepatic needle cholangiography had been performed for years prior to this, but therapeutic measures had remained outside the scope of radiologists. In the past 15 years the application of percutaneous measures has changed dramatically and continues to do so in the face of information gained by clinical trials, the development of endoscopy, and research on shock-wave and laser lithotripsy of gallstones.

Obstructive jaundice can now be diagnosed rapidly by ultrasonography, and the level and cause of obstruction are often evident. When full morphologic definition of the biliary tree is needed, retrograde opacification by way of endoscopic cannulation has become the first-line procedure. Percutaneous transhepatic cholangiography (PTHC) is now usually reserved for failed endoscopy or for patients with previous gastrointestinal surgery whose altered anatomy makes endoscopic examination difficult.

Although PBD was initially proposed as a routine preoperative measure for those with severe obstructive jaundice, its application in such patients has been limited by improvements in preoperative patient preparation, surgical technique, and endoscopic biliary drainage. It is now mainly employed as a palliative measure in patients with unresectable obstructing malignancies, but continues to play a useful role in those with biliary obstruction and suppurative cholangitis, in postoperative or posttraumatic biliary leakage, and in selected patients with cholelithiasis. Percutaneous interventions in biliary stone disease may see an increase with the dissemination of new techniques for stone removal.

PATIENT PREPARATION

As a matter of course, the indications for the requested procedure are reviewed, the patient's medical history is obtained, and any previous studies are examined. Coagulation tests and hematocrit are ordered if not already available. Medical anticoagulation must be stopped. Impaired coagulation must be reversed by administration of vitamin K, platelets, or fresh frozen plasma, as needed. In the face of irreversible coagulopathy, ascites, or advanced cirrhosis, alternative diagnostic or therapeutic measures must be considered. When percutaneous intervention is deemed appropriate, the procedure, its indications, risks, and alternatives are explained to the patient, and consent is obtained.

Antibiotics

For needle cholangiography, PBD and other manipulations in the biliary tree, patients take nothing by mouth for at least 8 hours, and broad-spectrum antibiotics should be started *at least 1 hour prior to* the procedure. Infected bile is present in about one third of all patients with obstruction from malignant disease and in over two thirds of those with stones or benign strictures.[2, 3] Previous biliary surgery predisposes to bacterial colonization. Despite the best of precautions, percutaneous biliary drains render bile colonized in virtually all cases by 2 to 3 weeks.[4, 5] Simple tube cholangiography or mere tube changes can provoke life-threatening cholangitis and sepsis.[6, 7] Antibiotics are therefore essential for adequate patient preparation!

One common antibiotic regimen is the combination of ampicillin and an aminoglycoside. *Escherichia coli* and *Klebsiella*, *Streptococcus* and *Staphylococcus* species are commonly found at the time of biliary drainage, and coverage for gram-negative bacteria is necessary.[4] A single cephalosporin, such as cefazolin (1.0 gm given intramuscularly or intravenously) may be used.[8] If acute cholangitis is present at the time of planned intervention, cefoperazone may be more appropriate because of its wider spectrum and its excretion in the bile.[8] Anaerobes are also commonly present in patients with acute cholangitis, making the addition of metronidazole or clindamycin advisable.

Anesthesia

A prerequisite for any percutaneous procedure is adequate anesthesia. Much of the pain produced from percutaneous biliary manipulations appears to be mediated primarily by intercostal somatic nerves.[9] Visceral neural receptors are implicated in acute nausea from distention of the bile ducts by injection of contrast media or by the mechanical effects of wires and catheters, but they otherwise do not seem to contribute substantially to the pain from transhepatic procedures.

Generous local anesthesia with lidocaine, and intravenous administration of benzodiazepines and opiates (such as midazolam and fentanyl) are usually sufficient for simple diagnostic fine-needle cholangiography. In some patients this combination of medications may allow the insertion of drainage catheters or dilatation of strictures without undue discomfort. However, many drainage procedures or extended manipulations can be quite painful, and other measures are called for. Vogelzang and Nemcek have described several alternatives, including intercostal nerve block, and epidural anesthesia.[9] Pleural block is another anesthetic technique of demonstrated value in percutaneous biliary drainage.[10] Epidural anesthesia and pleural block are best administered and monitored by an anesthesiologist.

Intercostal Nerve Block

Intercostal block is performed after the approach for biliary drainage is chosen, with injection of a mixture of bupivacaine 0.75% and epinephrine 1:200,000 in the intercostal space being traversed, as well as in the intercostal spaces immediately cephalad and caudad.[9] Puncture is made with a 25-gauge needle adjacent to the inferior rib margin about 10 cm posterior to the catheter entry site. Approximately 5 mL of the local anesthetic is infiltrated about the neurovascular bundle, carefully avoiding intravascular injection. Local effectiveness persists for 8 to 12 hours, and the nerve block procedure may be repeated later in the day and on succeeding days for residual pain.[9]

Pleural Block

Injection of bupivacaine 0.5% through an epidural needle or catheter into the right pleural space has been found quite effective in preventing pain from percutaneous biliary drainage. In the method described by Rosenblatt et al., after administration of local anesthesia and with the patient in a left lateral decubitus position, an 18-gauge epidural needle is attached to a 5-mL air-filled glass syringe and the needle is advanced into the pleural space.[10] When the pleural space is entered, the resistance at the needle tip drops, and the plunger of the syringe falls. A total of 30 mL of bupivacaine is then injected. After instillation of the medication, the patient is turned into a right lateral decubitus position for 10 minutes.

The anesthetic diffuses through the parietal pleura to produce multiple intercostal nerve blocks. As with intercostal nerve block, pleural block effectiveness is checked by pinprick. When an epidural-type catheter is left in place, additional bupivacaine can be administered at any time. The anesthetic effect lasts 4 to 8 hours.[10] Because of the potential for toxicity (cardiac arrhythmia, hypotension, or seizures), an anesthesiologist should be present.

Epidural Anesthesia

An epidural catheter is placed in the mid-thoracic spine, and it may be left in place for hours to days. Injection of an opiate produces a sensory blockade, with relatively little effect on motor fibers. However, if the anesthetic level reaches higher than anticipated, respiratory arrest is a risk. Only a certified anesthesiologist should perform this technique.

PERCUTANEOUS TRANSHEPATIC CHOLANGIOGRAPHY

Indications

As noted earlier, PTHC is now often a procedure of last resort. When a contrast study of the biliary ducts is desired, it is more often accomplished endoscopically. However, one large category of patients for whom PTHC remains quite valuable is those with previous gastrointestinal surgery and endoscopically inaccessible bilioenteric anastomoses. Fine-needle cholangiography can uncover the cause of jaundice or other symptoms related to stones, benign strictures, malignant tumors, or parasitic infection. Also PTHC is highly accurate in enabling the diagnosis of malignant obstruction, as well as for determining the resectability of hilar neoplasms.[11, 12]

Technique

With the patient supine on a tilting table, fluoroscopy is used to examine the depth of the lateral costophrenic sulcus. A needle placed through the eighth intercostal space (or below) at the midaxillary line is unlikely to puncture lung in most patients, although punctures above the tenth rib will traverse the pleural space.[13] The level chosen will depend on individual factors, such as presence of obstructive lung disease and size of the liver. When the appropriate entry point is chosen, anesthesia is given. Any intravenous sedation should be carefully administered to avoid oversedation. The patient must be able to cooperate by suspending respiration as directed.

Although ultrasonic guidance is feasible, it is not routinely needed for diagnostic cholangiography, especially if intrahepatic ductal dilatation is present. Rather, fluoroscopy is used to observe advancement into the liver of the long (15-cm or 20-cm) 21-gauge needle. For practical purposes, aiming toward the pedicle of the 12th thoracic vertebral body while staying in the midaxillary plane will place the needle through hepatic parenchyma near the hilus. The needle is advanced in a smooth motion during suspended respiration until it nearly reaches midline. Advancement should be stopped if a course toward the diaphragm or toward other extrahepatic structures is noted. When redirection or repeat puncture is required, the needle must be withdrawn to a point near to the hepatic capsule, without actually removing it from the liver.

With the needle in position, the stylet is removed, and a 20-mL syringe and connecting tubing filled with dilute water-soluble contrast medium are connected to the hub. Small intermittent injections are then made through the needle as it is withdrawn a few millimeters at a time under continuous fluoroscopic observation. If parenchyma is injected, a small, irregular stain results. Different flow patterns are evident if the needle tip encounters hepatic artery or vein, portal vein, or lymphatic channels. The arteries and veins will characteristically exhibit a rapid "washout" of the contrast agent. Injection of a bile duct, on the other hand, will show contrast material easily flowing from the needle tip to persistently fill branching tubular structures.

As soon as biliary filling is recognized, it is sometimes useful to measure biliary pressure or to allow 10 to 20 mL of bile to drain before further injection. However, if it appears that the needle position is precarious or if the intrahepatic ducts are not dilated, it is best to proceed directly with cholangiography. Full-strength (≥300 mg iodine/mL) contrast material is now injected, in order to maximize the opacification obtained while keeping the injected volume to a minimum. Undiluted contrast material is not used for needle placement, because intense parenchymal staining from multiple passes may prevent recognition of ductal filling during a later needle pass.

Because of the risk of bacteremia and septic shock, only enough contrast material to establish the diagnosis should be injected. Spot films are exposed in multiple projections. Because iodinated contrast material is heavier than bile, gravity can be used to fill the common hepatic and common bile ducts. For this reason, a tilt table should always be used for PTHC. If the left hepatic ducts fill poorly, the patient may be placed prone to assist in opacification of these vessels.

If the gallbladder is inadvertently entered during needle placement, its contents should be aspirated as completely as possible. The needle can then be withdrawn and redirected. However, it is also possible to inject contrast medium to obtain a diagnostic cholecystogram and (if the cystic duct is unobstructed) a full cholangiogram. If this maneuver is elected, the gallbladder must again be drained before removal of the needle.

The success rate of PTHC can be anticipated to be between 95% and 100% in the presence of intrahepatic ductal dilation, and from 70% to 80% in the absence of dilation.[14, 15] Success is proportional to the number of needle passes made, and at least 12 to 15 passes are recommended before one abandons the procedure. Fortunately, such a number of passes is not accompanied by a higher complication rate, as long as needles of 20-gauge or smaller are employed.

Findings

Stones

Stones appear as persistent filling defects on multiple films, usually distinguishable from air bubbles by their irregular or facetted shapes. The effects of gravity may be helpful in difficult cases, but it is not always possible to make the distinction. Blood clots, polypoid tumors, and biliary debris can also cause diagnostic problems. Unlike stones or air bubbles, polypoid tumors are fixed in location and show mural attachment in at least one projection.

Neoplasms
Most tumors affecting the biliary tree are not polypoid, but produce obstruction by ductal compression (as by hilar nodes or hepatic metastases) or infiltration. Typical of a pancreatic or biliary carcinoma is a short and irregular stricture, often producing a rat-tail appearance with gross dilation of the proximal ducts. Adenomas and other benign neoplasms are quite rare, and any tumor encountered should be considered malignant unless proved otherwise by histologic study.

Focal and Diffuse Strictures
Ampullary stenosis or inflammatory stricture from pancreatitis may be difficult to separate from malignant tumor, and endoscopic or percutaneous needle biopsy is indicated in such cases. Postoperative strictures tend to be longer and smoother than those of malignant neoplasms. Many such strictures are accompanied by ductal calculi.

Diffuse intrahepatic and extrahepatic stricture may represent advanced sclerosing cholangitis or infiltrating cholangiocarcinoma. Biopsy sampling of such lesions is also warranted, particularly because sclerosing cholangitis may be complicated by carcinoma.[16] When diffuse narrowing is found confined to the intrahepatic ducts, primary biliary cirrhosis also becomes a consideration.

Primary Sclerosing Cholangitis
Sclerosing cholangitis has been said to invariably involve the extrahepatic ducts, but cases have been described in which changes are confined within the liver.[17, 18] Most patients, however, will have diffuse disease. Focal strictures tend to be short and alternate with normal-sized duct to produce "beading." About half have "shaggy" mural irregularities, and ductal bands and diverticula are specific to primary sclerosing cholangitis.[17] Patients developing a superimposed cholangiocarcinoma commonly show marked and progressive ductal dilation, progressive stricture, and intraductal masses.[16] Some of these findings become evident only with time.

Other Forms of Cholangitis
In acute suppurative cholangitis the ducts are filled with debris and there may be abscesses of various size in communication with the ducts. Chronic recurrent cholangitis tends to produce multiple branch strictures with proximal areas of dilation, often containing stones. There are similar findings in those with Caroli's disease, congenital dilation of the intrahepatic ducts, which may be a variant of congenital hepatic fibrosis – polycystic liver disease. Parasitic worms and flukes within the bile ducts appear as linear defects on cholangiography.

Risks

Fine-needle transhepatic cholangiography is accompanied by a 3% rate of serious complications,[15] most due to sepsis or postprocedural bile leakage and peritonitis. Major bleeding arises in only 0.3% of patients, and overall mortality has been reported as 0.2%.[15] Other hazards include vasovagal reactions and pneumothorax. The risk of hypersensitivity reaction to contrast material is always present, but low. Although multiple passes are not correlated with a higher number of complications, the total amount of contrast medium injected during needle passes should be watched carefully in patients with renal insufficiency. Use of 18-gauge sheath needles for diagnostic cholangiography should be avoided, for the number of complications and mortality risk increase up to sixfold.[15]

PERCUTANEOUS BILIARY DRAINAGE

Indications and Cautions

Percutaneous biliary drainage is a primary palliation for patients with unresectable malignant obstruction producing pruritis or infection. Some forms of chemotherapy are contraindicated in the presence of jaundice, and drainage may be used to prepare for treatment in selected cases.[19] The procedure also provides a route of administration for internal radiation therapy.[20] With these specific exceptions, PBD should *not* be performed in those with otherwise asymptomatic jaundice, because of the risks involved. Also, drainage must be avoided in those with diffuse hepatic metastases, liver failure, and a life expectancy measured in only days to weeks. Complication rates are considerably higher in such patients, cancelling any potential benefits.[2] Other relative contraindications to PBD are the presence of significant ascites, advanced cirrhosis, and impaired coagulation. The first two conditions pose great technical difficulties for catheter introduction, and all three increase the possibility of serious complication.

Percutaneous drainage is the best nonoperative approach for treating postoperative strictures that cannot be cannulated endoscopically. Balloon dilatation and stent placement can follow drainage. The same holds true for patients suffering from primary sclerosing cholangitis, who may have focal strictures responsible for recurrent episodes of infection.[21] When the nature of a stricture is in doubt, the percutaneous tract provides access for brush or needle biopsy, if cytologic examination of collected bile is nondiagnostic.[22, 23] Postoperative biloma with continuing leakage calls for percutaneous diversion of bile flow, in addition to drainage of the fluid collection itself.[24]

The role of PBD in the preoperative correction of jaundice has been debated from the onset. At present, available data indicate that PBD is *not* routinely indicated before biliary surgery, but it can be a helpful adjunct in certain situations (see the following section). Acute suppurative cholangitis is one such situation, because emergency surgery carries a high mortality rate, and PBD can reliably stabilize patients with sepsis or shock.[25–27]

Advances in extracorporeal and contact lithotripsy, and in solvent therapy for biliary stones, may expand the indications for percutaneous biliary catheter placement in the future.[28–30]

Preoperative Drainage

It had long been taught that severe obstructive jaundice poses high risk for surgery. Those with serum bilirubin levels above 10 mg/dL have suffered 15% to 25% surgical mortality and morbidity as high as 40% to 60%.[31] Jaundice, malignancy, and anemia have been demonstrated to be independent risk factors.[3] Moreover, glomerular filtration rate and renal function are adversely affected by jaundice, and the risk of postoperative renal failure is increased.[3, 32]

In the late 1970s and early 1980s multiple reports suggested improved surgical results with routine preoperative drainage.[33–35] However, these studies were in part skewed by selection consequent to PBD and cholangiography; patients found to be unresectable or dying during the preoperative period were excluded from subsequent surgery. Historical controls were misleading, so prospective and randomized studies were especially important. Hatfield et al. and Pitt et al. randomized 57 and 79 patients, respectively, either to PBD followed 11 to 12 days later by operation or to surgery alone.[31, 36] Both groups found that when the complications of PBD were included in analysis, there was no clear benefit to preoperative drainage.

Although percutaneous drainage is no longer considered a routine preoperative measure, its benefit in the face of active cholangitis has already been mentioned. Huang and Ker treated 41 patients by PBD, including ten in septic shock, without a single death.[27] Immediate surgery in such circumstances is associated with mortality of 15% to 64%.[27] In Klatskin tumors (cholangiocarcinoma involving the confluence of ducts), a percutaneously placed drain can guide surgical resection. There are limited well-controlled reports suggesting that routine preoperative biliary decompression by endoscopic means *can* decrease overall mortality and morbidity, by virtue of a lower complication rate from endoscopic drainage.[37, 38] Problems with sepsis and bleeding are decreased, and the very low risk of tumor seeding of the transhepatic drainage tract is avoided.

A point needing emphasis is that preoperative cholangiography (whether percutaneous or endoscopic) can accurately predict which biliary tumors are unresectable.[11] Many patients found not to be candidates for curative surgery may be spared an operation as long as a drainage catheter can be placed. Up to 80% to 90% of patients submitting to surgery for malignant obstructive jaundice are treated by palliative bilioenteric bypass alone when the true extent of tumor is appreciated.[11, 34]

Palliative Drainage for Tumor

The established indications for nonsurgical palliative treatment of malignant neoplasms are in flux, with hepatic transplantation now being offered to some patients, and aggressive resection or drainage for even deeply infiltrating carcinomas to others. Some surgeons have railed against PTHC or PBD for hilar tumors, citing anecdotal evidence supporting peripheral bilioenteric surgical drainage.[39] Lacking in their appraisal is any unbiased statement of the short-term and long-term problems with such operations.[6]

Despite recent changes in approach, there remain patients suffering debilitating pruritis or infectious complications of biliary obstruction who are not considered suitable for surgical palliation. Among factors to be considered before choosing PBD are whether the risks of drainage outweigh the likely palliation, the ability of the patient or those supplying chronic care to deal adequately with an internal-external catheter, and the patient's psychologic capacity to adjust to such a tube. An endoprosthesis may be an alternative for those experiencing problems with tube acceptance and care (see "Endoprostheses," later in this chapter). If experienced operators are at hand, endoscopic endoprosthesis placement may be warranted as the first measure.[38, 40] Percutaneous drainage may be used alone or in collaboration with endoscopy when endoscopic methods fail, or it may be the initial intervention for hilar tumors.[40–42]

When PBD is used for palliation, problems with catheter plugging, displacement, external bile leakage, and cholangitis are to be anticipated. By the nature of the patients being treated, complications and early mortality are higher. Hilar tumors often cause separate obstruction of right and left hepatic ducts. Although drainage of only one side may reverse jaundice, undrained segments are likely to become infected, and bilateral drainage is recommended.[41, 43] Slow-growing tumors, such as cholangiocarcinoma, may require multiple tube changes over the course of months to years, and endoprostheses are best avoided in such cases.

Technique

Patient preparation with antibiotics and local, epidural, or other regional anesthesia is identical to that described for PTHC. Fine-needle cholangiography is performed in a standard fashion, but whenever PBD is planned, greater care should be taken to avoid traversing the pleural space. When a catheter is placed above the tenth rib at the midaxillary line, the pleura are almost invariably crossed.[13] If a catheter can be directed into the liver from a lower approach, uncommon but very serious pleural complications may be prevented. However, when the diaphragm is elevated or if the liver is small, crossing the right pleural reflection may be unavoidable.

Catheter Introduction Using a Fine Needle

In many cases the position of the cholangiographic needle may be suitable for placement of a fine mandril wire into the biliary system. The wire directs an exchange dilator mounted on a stiffening cannula into the ducts. The dilator and stiffener are advanced over the wire until the point of ductal entry is reached. The wire and cannula are then held stationary as the dilator is passed well into a major duct. A heavy-duty 0.035- or 0.038-inch guidewire is then used for further manipulations. Several coaxial introducers that permit a second guide-wire to be passed through an endhole sheath are available from various manufacturers. When the Cope catheter introduction system is used, the larger wire must exit a sidehole near the tip of the dilator, and a J-tip guidewire is required for this maneuver. If the wire supplied with this kit does not exit the sidehole, a 1.5-mm J Rosen wire is recommended.

If the cholangiogram needle enters a left-sided bile duct, makes an acute or awkward angle of entry, enters very near the point of obstruction, or has engaged an extrahepatic duct, it should *not* be used for drain placement. Instead, the needle should be left in place and contrast medium injected to direct a second needle toward a more appropriate spot. Should the gallbladder be inadvertently entered, contrast material may be injected for opacification of intrahepatic bile ducts, guiding conventional percutaneous transhepatic catheter placement.[44] The best method for directing a needle toward an opacified duct is to use a fluoroscopic C-arm to perfectly superimpose the needle (looking "down the barrel") upon the target. The C-arm is then rotated 90° and the needle advanced to the proper depth. With fluoroscopy one can often observe the duct move or recoil when entered by the needle.

Sheath Needles

A 5-F Teflon sheath needle can be used for biliary access instead of the 21-gauge needle in the technique just described. It has the advantage of allowing immediate placement of a heavy guidewire without dilator exchanges. A larger needle can also be more easily directed toward an opacified duct. In a hard, cirrhotic liver, a sheath needle may be the only alternative for successful percutaneous drainage. However, a larger needle increases the risk for significant bleeding or bile leakage, and its routine use is not recommended.

Passage of the Obstruction

Once a duct has been securely cannulated the obstruction may be gently probed with a straight or curved-tip catheter and soft-tipped guidewire (Fig 20–1). Steerable wires or "slippery" wires with a hydrophilic polymer coating are especially useful for passing obstructions. The lumen can usually be engaged in even extremely tight lesions. However, if the patient has long-standing distal common bile duct obstruction and the extrahepatic duct is markedly dilated, both wire and catheter may merely curl within the obstructed duct. For these patients a repeat attempt to pass the obstructing lesion frequently succeeds after several days of external drainage of bile.

When a guidewire passes through the common bile duct and into the duodenum, a straight-tapered 5- or 6-F catheter should be passed over it, well beyond the lesion. The initial guidewire is then replaced with a heavier one, such as an Amplatz Super-Stiff or Coons wire. Teflon dilators are advanced through the transhepatic tract and stricture or tumor. In very tight lesions, a tapered van Andel catheter may be introduced and then replaced with a balloon catheter for dilatation. Pigtail-tip, 8- to 10-F Teflon multisidehole catheters, such as the Ring biliary catheter, are easier to introduce at the time of initial passage. Sideholes should be present both proximal to the obstruction and distally within duodenum. There should be no sideholes in hepatic parenchyma. After the tract has been allowed to dilate and mature around this catheter for several days, a larger and softer tube can be introduced. Soft catheters are prone to "accordion" when they are advanced against heavy resistance.

Internal Drainage

Passage beyond the tumor or stenosis is desirable for several reasons. First, internal drainage of bile is made possible, eliminating the need for the patient to wear a bile bag, while preserving the enterohepatic circulation of bile salts and digestive physiology. However, if catheter passage is not possible, it is not normally necessary to replace bile salts orally. What is needed is fluid replacement, should the patient experience biliorrhea while on external drainage.

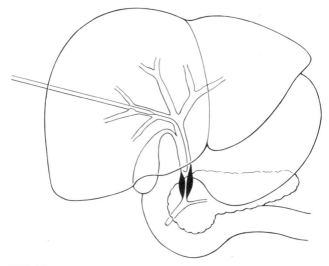

FIG 20–1.
Transhepatic catheter placement into a right intrahepatic duct with guidewire advancement to the point of distal common bile duct obstruction. (Modified from Cope C, Burke DR, Meranze S: *Atlas of Interventional Radiology.* New York, JB Lippincott, 1989.)

A catheter placed into duodenum is also less likely to be inadvertently displaced or removed than one with its tip in the biliary tree. A final advantage of internal drainage is that the full extent of an obstructing tumor is better appreciated by injection of a catheter traversing it.

Securing and Caring for the Catheter
Proper securing of the catheter is a critical step in percutaneous biliary drainage. Catheter displacement can be fatal![45, 46] Not only is emergency replacement mandatory, it is frequently more difficult than the initial procedure. Introduction of a wire through the tract is rarely successful after the drain has been removed, and the bile ducts are decompressed by free leakage into the peritoneum or pleural space. For this reason catheters are best secured both internally and externally.

Internal fixation is provided by a loop-retaining suture in a Cope drainage catheter. The loop will not allow withdrawal unless the suture is released, cut, or broken. Other catheter designs allowing positive retention are also available. For external fixation, two sutures are placed at the catheter entry site. An alternative to skin sutures is placement of an adhesive stoma ring over the site, with several sutures securing the catheter to the plastic ring. Wrapping a 6- to 8-cm length of cloth adhesive tape about a (meticulously dried) segment of the catheter, forming projecting tags or "wings" with the tape, allows sutures to be passed through the tape and looped about the catheter multiple times. This procedure makes catheter crimping, constriction, and other accidental injury less likely. Additional adhesive, such as a methylmethacrylate glue, prevents the tape from slipping when wet. Finally, a gauze dressing is placed over the skin entry site.

After initial catheter placement, drainage is best left to an external bag for at least a few days, even if the obstruction has been passed. Catheter output must be carefully monitored, and any fluid loss over 500 mL in 24 hours must be replaced with intravenous electrolytes. External drainage assures biliary decompression and resolution of any cholangitis, while tube patency is continually confirmed by maintained output. After several days an internal-external catheter may be capped for internal drainage as long as the patient is afebrile, the bile is clear and free of major debris, and there are no major abdominal symptoms to confuse clinical assessment of catheter-related problems.

A PBD catheter should be flushed daily with 5 to 10 mL of sterile saline. The patient or home-care provider must be instructed in clean technique and provided materials for this routine. A phone number arranging for emergency attention for any catheter-related problems should be given the patient before hospital discharge. Routine catheter changes are scheduled as an outpatient procedure at 4- to 6-week intervals. Patients may be prescribed oral antibiotics to be started 1 day before routine tube changes.

Problems to Watch For

As already mentioned, the catheter must be firmly secured to prevent displacement. Hemobilia often is noted in the first days after PBD, but clots in the bile ducts will lyse without intervention, as long as the catheter is maintained patent by regular flushing and drains externally. If hemobilia persists or recurs, catheter position must be checked to ensure that a sidehole is not located in parenchyma. Major bleeding is sometimes tamponaded by exchange for a larger catheter, but if it is refractory to conservative measures, arteriography is indicated. Mitchell et al. found severe hemobilia complicating 3% of percutaneous biliary drainages, and they were able to control most by catheter embolization of the affected vessel (see Chapter 12, Embolotherapy).[47] Arteriovenous fistulas and pseudoaneurysms are not rare after PBD, but they do not usually produce symptoms.

In the first days after drainage a small number of patients experience biliorrhea, with a daily output of 1 to 8 L. This is especially dangerous if it is not recognized early and fluid replacement falls short. Improper management has led to dehydration, renal failure, and death.[36, 45]

Patients showing signs of cholangitis should have their tubes opened to an external bag. If there is adequate flow of clear bile, a tube cholangiogram and ultrasound are indicated to uncover the source of any segmental biliary blockage or hepatic abscess. Such lesions will need separate drainage, except for the case of multiple segmental blocks in a patient suffering the terminal throes of cancer.

Leakage of bile or fluid from the skin entry site can present recurrent difficulties. If the patient has substantial ascites, leakage may be irremedial, and a stoma device and collection bag may need to be placed about the catheter. If bile leaks back along the catheter, tube patency must be checked. Placement of a larger catheter sometimes resolves the problem. However, there are certain patients with increased duodenal pressure leading to functional biliary obstruction or back-drainage of enteric contents. Internal catheter drainage may not be possible in this situation. Prior gastrointestinal surgery seems to predispose to duodenal pressure elevation.[45]

Left-Sided and Bilateral Drainage

Percutaneous drainage through the left lobe is indicated in the presence of separate ductal obstruction. However, there are advantages to a left-sided approach even for distal biliary obstructions. Subxiphoid catheter placement avoids the pleural space and its attendant risks. The left-sided approach is often less painful. In hilar obstructions, the left lobe drainage is more likely to remain intact, because the main hepatic duct is longer than its counterpart on the right. For drainage to succeed, the left lobe must be of sufficient size, it must be accessible, and at least 10 cm of dilated duct should be present.[9] The left ducts may be opacified by placing a fine needle from the right side in the standard fashion, or ultrasound may be used to direct subxiphoid needle placement.

Patients suffering hilar ductal obstruction may need bilateral percutaneous drainage. This generally means separate entry sites, but one may be able to pass a catheter across an obstructed ductal confluence, from right lobe to left, or vice versa. If this is possible, I have found that a coaxial tube arrangement—with a central 7-F catheter passing through a custom-cut sidehole to the opposite duct and with the larger 12-F outer catheter passing into duodenum in a conventional course—provides acceptable drainage. An O-ring and Y-adaptor allow each tube to be flushed separately.

Risks of Percutaneous Biliary Drainage

Among the possible complications of PBD are tube obstruction, displacement, cholangitis, sepsis, bile peritonitis, hemobilia or other bleeding. If the pleural space is crossed, patients are at risk for pneumothorax, malignant effusion, and a biliopleural or biliobronchial fistula.[48] Serum amylase elevation has been noted in up to 9% of patients, but clinical pancreatitis is, fortunately, rare.[47] Even so, tube-induced pancreatitis can be severe and fatal.[49] Tumor seeding of the drainage tract has been an uncommon problem, complicating fewer than 1% of drainages, but in individual patients it can produce serious morbidity.[50, 51]

As already mentioned, unusually high flow of externally draining bile can precipitate dehydration, hypotension, and renal failure.[36] Average tube output from a completely obstructed biliary system is under 600 mL/day.[33] Any volume exceeding this should be replaced with intravenous electrolytes. In properly selected patients, more than 80% will respond to drainage with a significant drop in serum bilirubin, and at least a 50% decrease in bilirubin is to be expected within 10 days.[43] However, supervening liver failure will negate any benefits of PBD.

Quoted incidence of complications and mortality from PBD vary widely, but it is clear that risks of the procedure are substantially higher in patients with malignant disease.[2, 7, 52] Whereas 30-day mortality in those treated with palliative drainage for cancer ranges from 22% to 43%, many series have reported no deaths when PBD was done for benign obstruction.[2, 5, 33, 52–54] Deaths directly attributable to percutaneous drainage have been cited in 0.7% to 4% of patients.[2, 26, 43, 50, 52, 53]

Taking both early and late complications into account, their incidence in patients with malignant tumors may be as high as 54% to 69%, mostly from minor and severe episodes of cholangitis.[2, 7, 50] Still, this is several times the incidence of cholangitis in patients with benign strictures or stones. The difference may seem paradoxic in that at the time of drainage, bacteria are more often cultured from the bile of those with benign lesions than from patients with carcinomas. However, external biliary catheters become uniformly colonized within several weeks, and patients with malignant tumors often have impaired immunologic function. Curiously, Carrasco et al. have reported more problems with cholangitis in those patients whose catheters were kept closed to internal drainage than those on external drainage.[45]

In order to put the complications of patients with PBD for malignant disease in perspective, those of surgical intervention must be taken into account. A recent review by Lai et al. of 97 patients with proximal biliary carcinoma showed that nearly 80% of tumors were grossly unresectable at operation.[55] Of all patients treated surgically, 26% did not survive their initial hospitalization, and the incidence of postoperative cholangitis varied from 15% to 37% depending on the procedure performed. Morbidity and mortality were considerably higher for patients undergoing hepatic resection. Excluding hospital deaths, mean patient survival was slightly longer than one year. Palliative intubation alone, without any attempt at resection, is recommended for those with tumor found unresectable at surgery.[55]

RELATED BILIARY INTERVENTIONS

Once percutaneous access to the biliary ducts could be established, it was not long before methods were developed for treating stones, strictures, and malignancy, as well as for obtaining additional diagnostic information through biopsies and perfusion studies.

Endoprostheses

Endoprostheses are designed to provide completely internal stenting of biliary strictures. They are most frequently applied in terminal palliative care. Because of the potential for stent occlusion over the long term, they are not indicated for benign strictures. The advantages of endoprostheses include absence of an external tube or drainage bag (an overriding psychological factor in some patients), avoidance of entry site pain and bile leakage, and perhaps some protection against recurrent cholangitis. Internal stents should be placed only if serum bilirubin levels fall substantially after PBD.

Although Szabo et al. did not observe any significant difference in the incidence of cholangitis between those with biliary endoprostheses and those with internal-extenal drains (33% and 38%, respectively), they did find that episodes arose much later after placement (at a mean of 4 months versus 6 weeks for conventional percutaneous tubes).[56] In the case of high common bile duct obstruction, suprapapillary placement of the distal tip has been advocated as a means of decreasing the risk of cholangitis. However, a protective effect has not been observed universally by those investigators leaving endoprostheses completely above the sphincter of Oddi.[57]

Tube patency can be related to diameter, and larger stents may be placed percutaneously than by unaided endoscopy. Stents up to 14 F in size can be introduced percutaneously, but this requires at least several days for tract maturation followed by serial dilations of the tract, a procedure which can be quite painful. Endoscopic stenting is generally limited to introduction of catheters no larger than 10 F, but Kerlan et al. have described a combined percutaneous-endoscopic technique allowing peroral placement of large-bore stents in a single session.[58] By grasping a long guidewire placed percutaneously into duodenum, an endoscope operator pulls the wire back through the mouth. The large stent and a pusher are loaded on from above, while a balloon catheter helps pull the stent into its final position.

Overall, percutaneous stent placement succeeds in over 90% of patients for whom it is attempted.[59, 60] Tubes 12 F or larger are usually patent at least 6 months.[19] Reported rates of obstruction have varied from 1% to 23%.[59–61] Still, one of the great disadvantages of internal stents is that their replacement requires *de novo* percutaneous biliary drainage or endoscopy. Endoprosthesis occlusion does not necessarily mean replacement is warranted. With renewed percutaneous access, a brush or dilator passed into the tube may reopen it, allowing months of subsequent effective drainage.[62]

Whenever a stent is placed, whether solely percutaneously or combined with endoscopy, a drainage catheter should be left above it for several days to ensure that blood or debris do not immediately occlude the endoprosthesis. A "safety" wire, a second guidewire within the biliary ducts, is advised for secure performance of the maneuvers necessary for stent insertion. A coaxial Lieberman sheath is advanced over the first wire, the inner dilator removed, and the second wire placed alongside the first before further manipulations are undertaken. If bilateral endoprostheses are desired for high biliary obstruction, both stents must be placed simultaneously, otherwise insertion of the second may displace the first!

Numerous different endoprosthesis designs have been tried. Some are straight catheters with a tapered leading end, and a trailing retaining suture, and plastic "button" to be affixed subcutaneously.[61] Others have Malecot "mushrooms," barbs, or other internal fixation strategies. Some are meant to remain completely above the papillary sphincter. Problems with migration have been described in 3% to 16%, depending on the type of endoprosthesis placed and the experience of the radiologist.[57, 59–61] If migration occurs early, it may be remedied by use of a deflecting wire or balloon catheter.[61] When a balloon catheter is used to reposition or remove an endoprothesis, the balloon is passed into the lumen of the device and inflated until the balloon catheter and stent move securely as a unit.

The application of internal stents depends on patient and physician preference, and the expected life-span of the patient. For unresectable pancreatic tumors, stents will often remain open for the patient's remaining few months. However, for slow-growing tumors, such as cholangiocarcinoma, it is best to maintain external access and regularly change an internal-external drain. It should be remembered that stent introduction may be a difficult and time-consuming procedure, and considerable fluoroscopic exposure is common.[60] Some of these problems may be overcome by the development of expandable metallic meshes, similar to the endoprostheses designed for vascular lesions. Such internal stents may attain a diameter of 1 cm (30 F).[63] If long-term patency is reliably maintained in clinical trials, expandable endoprostheses might be indicated for benign biliary strictures.

Biliary Cytology and Biopsy Procedures

When a biliary obstruction appears malignant but no diagnosis has been established, the percutaneous tract can be used for cytologic or histologic confirmation. Bile collected from an external drainage bag is more likely to yield positive cytologic evidence than fluid obtained at PTHC alone.[40] Cytologic study is positive in roughly 50% of patients with cholangiocarcinoma, although reported sensitivity varies widely.[11, 40, 64] The positive cytology rate is lower with pancreatic primary tumors.[43]

As with other diagnostic biopsy procedures, persistence produces greater accuracy. If cytologic examination is negative, percutaneous needle biopsy can be guided by tube cholangiography. Alternative approaches include brush, forceps, or needle biopsy through a transhepatic sheath.[23] Repeated biopsies by various methods can attain a sensitivity of 92% for biliary carcinoma and 73% for pancreatic carcinoma.[22]

Internal Irradiation of Biliary Neoplasms

Percutaneous biliary drainage provides a route for placement of radioisotopes within or adjacent to an obstructing tumor. Very high doses of local radiation can be administered by means of Iridium-192 seeds with little exposure of liver beyond the neoplasm. Survival of patients with cholangiocarcinoma has been extended to a mean of 17 months from diagnosis, compared with 3 months for patients with untreated tumors.[20] These data are rather difficult to interpret, because the simple act of palliative drainage can produce similar prolongation of survival. Also, patients treated with internal radiation therapy are subject to episodes of severe cholangitis, periductal abscess, and hemobilia.[65] A potential alternative to ionizing radiation for such tumors is the application of directed laser energy, a treatment deserving further investigation.[66]

Percutaneous Cholecystostomy

Percutaneous cholecystostomy is a means of treating patients with acute cholecystitis who present with high surgical risk. Needle puncture is directed by ultrasonography, and liver parenchyma should be traversed in order to enter the gallbladder at its hepatic attachment. An 8-F self-retaining catheter is then placed over a guidewire after suitable tract dilatation. Acute inflammation is safely resolved by this technique in well over 90% of patients.[44, 67, 68] Acalculous cholecystitis, which often arises in critically ill patients, can be effectively managed. Percutaneous cholecystostomy can even be achieved at the patient's bedside when needed. Catheters are removed after symptoms have resolved and the cystic duct is found to be unobstructed. All bile is aspirated from the gallbladder before catheter removal. When stones are present, various means for their removal can be applied through the tract, although stone removal is not absolutely necessary in the absence of persistent symptoms.

A limitation of percutaneous cholecystostomy is that gallbladder necrosis cannot be recognized, except by failure of prompt improvement in the patient's clinical status.[44] VanSonnenberg et al. reported a high incidence of severe vasovagal reactions during gallbladder drainage[69]; however, other investigators have not found this to be a problem. Accidental early removal of a catheter has led to fatal peritonitis in one reported case.[70]

Gallbladder puncture can also serve as an alternative to transhepatic PBD in selected patients, as long as the cystic duct is patent. Even if drainage is not planned by this route, contrast medium injection into the gallbladder (after initial evacuation of bile) often provides excellent opacification of intrahepatic and extrahepatic ducts. A conventional transhepatic drain can subsequently be placed under fluoroscopic guidance.

PERCUTANEOUS INTERVENTION IN BENIGN BILIARY DISEASE

Aside from cases of acute suppurative cholangitis, PBD is not normally indicated for patients with benign disease. There are exceptions, and each case must be assessed individually. Percutaneous access is invaluable in cases of retained common bile duct stones after cholecystectomy. Stones may be removed through a T-tube tract if they do not pass through an endoscopic sphincterotomy. Postoperative biliary strictures or fistulas can be treated nonoperatively by dilation or intubation. Sclerosing cholangitis presents its own set of chronic and recurrent challenges that may be addressed percutaneously. We have successfully dilated strictures and removed stones in patients with hepatic transplants, a whole new population for which percutaneous methods are preferable to reoperation. Beyond this, the place of nonsurgical stone removal, in any or all of its novel manifestations, remains to be established. Percutaneous access may play an important adjunct role for at least some of these patients.

Strictures

There is no strict consensus on which benign strictures should undergo balloon dilatation. When strictures are dilated, there is no standard on the number of balloon dilations, the duration of balloon inflation, or the duration of postprocedural stenting. Because most common duct strictures are iatrogenic and respond to operative repair, it is prudent to reserve percutaneous methods for those with contraindications to surgery or failed repeat operation.

The response of any individual lesion is difficult to predict. Choledochoenteric anastomotic strictures have shown the best results in some series.[19, 71] This has not held true for others.[72] Nevertheless, it is evident that focal lesions in primary sclerosing cholangitis are more likely to recur (see the following section). Balloon size should match that expected for a normal duct. Although very long and repeated inflations (up to an hour at a time) have been used, immediate success can be expected with shorter duration inflations in most lesions.[72, 73] Very tough strictures may need passage of coaxial Teflon dilators to 12 or 14 F. Stents are left across treated lesions from weeks to months. Bret et al. advocate soft silicone stents up to 18 F in size.[74] Expandable endoprostheses, although clinically unproved, may prevent stricture recurrence in the future.[63]

Success of balloon dilatation for benign strictures is about 70% at 1 to 2 years of follow-up.[72] Longer term follow-up is essential, because surgical experience has shown that strictures can recur many years after treatment.

Interventions in Sclerosing Cholangitis

Primary sclerosing cholangitis, a condition associated with ulcerative colitis and most commonly affecting young and middle-aged men, is characterized by extensive submucosal fibrosis. This progressive disease leads to biliary cirrhosis and hepatic failure, with over one third of patients dying within 7 years of diagnosis.[75] Conventional surgical bypass results have been generally disappointing, and hepatic transplantation is being aggressively performed in some centers.

Percutaneous drainage and dilatation can provide palliation for many with this disease. Focal strictures producing jaundice, pruritis, and recurrent cholangitis can be ameliorated. Those who have had such symptoms longer than 6 months are less likely to respond to balloon dilatation and stenting.[21] Even patients with a good response usually have recurrent symptoms within 2 years.[75]

Because of the long-term problems of sclerosing cholangitis, Russell et al. have formulated a combined surgical-radiologic approach.[76] A side-to-side biliary jejunal anastomosis is created by a Roux-en-Y limb, with jejunum tacked subcutaneously and marked by surgical clips. Access to the limb and biliary tree is obtained by percutaneous puncture at this site. Multiple balloon dilatations of biliary strictures can be performed at 2- and 6-month intervals in this fashion.[76] An unresolved issue is how such surgery might affect future transplantation should the patient's disease progress to hepatic failure or cholangiocarcinoma.[75]

Biliary Perfusion Studies

Occasionally, the significance of ductal dilation may not be clear, or partial biliary obstruction may be suspected in the absence of dilated ducts. In analogy to the Whitaker uroperfusion test for hydronephrosis, vanSonnenberg et al. developed a perfusion test for detecting partial biliary obstruction.[77] With the use of a manometer, perfusion pump, and a three-way stopcock, biliary pressure is measured after 5 minutes of dilute contrast medium infusion through a biliary drain or needle. The initial infusion rate is 2 mL/min, increased to 4 mL/min and 8 mL/min if obstruction is not documented at the lower rates. If pressure readings remain normal, subsequent measurements are obtained after 3 minutes of 15 mL/min and 2 minutes of 19 mL/min as tolerated. A study is considered abnormal if biliary pressure exceeds 20 cm of normal saline at any infusion rate. Biliary manometry and perfusion are helpful for uncovering subtle ampullary stenosis, as well as for assessing the results of stricture dilatation.[77]

Biliary Stones

Cholesterol gallstones form as a consequence of stasis and supersaturation of bile by cholesterol. "Pigment" stones contain calcium salts of carbonate, bilirubin, phosphate, and alkanoate.[78] Standard treatment for stone disease has been cholecystectomy and common bile duct exploration. Development of catheter and basket techniques of extracting stones through T-tube tracts have removed the need for a second operation in patients found to have retained ductal stones.[79] Endoscopic sphincterotomy and manipulations are also used to treat such patients with a high degree of success.

Recently, great interest has been generated by extracorporeal shockwave lithotripsy (ESWL) treatment of gallstones, alone or in combination with oral medical therapy.[28, 80, 81] Mechanical, ultrasonic, and laser lithotripsy, as well as solvent infusions, have been performed through cholecystostomies.[29, 30, 82–84] However, one third of patients will have recurrent stones within 4 years if the gallbladder is not removed.[85] Ablation of the gallbladder by sclerosing agents or other means, or long-term adjuvant medical therapy, may be needed before any nonsurgical treatment of gallbladder stones can become the standard.

T-Tube Manipulations

Burhenne pioneered the removal of retained stones through T-tube tracts.[79] The tract must be allowed to mature at least 5 weeks after surgery, and longer if the tube is smaller than 14 F. Cholangiography is performed after removal of the T-tube, and a sheath is introduced. A stone basket is advanced beyond the stone before being allowed to open. The basket is withdrawn and rotated to trap the stone within. If the stone is 8 mm in diameter or smaller, the basket and sheath are completely withdrawn in a smooth motion *without* closing the basket. Closing the basket often causes fragmentation, a desirable result if a large stone is snared, but undesirable for small calculi. The procedure is repeated as many times as necessary. Stones less than 3 mm in diameter should pass spontaneously into the duodenum if no stricture is present. In difficult cases steerable catheters, balloon catheters, safety wires, and flushing are employed.

Multiple sessions are often needed to render patients free of stone. A catheter the size of the tract should be left in place after each session, with removal only after follow-up cholangiography shows no residual stone or obstruction. With experience, success of T-tube tract extraction can be as high as 95%, with complications arising in only 4%[79].

Percutaneous Biliary Drainage for Stone Removal

Berkman et al. have treated common bile duct stones by percutaneous transhepatic catheter placement, followed several days later by balloon dilatation of the ampulla of Vater.[86] They resorted to balloons only slightly larger than the largest stone present. After the dilation procedure, stones were pushed through the ampulla into the duodenum by a balloon catheter. Of the 17 patients so treated, 11 had stones larger than 10 mm in diameter, but all were successfully removed without mortality or significant complications.[86]

Perfusion of Stones With Solvents

While oral chenodeoxycholic and ursodeoxycholic acid may dissolve gallstones, months to years of treatment are required for complete removal of stones, and only a minority of patients are suitable for such treatment.[78] Solvents perfused directly into the biliary ducts have been applied as a more rapid alternative for selected patients. The agents most commonly studied, glycerol-mono-octanoin (monooctanoin) and methyltertiary butyl ether (MTBE), both remain investigational and require an Investigation Device Exemption for use in the United States. No matter which solvent is used, pigment stones and stones showing calcification are refractory to dissolution attempts.

Monooctanoin is a very viscous fluid that must be gently heated during infusion. Outflow through a patent duct or sump catheter must be documented before starting treatment, and an overflow valve must be included in the line to prevent a pressure of 30 cm of saline from being exceeded.[87] Infusion for days to weeks is required for even modest success.

The agent MTBE represents a less tedious alternative but has its own peculiar hazards. It has a fairly high boiling point for an ether (55°C), but like all ethers it must be used with meticulous ventilation and with respect for its explosive potential![88] Percutaneous cholecystostomy has been used to deliver the agent for treatment of gallbladder stones in patients refusing or at high risk for surgery. In the study of Thistle et al., 32 of 75 patients were rendered stone-free at last follow-up, and mean duration of infusion was nearly 13 hours.[30] Disadvantages of MTBE infusion include the necessity for manual instillation and aspiration of fluid over such a protracted period, sedation occurring with systemic absorption, and the risk of duodenitis and intravascular hemolysis. The efficacy of MTBE treatment may be enhanced with stone fragmentation, which increases the surface to volume ratio.[29]

Biliary Lithotripsy

Presently the hottest topic in gallstone therapy, ESWL has been applied to both gallbladder stones and common duct calculi.[28, 80] When the gallbladder is treated, common duct patency must be demonstrated by opacification on oral cholecystogram. In a series of 175 carefully selected patients, Sackmann et al. were able to reduce stones to fragments no larger than 3 mm in all but 20%, and with adjuvant oral medication, 91% of patients had no demonstrable stones at 18 months.[28] Stones with calcified rims are amenable to ESWL therapy. Second-generation lithotripsy devices can treat patients without the need for anesthesia or sedation.[89] Bland et al. have had 74% success in clearing common duct stones by ESWL combined with endoscopic or percutaneous drainage.[28]

At the University of Wisconsin-Madison, we have used electrohydraulic and ultrasonic lithotripsy to fragment stones in selected patients, including one woman with a transplanted liver.[84] However, the best prospect for future contact lithotripsy is with lasers, and preliminary studies are promising.[29, 90] A successful system should include a small-caliber steerable catheter that can be passed transhepatically through a sheath.

REFERENCES

1. Molnar W, Stockum AE: Relief of obstructive jaundice through percutaneous transhepatic catheter: A new therapeutic method. *AJR* 1974; 122:356–367.
2. Yee ACN, Ho C-S: Complications of percutaneous biliary drainage: Benign versus malignant diseases. *AJR* 1987; 148:1207–1209.
3. Greig JD, Krukowski ZH, Matheson NA: Surgical morbidity and mortality in one hundred and twenty-nine patients with obstructive jaundice. *Br J Surg* 1988; 75:216–219.
4. Audisio RA, Bozzetti F, Severini A, et al: The occurrence of cholangitis after percutaneous biliary drainage: Evaluation of some risk factors. *Surgery* 1988; 103:507–512.
5. Olak J, Stein LA, Meakins JL: Palliative transhepatic biliary drainage: Assessment of morbidity and mortality. *Can J Surg* 1986; 20:243–246.
6. Choi TK, Fan ST, Lai ECS, et al: Malignant hilar biliary obstruction treated by segmental bilioenteric anastomosis. *Surgery* 1988; 104:525–529.

7. Cohan RH, Illescas FF, Saeed M, et al: Infectious complications of percutaneous biliary drainage. *Invest Radiol* 1986; 21:705–709.

8. Spies JB, Rosen RJ, Lebowitz AS: Antibiotic prophylaxis in vascular and interventional radiology: A rational approach. *Radiology* 1988; 166:381–387.

9. Vogelzang RL, Nemcek AA Jr: Toward painless percutaneous biliary procedures: New strategies and alternatives. *J Intervent Radiol* 1988; 3:131–134.

10. Rosenblatt M, Robalino J, Bergman A, et al: Pleural block: Technique for regional anesthesia during percutaneous hepatobiliary drainage. *Radiology* 1989; 172:279–280.

11. Lyn RB, Wilson JAP, Cho KJ: Cholangiocarcinoma: Role of percutaneous transhepatic cholangiography in determination of resectability. *Dig Dis Sci* 1988; 33:587–591.

12. Gibby DG, Hanks JB, Wanebo HJ, et al: Bile duct carcinoma: Diagnosis and treatment. *Ann Surg* 1985; 202:139–144.

13. Neff CC, Mueller PR, Ferrucci JT Jr, et al: Serious complications following transgression of the pleural space in drainage procedures. *Radiology* 1984; 152:335–341.

14. Keller FS, Katon RM, Dotter CT, et al: Fine needle cholangiography (FNC) in the nonjaundiced patient. *J Clin Gastroenterol* 1979; 1:125–129.

15. Harbin WP, Mueller PR, Ferrucci JT Jr: Transhepatic cholangiography: Complications and use patterns of the fine-needle technique. *Radiology* 1980; 135:15–22.

16. MacCarty RL, LaRusso NF, May GR, et al: Cholangiocarcinoma complicating primary sclerosing cholangitis: Cholangiographic appearances. *Radiology* 1985; 156:43–46.

17. MacCarty RL, LaRusso NF, Wiesner RH, et al: Primary sclerosing cholangitis: Findings on cholangiography and pancreatography. *Radiology* 1983; 149:39–44.

18. Li-Yeng C, Goldberg HI: Sclerosing cholangitis: Broad spectrum of radiographic features. *Gastrointest Radiol* 1984; 9:39–47.

19. Ring EJ, Kerlan RK Jr: Interventional biliary radiology. *AJR* 1984; 142:31–34.

20. Karani J, Fletcher M, Brinkley D, et al: Internal biliary drainage and local radiotherapy with iridium-192 wire in treatment of hilar cholangiocarcinoma. *Clin Radiol* 1985; 36:603–606.

21. May GR, Bender CE, LaRusso NF, et al: Nonoperative dilatation of dominant strictures in primary sclerosing cholangitis. *AJR* 1985; 145:1061–1064.

22. Cohan RH, Illescas FF, Braun SD, et al: Fine needle aspiration biopsy in malignant obstructive jaundice. *Gastrointest Radiol* 1986; 11:145–150.

23. Kuroda C, Yoshioka H, Tokunaga K, et al: Fine-needle aspiration biopsy via percutaneous transhepatic catheterization: Technique and clinical results. *Gastrointest Radiol* 1986; 11:81–84.

24. Berger H, Winter T, Pratschke E, et al; Perkutane drainagebehandlung fistelassoziierter abszess und biliärer fisteln. *ROFO* 1989; 150:342–345.

25. Nunez D Jr, Guerra JJ Jr, Al-Sheikh WA, et al: Percutaneous biliary drainage in acute suppurative cholangitis. *Gastrointest Radiol* 1986; 11:85–89.

26. Lois JF, Gomes AS, Grace PA, et al: The risks of percutaneous transhepatic drainage in patients with cholangitis. *AJR* 1987; 148:367–371.

27. Huang MH, Ker CG: Ultrasonic guided percutaneous transhepatic bile drainage for cholangitis due to intrahepatic stones. *Arch Surg* 1988; 123:106–109.

28. Bland KI, Jones RS, Maher JW, et al: Extracorporeal shockwave lithotripsy of bile duct calculi: An interim report of the Dornier US bile duct lithotripsy prospective study. *Ann Surg* 1989; 209:743–753.

29. Faulkner DJ, Kozarek RA: Gallstones: Fragmentation with a tunable dye laser and dissolution with methyl tert-butyl ether in vitro. *Radiology* 1989; 170:185–189.

30. Thistle JL, May GR, Bender CE, et al: Dissolution of cholesterol gallbladder stones by methyl-tert-butyl ether administered by percutaneous transhepatic catheter. *N Engl J Med* 1989; 320:633–639.

31. Pitt HA, Gomes AS, Lois JF, et al: Does preoperative percutaneous biliary drainage reduce operative risk or increase hospital cost? *Ann Surg* 1985; 201:545–553.

32. Smith RC, Pooley M, George CR, et al: Preoperative percutaneous transhepatic internal drainage in obstructive jaundice: A randomized, controlled trial examining renal function. *Surgery* 1985; 97:641–648.

33. Norlander A, Kalin B, Sundblad R: Effect of percutaneous transhepatic drainage upon liver function and postoperative mortality. *Surg Gynecol Obstet* 1982; 155:161–166.

34. Denning DA, Ellison EC, Carey LC: Preoperative percutaneous transhepatic biliary decompression lowers operative morbidity in patients with obstructive jaundice. *Am J Surg* 1981; 141:61–65.

35. Gobien RP, Stanley JH, Soucek C, et al: Routine preoperative biliary drainage: Effect on management of obstructive jaundice. *Radiology* 1984; 152:353–356.

36. Hatfield, ARW, Tobias R, Terblanche J, et al: Preoperative external biliary drainage in obstructive jaundice: A prospective controlled clinical trial. *Lancet* 1982; 2:896–899.

37. Lygidakis NJ, van der Heyde MN, Lubbers MJ: Evaluation of preoperative biliary drainage in the surgical management of pancreatic head carcinoma. *Acta Chir Scand* 1987; 153:665–668.

38. Stanley J, Govien RP, Cunningham J, et al: Biliary decompression: An institutional comparison of percutaneous and endoscopic methods. *Radiology* 1986; 158:195–197.

39. Lewis WD, Cady B, Rohrer RJ, et al: Avoidance of transhepatic drainage prior to hepaticojejunostomy for obstruction of the biliary tract. *Surg Gynecol Obstet* 1987; 165:381–386.

40. Okuda K, Ohto M, Tsuchiya Y: The role of ultrasound, percutaneous transhepatic cholangiography, computed tomographic scanning, and magnetic resonance imaging in the preoperative assessment of bile duct cancer. *World J Surg* 1988; 12:18–26.

41. Deviere J, Baize M, de Toeuf J, et al: Long-term follow-up of patients with hilar malignant stricture treated by endoscopic internal biliary drainage. *Gastrointest Endosc* 1988; 32:95–101.

42. Lameris JS, Stoker J, Dees J, et al: Non-surgical palliative treatment of patients with malignant biliary obstruction: The place of endoscopic and percutaneous drainage. *Clin Radiol* 1987; 38:603–608.

43. Günther RW, Schild H, Thelen M: Percutaneous transhepatic biliary drainage: Experience with 311 procedures. *Cardiovasc Intervent Radiol* 1988; 11:65–71.

44. Vogelzang RL, Nemcek AA Jr: Percutaneous cholecystostomy: Diagnostic and therapeutic efficacy. *Radiology* 1988; 168:29–34.

45. Carrasco CH, Zornoza J, Bechtel WJ: Malignant biliary obstruction: Complications of percutaneous biliary drainage. *Radiology* 1984; 152:353–346.

46. Nichols DM, Cooperberg PL, Golding RH, et al: The safe intercostal approach? Pleural complications in abdominal interventional radiology. *AJR* 1984; 141:1013–1018.

47. Mitchell SE, Shuman LS, Kaufman SL, et al: Biliary catheter drainage complicated by hemobilia: Treatment by balloon embolotherapy. *Radiology* 1985; 157:645–652.

48. Joyce FS, Thorup J, Burcharth F: Bilio-bronchial fistula: A rare complication to biliary endoprosthesis. *ROFO* 1988; 148:723–724.

49. Rypins EB, Bitzer LG, Sarfeh IJ, et al: The role of percutaneous transhepatic internal biliary drainage in preoperative patients. *Am Surg* 1987; 53:562–564.

50. Hamlin JA, Friedman M, Stein MG, et al: Percutaneous biliary drainage: Complications of 118 consecutive catheterizations. *Radiology* 1986; 158:199–202.

51. Chapman WC, Sharp KW, Weaver F, et al: Tumor seeding from percutaneous biliary catheters. *Ann Surg* 1989; 209:708–713.

52. Stambuk EC, Pitt HA, Pais SO, et al: Percutaneous transhepatic drainage: Risks and benefits. *Arch Surg* 1983; 118:1388–1394.

53. Schoenemann J, Willems M, Wolf G, et al: Ergebnisse der perkutanen transhepatischen gallengangsdrainage. *ROFO* 1987; 147:619–623.

54. Bonnel D, Ferrucci JT Jr, Mueller PR, et al: Surgical and radiological decompression in malignant biliary obstruction: A retrospective study using multivariate risk factor analysis. *Radiology* 1984; 152:347–351.

55. Lai ECS, Tompkins RK, Roslyn JJ, et al: Proximal bile duct cancer: Quality of survival. *Ann Surg* 1987; 205:111–118.

56. Szabo S, Mendelson MH, Mitty HA, et al: Infections associated with transhepatic biliary drainage devices. *Am J Med* 1987; 82:921–926.

57. Lammer J: Perkutane transhepatische gallengangsendoprothese. *ROFO* 1985; 142:243–253.

58. Kerlan RK Jr, Ring EJ, Pogany AC, et al: Biliary endoprostheses: Insertion using a combined peroral-transhepatic method. *Radiology* 1984; 150:828–830.

59. Mueller PR, Ferrucci JT Jr, Teplick SK, et al: Biliary stent endoprosthesis: Analysis of complications in 113 patients. *Radiology* 1985; 156:637–639.

60. Mendez G, Russell E, LePage JR, et al: Abandonment of endoprosthetic drainage technique in malignant biliary obstruction. *AJR* 1984; 143:617–622.

61. Coons HG, Carey PH: Large-bore, long biliary endoprostheses (biliary stents) for improved drainage. *Radiology* 1983; 148:89–94.

62. Teplik SK, Haskin PH, Pavlides CA, et al: Management of obstructed biliary endoprostheses. *Cardiovasc Intervent Radiol* 1985; 8:164–167.

63. Dick R, Gillams A, Dooley JS, et al: Stainless steel mesh stents for biliary strictures. *J Intervent Radiol* 1989; 4:95–98.

64. Harell GS, Anderson MF, Berry PF: Cytologic bile examination in the diagnosis of biliary duct neoplastic strictures. *AJR* 1981; 137:1123–1126.

65. Meyers WC, Jones RS: Internal radiation for bile duct cancer. *World J Surg* 1988; 12:99–104.

66. Kubota Y, Seki T, Nakano T, et al: A case of bile duct cancer treated by laser via percutaneous transhepatic choledochoscopy. *Hepatogastroenterology* 1988; 35:213–214.

67. Larssen TB, Gothlin JH, Jensen D, et al: Ultrasonically and fluoroscopically guided therapeutic percutaneous catheter drainage of the gallbladder. *Gastrointest Radiol* 1988; 13:37–40.

68. Eggermont AM, Laméris, Jeekel J: Ultrasound-guided percutaneous transhepatic cholecystostomy for acute acalculus cholecystitis. *Arch Surg* 1985; 120:1354–1356.

69. vanSonnenberg E, Wing VW, Pollard JW, et al: Life-threatening vagal reactions associated with percutaneous cholecystostomy. *Radiology* 1984; 151:377–380.

70. Shaver RW, Hawkins IF Jr, Soong J: Percutaneous cholecystostomy. *AJR* 1982; 138:1133–1136.

71. Martin EC, Fankuchen EI, Laffey KJ, et al: Percutaneous management of benign biliary disease. *Gastrointest Radiol* 1984; 9:207–212.

72. Gibson RN, Yeung AE, Savage A, et al: Percutaneous techniques in benign hilar and intrahepatic strictures. *J Intervent Radiol* 1988; 3:125–130.

73. Salomonowitz E, Casteñeda-Zuñiga WR, Lund G, et al: Balloon dilatation of benign biliary strictures. *Radiology* 1984; 151:613–616.

74. Bret PM, Bretagnolle M, Fond A, et al: Use of large silicone catheters in patients with long-term percutaneous transhepatic biliary drainage. *Cardiovasc Intervent Radiol* 1986; 9:57–58.

75. Skolkin MD, Alspaugh JP, Casarella WJ, et al: Sclerosing cholangitis: Palliation with percutaneous cholangioplasty. *Radiology* 1989; 170:199–206.

76. Russell E, Yrizarry JM, Huber JS, et al: Percutaneous transjejunal biliary dilatation: Alternate management for benign strictures. *Radiology* 1986; 159:209–214.

77. vanSonnenberg E, Ferrucci JT Jr, Neff CC, et al: Biliary pressure: Manometric and perfusion studies at percutaneous transhepatic cholangiography and percutaneous biliary drainage. *Radiology* 1983; 148:41–50.

78. Hofmann AF: Bile, bile acids, and gallstones: Will new knowledge bring new power? *AJR* 1988; 151:5–12.

79. Burhenne HJ: Percutaneous extraction of retained biliary tract stones: 661 patients. *AJR* 1980; 134:888–898.

80. Sackmann M, Delius M, Sauerbruch T, et al: Shock-wave lithotripsy of gallbladder stones: The first 175 patients. *N Engl J Med* 1988; 318:393–397.

81. Bühler H, Jaeger P, Ammann R, et al: Extrakorporale stosswellenlithotripsie (ESWL) zur behandlung von gallengangssteinen. *Schweiz Med Wochenschr* 1988; 118:113–117.

82. Tadavarthy SM, Klugman J, Castañeda-Zuñiga WR, et al: Removal of large and small biliary duct stones. *Cardiovasc Intervent Radiol* 1981; 4:93–96.

83. Laffey KJ, Martin EC: Percutaneous removal of large gallstones. *Gastrointest Radiol* 1986; 11:165–168.

84. Gacetta DJ, Cohen MJ, Crummy AB, et al: Ultrasonic lithotripsy of gallstones after cholecystostomy. *AJR* 1984; 143:1088–1089.

85. Gibney RG, Chow K, So CB, et al: Gallstone recurrence after cholecystolithotomy. *AJR* 1989; 153:287–289.

86. Berkman WA, Bishop AF, Palagallo GL, et al: Transhepatic balloon dilation of the distal common bile duct and ampulla of Vater for removal of calculi. *Radiology* 1988; 167:453–455.

87. Gadacz TR: The effect of monooctanoin on retained common duct stones. *Surgery* 1981; 89:527–531.
88. Lee LL, McGahan JP: Dissolution of cholesterol gallstones: Comparison of solvents. *Gastrointest Radiol* 1986; 11:169–171.
89. Ackermann C, Meyer B, Rothenbühler JM, et al: Schmerzfreie piezoelektrische extrakorporelle stosswellenlithotripsie bei gallenblasensteinen. *Schweiz Med Wochenschr* 1989; 119:720–723.
90. Wenk H, Thomas S, Baretton G, et al: Die percutane transhepatische laserlithotripsie von gallenblasensteinen: Tierexperimentelle ergebnisse. *Langenbecks Arch Chir* 1989; 374:169–174.

21

Genitourinary Interventions

KEY CONCEPTS

1. Percutaneous nephrostomy is indicated for most cases of obstructive hydronephrosis, if a retrograde ureteral stent cannot be placed.
2. A posterolateral approach below the 12th rib, with entry into a posterior calyx or infundibulum, is desirable for most nephrostomies.
3. Percutaneous nephrolithotomy and lithotripsy continue to be indicated for large (over 3 cm), infected, or cystine stones, as well as those associated with ureteral obstruction.
4. Response of ureteral strictures to dilation and stenting is difficult to predict, but anastomotic ileal conduit strictures have a poor prognosis.
5. The Whitaker test, measurement of a pressure gradient between renal pelvis and bladder after a perfusion challenge, establishes the significance of questionable obstructions.

Percutaneous nephrostomy, first used solely for the decompression of obstructed upper urinary tracts, has fostered the development of an entirely new approach to urologic disease. The field of endourology, which makes use of both percutaneous and retrograde access to the genitourinary system, permits conditions previously needing open surgery and prolonged postoperative recuperation to be treated with decreased burden to the patient. The introduction of extracorporeal shockwave lithotripsy (ESWL) in the past decade has not eliminated the

need for percutaneous access to the collecting system in those with
urinary calculi, but has modified the application of percutaneous nephrostomy for renal stone disease.[1]

Percutaneous nephrostomy clearly remains the procedure of choice
for sepsis consequent to ureteral obstruction, although retrograde ureteral stent placement by way of cystoscopy may be attempted first in
some patients. Percutaneous nephrostomy is also a prime treatment
for uremia or flank pain due to chronic obstruction, whether of benign
or malignant cause. However, percutaneous drainage is of dubious
value in patients with advanced and disseminated cancer.[2]

The access provided by nephrostomy permits ureteral strictures to
be dilated and stented, flow through urinary fistulas to be diverted,
and traumatic ureteral or renal pelvic perforations to heal.[3, 4] Nephrostomy is a route for renal endoscopy, performance of endopyelotomy
of ureteropelvic junction strictures, and topical treatment for upper
tract transitional cell carcinoma.[5, 6] Percutaneous methods are especially valuable in patients who have received renal transplants and
experience immediate or delayed postoperative problems.[7] Diabetics
or others developing fungal pyelonephritis may need urinary decompression and direct perfusion of amphotericin to resolve infection.[8]

Percutaneous lithotripsy achieved widespread application in the early
1980s, only to be superseded by ESWL for the treatment of renal
stones. However, ESWL alone cannot remove staghorn calculi or other
large stones, and nephrostomy is often needed in preparation for complicated stone extractions.[9] Stone dissolution therapy relies on infusion
and drainage through percutaneous tubes.[10] Retrograde transvesical
nephrostomy and other retrograde manipulations have been developed
and present an alternative for selected patients.[11, 12] No matter what
the clinical problem being addressed, the patient's interests are best
served by cooperation between radiologist and urologist, whose technical skills are exercised optimally in a complementary manner.

PATIENT PREPARATION

As before any invasive procedure, the patient's history and previous
studies must be reviewed, and clear objectives formulated. When multiple interventions are likely, or in the case of complicated stone disease, the treatment plan and percutaneous approach should be discussed
directly with the attending urologist prior to nephrostomy placement.
For example, the removal of multiple stones in the renal collecting
system depends critically on the point of entry of the nephrostomy

tube.[13] If a patient has bilateral hydronephrosis, the bladder must be decompressed by a Foley catheter (or suprapubic cystostomy in rare instances) to eliminate the possibility of placing bilateral nephrostomies for bladder outlet obstruction! This unlikely scenario typically arises when a patient is directly referred from a nursing home or a medical ward where staff are unaccustomed to managing urologic problems.

The patient is then seen; the procedure, its indications, alternatives, and risks are explained; and consent is obtained. Any anticoagulant medication is stopped. Prothrombin time, partial thromboplastin time, and platelet count are checked, and any evident coagulation defects are corrected, as possible. Any history of hypersensitivity to iodinated contrast medium or other medication is elicited. Note must be made of splenomegaly (for left-sided nephrostomy), severe scoliosis, or other anatomic anomalies that could affect approach. Percutaneous nephrostomy is possible in cases of pelvic kidney or horseshoe kidney.[14] However, when renal anatomy is grossly distorted, computed tomography (CT) may be needed for planning. Scanning with CT or ultrasound (US) can also be used to estimate the likelihood of restoring function to a chronically obstructed kidney, by demonstrating renal size and cortical thickness.

Patients with signs of urinary tract infection need administration of intravenous antibiotics at least 1 hour prior to intervention.[15] Those undergoing drainage of pyonephrosis have a 7% incidence of septic shock, despite aminoglycoside prophylaxis.[15] Because renal stones are commonly associated with infection, premedication with antibiotics is also used routinely before nephrostomy for stone removal.[16] Antibiotics need not be given universally to otherwise asymptomatic patients, but if such a policy is pursued, urinalysis and urine culture are recommended before elective procedures. Simple changes of a nephrostomy tube do not require medication as long as the tube has not been obstructed.

Urinary tract pathogens are predominantly gram-negative bacteria, such as *Escherichia coli*, *Klebsiella*, *Enterococcus*, or *Proteus* species. For prophylaxis in the absence of any overt infection, cefazolin or cefoperazone may be given and continued for 48 hours.[15] When infection is present, therapy is directed toward the isolated organisms, and ticarcillin, piperacillin, and an aminoglycoside can be used to provide a broader range of antibiotic coverage.[15]

Simple nephrostomy tube placement in adults can be performed with local anesthesia, supplemented by an intravenous benzodiazepine (midazolam or diazepam) and an opiate for pain (morphine or fentanyl).[17] Only in special cases is epidural or general anesthesia needed. However, when extensive tract dilatation is planned, as for renal stone removal, regional or general anesthesia is indicated.

PERCUTANEOUS NEPHROSTOMY

Anatomic Considerations

The kidneys are normally obliquely oriented in the retroperitoneum, with their upper poles medial and posterior in respect to their lower poles. They have a medial-to-lateral posterior angulation of about 30° as measured from the coronal plane.[18] Therefore, on an anteroposterior projection, the calyces appearing to be in profile and laterally placed are the anterior calyces. Neighboring structures to be respected include the pleural space, liver, spleen, and colon. Although the pleural space may be safely traversed when necessary, it is best avoided for simple drainage procedures.[19] Tubes placed posteriorly through the 11th intercostal space will pass through pleura, and those placed above the 11th rib may injure the lung.[20] The inferior margins of the ribs must be avoided to prevent bleeding from the subcostal artery, as well as pain from periosteal irritation.

Most kidneys have their dominant segmental arterial supply distributed anteriorly, and their only posterior segmental artery passes behind the upper pole infundibulum.[13] The posterolateral watershed of these vessels, described by Brödel as the "bloodless line of incision," provides the safest approach into the kidney.[13] However, the site of skin entry should not be placed too laterally, for as one approaches the posterior axillary line, the chances for injuring a posteriorly positioned colon or spleen increase.[21 22] Prevention of injury to an adjacent organ requires careful US or fluoroscopic observation during needle placement, tract dilatation, and tube introduction.

Technique

Nephrostomy placement can be guided by US, fluoroscopy or both. If the kidney is unobstructed or partially obstructed and capable of excretion, intravenous iodinated contrast material is administered. In certain cases, such as for removal of nonobstructing stones, a retrograde ureteral catheter may have been placed during cystoscopy. If such a catheter is in position, it can be used to inject the contrast medium. Not only iodinated contrast medium but also carbon dioxide (CO_2) may be injected through retrograde catheters. In a prone patient, CO_2 accumulates in posterior calyces, aiding in their recognition.

Both intravenous and directly injected contrast medium are valuable

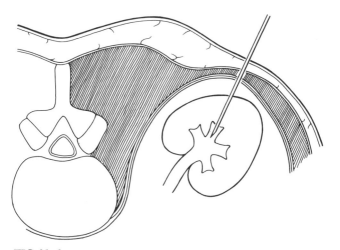

FIG 21–1.
Needle entry into a posterolateral calyx: the first step in percutaneous nephrostomy.

for defining anatomy and position, as well as for distending the renal collecting system. Distention produces fewer complications of perforation and excessive bleeding.[20] Severe hydronephrosis is easily defined by US scanning in the great majority of cases, and needle placement is relatively simple, except in grossly obese patients.

Initial entry with a fine (21-gauge) needle is preferable. Routine use of a large sheath needle places the patient at greater risk for pseudoaneurysm, arteriovenous fistula, and major bleeding.[23] There may be little practical alternative to a sheath needle, however, if there is heavy perinephric scarring from inflammation or previous surgery or if the patient has a thick layer of perinephric fat.

Except when a particular calyx must be entered in conjunction with stone removal, the needle is best directed toward a lower pole or midnephric posterior calyx from a posterolateral approach (Fig 21–1). Puncture into an anterior calyx makes guidewire and catheter placement extremely difficult, if not impossible, because of the acute angle that must be negotiated between the entry site and renal pelvis. Ultrasound study is helpful in choosing an approach even if the collecting system is opacified, because intervening spleen or colon can be recognized. The collecting system should be entered as peripherally as possible, avoiding the larger vessels near the renal hilus. The route taken must be made through parenchyma; otherwise, perinephric leakage of urine may result. In difficult cases it may be worthwhile to place a fine needle directly into the renal pelvis from a posterior approach. Contrast medium is injected to opacify the intrarenal collecting system in order to guide a subsequent more peripheral needle placement

When US is used to direct the needle, a biopsy guide is helpful. The needle is inserted to the measured depth, and its tip may be visible sonographically. Tapping the needle gently or sliding the stylet in and out may make the tip position more evident.

Needle placement with fluoroscopic guidance is greatly aided by use of a C-arm. After the entry site is selected, the calyx, fluoroscope, and needle are aligned, and the needle then is precisely superimposed over its target (looking "down the barrel"). With the patient's breathing suspended at the proper moment, fluoroscopy is interrupted and the needle is quickly and smoothly advanced into the kidney. An increase in resistance is usually felt as the needle encounters the renal capsule. Rotation of the C-arm 90° will then show the depth of placement. To prevent major arterial injury, the needle tip should not be passed deeper than midpelvis.

Alternatively, the C-arm may be rotated after the needle is aligned but before it is advanced. In this configuration, continuous fluoroscopy during needle advancement may guide the depth of insertion, often allowing the moment of entry to be recognized. If it is clear that the needle has deviated in an undesired direction, it may be left in position while a second one is placed (the "tandem needle" technique). The direction of the first needle is a gauge by which corrections in angle can be made with the second needle.

Once the needle has been inserted, the stylet is removed. If hydronephrosis is present and the collecting system has been entered, urine will return spontaneously; otherwise, a connecting tube is used to attach a contrast material–filled syringe to the needle. Gentle intermittent aspiration is applied as the needle is slowly withdrawn. When urine returns, a small amount of contrast medium is injected to confirm proper entry. If no urine is aspirated, the needle placement procedure is repeated. Contrast medium should not be introduced in the absence of fluid return, because repeated injection into the soft tissues will rapidly obscure the kidney. In an obstructed collecting system, a larger volume of contrast medium is injected only after

removal of a similar volume of urine. Little if any ⏐
should be instilled if grossly infected urine returns, fc_
sepsis.

A fine (0.018-inch) guidewire is passed through the needle and should
advance freely into the renal pelvis. If difficulties are encountered,
repeat injection of contrast medium must confirm that the needle tip
is still within the collecting system. With successful guidewire passage,
a tapered dilator-stiffener combination is then passed over the wire to
allow a heavier, standard (0.038-inch) guidewire to be introduced. In
the Cope catheter introduction system the standard-size J-tip wire exits
a sidehole near the tip. Other systems employ a coaxial design, which
provides access through an endhole. Puncture with an 18-gauge sheath
needle permits introduction of a standard guidewire without the need
for an intermediate step. Teflon dilators are passed serially over the
larger wire before final placement of the nephrostomy catheter.

If the course of the soft tissue tract is difficult to traverse because
of local scarring or because of coiling in retroperitoneal fat, a dilator
must be gingerly reintroduced into the collecting system. The wire is
exchanged for another extremely stiff wire for further manipulations.
At times it is prudent to employ a sheath to introduce a second "safety"
wire.

A nephrostomy catheter left in place should be self-retaining, such
as the Cope loop catheter. Simple pigtail catheters are displaced by
respiratory motion with disconcerting frequency. An exception is made
when the nephrostomy is for percutaneous nephrolithotomy. In this
situation a long, straight catheter with multiple sideholes directed down
the ureter is preferred. Wire introduction is easier and access is more
secure for subsequent tract dilatation and insertion of the working
sheath. If tract dilatation is not performed in the same session, sideholes
must be present in the renal pelvis to ensure adequate drainage.

Tube Management

The catheter is sutured to the patient's skin or to a stoma ring (see
Chapter 20, Biliary Interventions). Vital signs must be monitored frequently for several hours to detect any possible retroperitoneal bleeding. No anticoagulation should be instituted for at least 1 day following
percutaneous nephrostomy (a point to remember, if the patient is
undergoing hemodialysis). Some degree of hematuria is to be expected
after catheter placement, but even large pelvic clots lyse rapidly and
the urine normally clears within 1 to 2 days.

The nephrostomy is left to gravity drainage. Tube output should be
charted, and note must be made of any postobstructive diuresis. Reg-

ular catheter flushing is not ordinarily needed. Patients may be discharged with nephrostomy catheters in place. Tube changes may be scheduled as outpatient procedures every 4 to 6 weeks, although a few patients may need more frequent attention because of particularly high rates of catheter encrustation.

Risks

Unlike biliary drainage procedures, percutaneous nephrostomies are not often complicated by sepsis, except when pyonephrosis is present.[15] Even renal transplant patients, with their immunosuppressive medications, do not often suffer infections due to nephrostomy. However, sepsis arising in transplant patients can be fatal.[7] In his multi-institutional review of complications of percutaneous nephrolithotomy and lithotripsy (perhaps the most invasive form of percutaneous nephrostomy), Lang found that 1% of patients developed perinephric abscess and 0.1% experienced septic shock.[20]

Colonic perforation has been reported in 0.2%, but it does not usually need surgical repair.[20, 21] In cases of stone removal, nephrostomy is complicated by hydrothorax or hydropneumothorax in up to 12% of patients when the puncture is made above the 12th rib.[19] Ureteral or pelvic perforation is not a major problem, because urothelial tears will heal quickly as long as the kidney is drained. However, complete ureteral transection must be treated by stent placement or surgery.

The major risk of percutaneous nephrostomy is hemorrhage. Cope and Zeit found that nephrostomy with an 18-gauge sheath needle led to life-threatening bleeding in 1% of patients.[23] Use of the fine-needle technique has decreased this risk substantially.[16] The rare deaths from percutaneous nephrolithotomy are due to hemorrhage.[20] If gross blood persists longer than 1 to 2 days in drained urine, cold saline irrigation may be helpful. Active bleeding arising from the nephrostomy tract should undergo tamponade with a larger catheter or a balloon. If bleeding does not respond to conservative measures, angiography is indicated to identify a pseudoaneurysm or arteriovenous fistula, which can then be treated by embolization.[23]

STRICTURE DILATATION AND STENT PLACEMENT

Aside from malignant obstruction and renal stone disease, one of the most common problems addressed by endourologic intervention is ureteral stricture. Stricture is commonly caused by ischemia, which may be produced by a variety of factors: radiation therapy, surgical stripping of the ureter, pressure necrosis from ureteral stone, trauma, inflammatory disease, and chronic transplant rejection.[3] A vicious cycle of ischemia, scarring, and progressive ischemia is postulated as the responsible mechanism. The observed great variability in response of strictures to balloon dilatation and ureteral stent placement may be related to the residual vascular supply to the involved segment.[3]

Prognostic Factors

Devitalized ureter does not respond to dilation, but devitalized segments are difficult to recognize *a priori*. Lang has described a smooth pipestem appearance of the stricture with proximal ureteral dilatation as indicative of devitalization.[3] Short strictures and those involving the upper ureter are more amenable to dilatation. If narrowing has been present for less than 3 months and there is no reason to suspect devitalization, balloon dilatation has produced lasting improvement in over 90% of attempts.[3] Renal transplant ureteral stenoses also can be treated effectively by dilatation, and an attempt is warranted no matter how long after transplantation a lesion becomes evident.[24, 25]

On the other hand, benign anastomotic strictures in patients with ureteral diversion to ileal loops are quite refractory to nonoperative management. About 10% of all patients having such diversions experience obstruction.[26] Although short-term positive response after balloon dilatation is found in 40% to 60% of anastomotic strictures, Shapiro et al. have noted only 16% patency at 1 year, and even later recurrences are observed.[26] Anastomotic strictures in patients treated by radiation have an especially poor prognosis. Ureteropelvic junction obstructions are also poorly responsive to balloon dilatation, and percutaneous endopyelotomy with a mechanical or electrical pyelotome is a more appropriate intervention.[27]

Technical Points

For treating strictures, a mid- to upper-pole nephrostomy lends a mechanical advantage by allowing a more straight-line approach. Tight strictures may be traversed by a hydrophilic polymer-coated guidewire, or a Ring-Lunderquist torque guidewire. If the character of the stricture is in doubt, cytologic material can be obtained by brush biopsy through a sheath.

Initial dilatation with a tapered Teflon catheter, such as a van Andel catheter, facilitates introduction of the balloon. If necessary, a stiffer wire can be introduced through the stricture after predilatation in this manner. Especially tough strictures may respond only to coaxial Teflon dilators, which are available in sizes through 18 F.[28] If problems are encountered with catheter coiling in renal pelvis or in the retroperitoneal soft tissues, long Teflon sheaths allow more effective transmission of force against resistant lesions. In the future, rotational wires or catheters developed for treating vascular obstruction may be modified for recanalizing occluded ureters.[29]

In cases in which a wire successfully passes the lesion but no catheter will follow, the wire may be snared in the bladder or ileal conduit by a retrograde basket or snare (with or without the aid of cystoscopy). With both ends of a guidewire in hand and held taut, almost any stricture can be passed, dilated, and stented. If the necessity of a "through-and-through" technique can be anticipated, a long exchange guidewire is used to pass the stricture. After the wire tip is retrieved, care must be taken to protect the urethra from injury by placement of a sheath over that portion of wire traversing it.

Balloons

High-pressure balloons 4 to 8 mm in diameter are best for treating ureteral strictures. Balloons up to 10 mm in size may be applied in ureteroileal anastomoses.[3] Inflations are made for 30 to 60 seconds and are repeated until residual deformity disappears. Sometimes, repeat dilatation procedures several days apart with successively larger balloons are useful.[24] At the end of the dilatation, the site must be stented.

Stents

The size of the stent and its duration of placement are factors not well defined for optimal prevention of stricture recurrence. Stents of 6 to 8 F in diameter are commonly used. Much larger stents may in themselves promote ureteral ischemia.[30] The length between renal pelvis and bladder can be measured by the "bent-wire" technique: passing a wire through a catheter placed antegrade into bladder, making one crimp in the guidewire at the external hub when its tip is at the ureterovesical junction, and making a second bend at the hub when the wire tip has been withdrawn to the renal pelvis.

Stenting catheters are either internal-external or completely internal. The simplest internal-external device is a pigtail catheter, with its tip in the bladder and sideholes in the upper collecting system. Such stents have the advantage that they can easily be removed or replaced. Other designs involve a midcatheter Cope loop or other internal fixation device meant to be positioned in the renal pelvis.

Internal stents are inserted antegrade over a guidewire with a pushing catheter. Soft polyurethane or silicone stents are best introduced with an inner stiffener or through a sheath; otherwise, they will buckle or "accordion" when pushed against resistance. Some are supplied with a suture placed through the trailing end as a means of pulling the tube back if it has been advanced too far. When the leading J-tip or pigtail is in proper position, the inner stiffener is removed but the guidewire is left in place. The suture is then cut and pulled out of the nephrostomy tract. The wire is then removed from the stent, while the pusher keeps the trailing end of the stent within the upper collecting system. As the last step, a nephrostomy tube is replaced over a second wire that is introduced either at the start of manipulations or through the pushing catheter.

The nephrostomy should be left in place for 1 to 2 days after internal stent placement to ensure that blood or debris does not cause early occlusion. Stents should not be introduced in the face of unresolved bleeding or infection. When a follow-up nephrostogram shows absence of debris and free flow of contrast medium through the stent, the nephrostomy can be removed. Although internal stents are a lesser day-to-day burden for the patient, removal or exchange requires cystoscopy or repeat nephrostomy.

Ureteroenteric anastomotic strictures may be chronically stented by passing a wire antegrade through the stricture and out the conduit. A stenting catheter is inserted retrograde, and the end of the catheter is left in the collection bag of the conduit. In this fashion the patient's nephrostomy can be removed later, but future stent changes are easily performed over a guidewire. One must beware that repeated balloon dilatations and prolonged stent placement across an anastomotic stricture place a patient at risk for uretero-arterial fistula and massive bleeding.[31]

When used after stricture dilatation, stents are left in place for weeks to months, with 6 to 8 weeks a period often employed.[24, 27] Any residual narrowing found after stent removal can be evaluated by a Whitaker test (see the following section). Stents placed for palliation of malignant obstruction are regularly changed for the duration of the patient's life. Exchanges must be performed relatively frequently (about every 4 weeks) in patients with a history of stones, to prevent heavy encrustation of the stent.

Of the various materials used for stents, Teflon has a low tendency for encrustation, but it is quite stiff.[32] Polyurethane is softer than polyethylene, but both materials tend to become brittle with time.[30] Silicone and Silastic stents may be difficult to introduce because of their softness and high coefficient of friction, but their long-term patency is good.[32]

URETERAL PERFUSION CHALLENGE (THE WHITAKER TEST)

The significance of urinary tract dilation in a patient passing urine may be quite difficult to interpret, particularly if the patient has had previous pyeloplasty, ureteral reimplantation, reflux, or renal transplantation. Subtle obstruction may be difficult to establish, even when antegrade pyelography is performed. Residual narrowing after treatment of a ureteral stricture may or may not impede urine flow. For such situations, a urinary flow-pressure test is invaluable.

In the test formulated by Whitaker, the renal pelvis is perfused with up to 10 mL/min of fluid for 3 to 5 minutes and the pressure gradient between kidney and bladder is measured.[33] A healthy individual will show an absolute renal pelvic pressure of less than 25 cm of water, and a pressure gradient of less than 15 cm of water at maximal flow challenge. A Foley catheter should be in place, because high absolute intrarenal pressure may simply reflect lower tract obstruction or a hypertonic bladder.[33]

Before ureteral perfusion is begun, baseline manometry is performed. If a large gradient is found at rest, or if the resting pressure within kidney approaches 30 cm of water, no perfusion test is needed. Normal ureteral flow is between 0.25 and 0.50 mL/min, but the unobstructed ureter can accommodate a much higher flow when necessary.[34] Infusion by automatic injector may be started at a low rate, such as 4 mL/min for 5 minutes, with repeat infusion at a higher rate only if abnormal pressures are not elicited. The fluid infused is 30% iodinated contrast medium, which allows fluoroscopic observation of the amount of collecting system distention present. The kidney should not be exposed to pressures above 30 cm of water, because of the risk of pyelotubular backflow and sepsis.[34] A gradient of 15 cm of water or more indicates significant obstruction.

It should be noted that, just as not all dilated renal collecting systems are obstructed, not all obstructed kidneys are hydronephrotic. Perhaps as many as 5% of patients with urinary tract obstruction will show no dilation.[35] Lack of dilation may indicate tumor encasement or retroperitoneal fibrosis, but in some cases there is no obvious cause. A Whitaker test may be indicated when nondilated obstructive uropathy is suspected, followed by placement of a nephrostomy tube if the test is positive.

RENAL STONE REMOVAL

About one patient in every thousand hospitalized is suffering from renal stone disease, but 60% of stones will pass spontaneously.[36] The other 40% require intervention, which until fairly recently meant open pyelolithotomy, a major surgical procedure requiring prolonged convalescence. The first percutaneous nephrostomy for stone removal was performed in 1976.[37] Since that time, open surgery has been largely displaced by percutaneous, retrograde ureteral, and more recently by ESWL treatments. Not only are the newer methods less invasive, they also reduce the hospitalization and recovery periods.

The ESWL procedure may now be applied to over 80% of renal and upper ureteral stones.[37] However, percutaneous nephrolithotomy and lithotripsy are still indicated for large (over 3.0 cm in diameter), staghorn, infected, or cystine stones; for patients with obstructed outflow; for children; and for massively obese patients.[37] Some patients with staghorn calculi and little residual function in the affected kidney are best treated by nephrectomy.

Calcium oxalate is the sole or major component of over 80% of urinary stones.[36, 38] Other calculi are composed of apatite, struvite (magnesium ammonium phosphate), urate, cystine, or organic matrix.[38] As noted, cystine stones respond poorly to ESWL. Most staghorn calculi are composed of struvite, which forms in the presence of urea-splitting bacteria such as *Proteus mirabilis*.[39] Struvite stones are more difficult to treat and, being infected, are more likely to cause septic complications. Because percutaneous stone removal yields up to 84% stone-free rates at hospital discharge for infected calculi (a rate much better than ESWL), it is the procedure of choice for such patients.[37] Complete stone removal is important, for only 14% of patients with no residual fragments after open nephrolithotomy have recurrence, compared with a rate of 70% for those with some stone material left behind.[40] Chemolysis can sometimes be used to dissolve retained fragments (see "Chemodissolution," later in this chapter).

The tools of percutaneous extraction include various types of wire baskets; steerable catheters; Fogarty, occlusion, and other balloon catheters; US and electromechanical lithotriptors; Randall forceps; nephroscopes; and ureteroscopes.[41] Ureteral occlusion balloons are helpful for preventing migration and impaction of intrarenal stones during manipulations. Most percutaneous procedures require dilation of the tract to 24 to 30 F, which can be accomplished either by passage of a set of fascial dilators or use of a balloon catheter designed for the purpose. A large Teflon sheath is introduced, through which endoscopes and other instruments are passed. Stones nearly 1 cm in diameter can be removed through a 30-F sheath. Tract dilation may be performed immediately after percutaneous nephrostomy or as a separate delayed procedure. Dilatation is performed with the patient under general or epidural anesthesia. After a large nephrostomy tract has been created, a catheter of appropriate size must be left in place for at least several days.

If there is a tear in the urothelium, stone fragments may be lost into the perinephric soft tissues. This occurrence is usually innocuous, but an infected fragment may cause an abscess.[20] Percutaneous nephrolithotomy and lithotripsy have a major complication rate of only 1.5% to 3% in experienced hands.[20, 42]

Ureteral Stones

Just as methods for percutaneous extraction of ureteral stones were being perfected, with removal rates of 90% to 100% attained, ESWL came into use.[43, 44] Retrograde ureteroscopic techniques also developed apace, further reducing the hospitalization and recovery periods after ureteral stone removal.[45] Presently, percutaneous access is rarely needed.

When nephrostomy is performed for ureteral stones, it often must be through an intercostal approach into an upper or mid calyx.[19] This provides the most straight-line approach possible to the ureter. The potential for pleural complications from high puncture must be appreciated. Because retrograde methods alone are almost uniformly effective for removing ureteral stones within the pelvis, percutaneous nephrostomy is typically applied only to mid- and upper-ureteral calculi.

The presence of a retrograde ureteral catheter greatly facilitates stone removal. The retrograde catheter can be used to push or flush the stone back into renal pelvis. Flushing is most effective when it is performed through an occlusion balloon catheter. With a nephrostomy in place, Hunter et al. have even power-injected dilute contrast medium retrograde at rates up to 25 mL/sec for 2 seconds with no complications and good results in 31 patients![44] If flushing does not budge the stone, a flexible ureteroscope is introduced to guide the application of forceps or stone baskets.[44] Some stones become embedded in mucosa and may be very difficult to extract, but a steerable catheter or soft wire may sometimes be used to pry off such a calculus.[44] A safety wire or catheter placed past the stone will interfere with retrieval attempts by means of a basket.[9] Ureteral spasm can be a difficult problem, and spasm does not respond reliably to instillation of nitroglycerin or other medications. Complications needing surgical intervention are rare.[43, 44] If edema causes obstruction or if ureteral injury is suspected, a stent should be left in place.

Contact Lithotripsy

When stones are too large to be removed through a dilated percutaneous tract, they must be shattered. Shock waves may be directly applied by a variety of generators, from electrohydraulic and US lithotriptors to laser contact fibers.[36] Ultrasonic lithotriptors have the disadvantage of being rigid devices, which cannot be directed around curves. Despite this drawback, they have gained wide acceptance and can be used with great success.

Because of the inflexibility of US lithotriptors and the limited flexibility of conventional nephroscopes, the entry site of the percutaneous tract may be critical to success. The approach to a simple renal pelvic stone is straightforward, but in many cases a calyceal stone can be removed only when that calyx is entered directly. Although some investigators prefer to "trap" a calyceal stone by entering at the infundibular junction central to it, others insist that the point of entry must be within the calyx, peripheral to the stone itself.[13, 16] Upper pole stones, multiple stones, and staghorn calculi may be best approached through a lower pole calyx.[13] Whatever the chosen approach, a catheter should be placed down ureter, because access is much more likely to be lost during tract dilation if the wire cannot be advanced beyond the upper collecting system.

Because a large volume of fluid is perfused during lithotripsy, use of a sheath is imperative. A sheath allows the flush solution to be vented efficiently and keeps intrarenal pressure down.[46] It also makes hydrothorax less likely when the approach is intercostal.[19] The input-output balance of infusate must be carefully watched to prevent fluid intoxication of the patient from an unrecognized urothelial tear.[20] Rarely, decreased return of infusate may be the first indication of colonic perforation.[21]

Staghorn calculus, a prime remaining indication for percutaneous lithotripsy, presents a special challenge. Multiple nephrostomies in the treated kidney are commonly needed for effective stone removal.[39] As already mentioned, staghorn calculi are typically infected and present a higher risk of complication. Transfusion is often needed, although only 2.5% of patients have had bleeding severe enough to warrant angiography.[39] Coleman et al. have recommended that all patients undergoing percutaneous lithotripsy have at least 2 units of blood typed and cross-matched prior to the procedure.[13]

Contact lithotripsy can be extremely successful when performed by an experienced team, either as a one- or two-stage procedure. Stone-free hospital discharge rates of 96% or greater have been achieved.[37, 38] Serious vascular complications may arise in up to 3%, and represent the primary risk of the procedure.[42] Large catheters not only allow residual fragments to be flushed from the renal collecting system, but also tamponade the tract and promote healing. Because major bleeding can occur on removal of the catheter, removal should only be performed in the hospital when there is a safety wire in place. A balloon catheter can be expeditiously inserted over the guidewire and inflated to control hemorrhage. When such bleeding does occur, it usually resolves with replacement of a large nephrostomy catheter for several more days. If this course of action fails, arteriography and possible embolization are indicated.

Extracorporeal Shock-Wave Lithotripsy

Developed in West Germany in the early 1980's, ESWL sharply focuses electrically generated shock waves onto a renal or ureteral stone. The change in impedance between soft tissue and stone results in shearing and tearing forces, which disintegrate the stone through repeated application. A maximum of 1,500 to 2,400 separate shocks are administered in a single treatment session.[1, 47] First-generation instruments produce energy in a water bath, with a condenser discharge vaporizing water to generate the requisite shock wave.[48] Patients are immersed in the bath and are heavily anesthetized for treatment by these devices. The stone is placed in the shock wave focus under the observation of orthogonally oriented fluoroscopes, and treatment is administered with electrocardiographic (ECG) gating. Newer designs employ piezoelectrical ceramic elements, with stone positioning provided by US. Shock-wave generation may be less efficient with the latter technique, but it is also less painful and freed from the constraints of a large water bath and ECG-triggering.[6]

The ESWL procedure has been applied with excellent results in solitary small pelvic stones, requiring an average 4-day hospitalization and yielding a stone-free rate of over 90%.[48] In less highly-selected patients with intrarenal calculi, long-term stone-free rates of 73% are typical, and many of the remaining patients have only very small residual fragments.[40] Apatite and cystine stones, as well as calculi in the lower pole calyces, show the worst stone-free rates after ESWL, 60%

or less.[40] About one of every ten ESWL patients requires retrograde ureteral manipulations, stent placement, or percutaneous nephrostomy to complete treatment.[1, 9] Larger stones are prone to cause acute obstruction by forming a sandy cast of the ureter (steinstrasse). Balloon dilatation of the ureterovesical junction followed by forceful antegrade flushing has been effective for alleviating such obstruction.[9] Prophylactic stent placement may be warranted for bulky stones. As noted earlier, ESWL should not be used as the first treatment for staghorn or infected calculi, but it may be indicated after a debulking session of percutaneous lithotripsy.

Early results in extracorporeal treatment of ureteral stones were somewhat disappointing. However, it is now recognized that passage of a stent beyond the stone enhances the effects of ESWL, possibly by increasing the fluid-stone interface. Success rates after stent placement approach 100%, so that ESWL can be recommended as the first treatment for ureteral stones above the iliac crest.[49]

An issue of some debate is the long-term effect of ESWL on renal parenchyma. Until definitive information is available, ESWL is not recommended in children.[37] Therapy with ESWL has resulted in extremely low mortality, and large series of patients have been treated without the loss of a single kidney.[40, 48] Interstitial edema and transient gross hematuria are common side effects of ESWL. Although the incidence of subcapsular hematoma has been cited as low as 0.6%, others have detected subcapsular fluid in 24% to 31% of patients.[47, 50] A small persistent drop in the percentage of effective renal plasma flow to the treated kidney has been found in patients 17 to 21 months after ESWL.[51]

The most disturbing clinical finding has been new hypertension in 8% of patients within 1 year of stone treatment.[51] Liedl et al. have failed to find such an effect in hundreds of patients followed a mean of 3.6 years, with the observed incidence of new hypertension in their patients matching that expected with aging alone.[47] A possible reason for this discrepancy is the lower number of shocks administered to their patients, and they recommend that 1,500 shocks be the maximum administered at a single session. In animal studies a positive correlation between the number of shocks and microscopic renal damage has been documented.[52] It should also be noted that patients with renal stone disease tend to have a higher baseline incidence of hypertension than the general population.[47]

Chemodissolution

Only a small number of renal stones can be dissolved by local perfusion of a solvent. Urate stones are amenable to treatment by sodium bicarbonate, and cystine stones have been perfused with acetylcysteine.[10] Struvite stones may respond to hemiacidrin, an electrolyte solution. Adequate outflow is essential for all forms of chemodissolution therapy. If ureteral obstruction is present, the nephrostomy tube should be a sump catheter, or a second percutaneous nephrostomy must be placed as a vent.[10] Infusion begins with a saline challenge to establish a maximum flow tolerance before pain or pressure elevation above 30 cm of water supervenes. Solvent infusion is then started at half the maximum tolerated rate and slowly increased. Dretler and Pfister have set an arbitrary goal of 120 mL/h for hemiacidrin infusion.[10] The great majority of properly selected stones will dissolve, but infusions of 20 to 30 days are typical! For this reason, chemodissolution is now limited to resolution of residual fragments after ESWL or percutaneous treatment.

URETERAL FISTULAS AND URINOMAS

Percutaneous methods are well suited to the treatment of many ureterocutaneous or ureterovisceral fistulae, as well as urinomas. Fistulas may result from trauma or surgery, and surgical repair leads to nephrectomy in 20% and mortality of 10%.[4] Simple percutaneous decompression of the upper collecting system may leave a ureteral stricture, so an internal-external stenting catheter is recommended, as long as no sideholes are located near the fistula. Long-term success had been reported as 70% in resolving fistulas of benign origin, with even better results when fistulas are treated promptly.[4] Urine leakage from renal transplants is somewhat less likely to respond, perhaps because of rejection and the possibility of ureteral necrosis.[4, 7]

Leakage caused by malignant disease or related to radiation therapy is harder to treat. Balloon occlusion of the ureter has been helpful in some cases, but inflation pressure must be kept low to prevent urothelial necrosis.[9] Electrofulguration or special methods of percutaneous clip placement for permanent ureteral occlusion hold more promise for such difficult clinical problems.[53, 54]

RETROGRADE INTERVENTIONAL METHODS

Hawkins et al. have developed a retrograde method for nephrostomy creation, using a coaxial Teflon catheter system and a long, sheathed 20- or 21-gauge needle.[11] The system is directed into a posterior calyx, and the needle—or a very sharp, stiff "rocket" wire (of an alloy developed by the National Aeronautics and Space Administration)—is advanced until it punctures skin. Caudad angulation can be created and controlled by pushing the guiding catheter cephalad. The technique is used for percutaneous stone removal and is particularly helpful in obese patients and nondilated collecting systems.[11]

Effectiveness of retrograde transvesical ureteral catheterization can be greatly enhanced through the collaboration of radiologist and urologist in difficult cases.[12] Amendola et al. have successfully assisted 168 of 180 attempted retrograde ureteral stent placements, ureteral dilatations, and brush biopsies by employing fluoroscopy, guidewires, and long sheaths to traverse obstructions and prevent wire or catheter coiling within the bladder.[12] The patient is brought to the radiologic interventional suite directly from cystoscopy with a partially placed ureteral catheter left in situ. Fluoroscopically guided retrograde interventions have been used in 5% of all patients undergoing ESWL at the University of Pennsylvania, and have obviated the need for percutaneous nephrostomy in many cases.[12] Failure of such intervention is more likely in the face of distal ureteral stenosis.

The same group has applied similar techniques for the cannulation of ureters anastomosed to an ileal conduit.[55] Only those ureters showing reflux on a loop injection of contrast medium are candidates for retrograde catheterization. A curved angiographic catheter with a guidewire advanced slightly beyond its tip is placed as far into the conduit as possible. While contrast medium is injected through a Foley catheter occluding the distal loop, the catheter is slowly withdrawn. When the ureteral anastomosis is engaged, catheter and guidewire are advanced. By means of a stiff wire, such as the Ring-Lunderquist torque guidewire, and Teflon sheaths, even stenotic ureters may be cannulated. Retrograde cannulation under fluoroscopy tends to be less tedious and frustrating than endoscopic catheterization.[55]

RENAL CYST PUNCTURE

When US and CT first provided a cross-sectional view of the kidney, there was great concern about the possible association of malignant neoplasm with renal cysts. Between 2% and 7% of cysts had been described as coexistent with carcinoma.[56] Guided percutaneous cyst puncture became a popular procedure to determine the character of a given cystic lesion. In the past decade, the imaging characteristics of uncomplicated benign renal cysts have been defined with high diagnostic accuracy, and the high incidence of cysts in the normal elderly population has become better appreciated. However, 5% to 8% of renal masses will still have an indeterminate nature by noninvasive studies alone.[57] It is for such lesions, as well as for cysts causing symptoms, that percutaneous puncture is indicated.

A fine needle is guided into the lesion under US or fluoroscopic guidance (when the lesion is larger than 3 cm). Aspirated fluid should be straw-colored and crystal clear, reflecting the proximal tubular origin of a simple cyst.[57] Laboratory studies to be performed include cytologic study, culture, and lactic dehydrogenose, protein, and fat content determinations. High levels of fat are characteristic of neoplasm, while elevated fluid protein may be the result of tumor or inflammation.[57] After fluid aspiration, iodinated contrast medium and air may be injected, with filming in cross-table prone, supine, decubitus, and upright positions for detection of any mural nodules or irregularities. Large, simple cysts will not resolve by simple aspiration, and they may be treated by injection of a sclerosing agent. One quarter of the original cyst volume may be injected with alcohol. The patient is rolled to expose as much of the cyst wall as possible before the alcohol is aspirated after 5 minutes.[57]

BALLOON DILATATION OF THE PROSTATIC URETHRA

Most men over 50 years of age have benign prostatic hypertrophy (BPH). The conventional mode of treatment has been transurethral prostatic resection. Castañeda et al. have used intraurethral balloons to treat urethral obstruction from prostatic hypertrophy.[58] Dilation is performed only after cystoscopy, voiding cystourethrography, and transrectal sonography. Retrograde urethrography is used to mark the site of the external urinary sphincter, and a 25-mm balloon dilates the urethra proximal to the external sphincter. The balloon is left inflated for 10 minutes, followed by repeat urethrography. Transient hematuria and dysuria are common, and a Council catheter is left in place for 24 hours. Early success has been good, with median lobe hypertrophy responsible for cases of failure.[58] The method does not allow histologic examination of prostatic fragments for possible carcinoma in situ, and longer term follow-up is needed to determine the place of balloon dilatation in the treatment of BPH.

SELECTED RADIOLOGIC METHODS IN MALE INFERTILITY

Vasogenic Impotence

Vascular diseases have been increasingly recognized as a cause of impotence. Noninvasive tests, such as the penile-brachial systolic index; nocturnal penile tumescence; and penile rigidity have been used to provide a reliable screen for those with arterial insufficiency. Most men with significant aortoiliac occlusive disease are impotent, and standard pelvic angiography will document this.[59] More distal occlusive disease is more difficult to demonstrate, requiring selective internal iliac arteriography.

The use of low-osmolarity contrast media and vasodilatation with intra-arterial nitroglycerin (300 µg in 10 mL of saline injected slowly immediately before arteriography) with or without papaverine (30 to 50 mg) have produced much better studies than previously possible.[59] The internal pudendal and penile arteries are best visualized with the opposite posterior oblique projection of the internal iliac being injected (i. e., left posterior oblique projection for the right internal iliac study, with the penis draped over the left thigh), using magnification technique, injection of 4 to 6 mL/sec of contrast material for 36 mL, and filming over 32 seconds.[59] Filling of the penile, cavernosal, and dorsal penile arteries should be observed. If one side is found to be normal, examination of the opposite side is unnecessary. Only one side need be revascularized for the relief of arterial impotence.

Dynamic cavernosography is a means of evaluating impotence resulting from venous leakage. The corpora cavernosa are punctured with 19-gauge needles about midshaft, halfway between the dorsal and ventral surfaces.[60] Because there are communications between the corpora through the septum, only one needle need be perfused, and pressure measurements may be obtained through the other. Infusion of dilute low-osmolality contrast begins at 40 mL/min and is increased until the penis appears erect.[60] Erection is produced in a normal penis with a flow of 80 to 120 mL/min, and is maintained by less than half that flow. Measured pressure should attain at least 80 mm Hg. Venous leakage is seen as filling of the veins of the prostatic plexus during erection or high infusion rates.

Direct injection of papaverine (30 to 60 mg) into the corpora cavernosa stimulates normal erectile physiology, decreasing the infusion rates needed to elicit erection.[61] Corpora cavernosography is repeated 10 minutes after injection of papaverine. Intracavernous papaverine is not absolutely necessary for the diagnosis of impotence due to venous leakage, and use of the drug does present a risk for thrombosis and priapism.[60]

Varicoceles

Varicocele is present in 4% of adult males and in nearly 40% of men with complaints of infertility.[62] It may adversely affect fertility by elevating scrotal temperature or by poorly understood effects of stasis and hypoxia. Although subclinical varicocele may be suspected by sonography or thermography, the diagnosis is made by selective internal spermatic vein venography. The left internal spermatic vein, which empties into the left renal vein, is more likely to be incompetent than the right, which is a direct tributary of the inferior vena cava. Reflux of contrast material beyond the level of the mid-lumbar spine is abnormal, and unilateral varicocele may be responsible for diminished fertility.[63]

Definition of the venous anatomy in varicocele is important because two-thirds of patients show venous duplications or communicating retroperitoneal veins.[63] Anatomic anomalies are responsible for the 15% to 20% incidence of recurrence after surgical venous ligation. Full venographic delineation of anatomy permits rational planning of treatment, including selective catheter embolization or venous sclerotherapy. Zeitler et al. have found a low (5%) incidence of recurrent varicocele after transcatheter injection of a sclerosing agent.[63]

REFERENCES

1. Hulbert JC: The role of endourologic procedures in relation to extracorporeal shock wave lithotripsy. *Semin Intervent Radiol* 1987; 4:50–52.

2. Keidan RD, Greenberg RE, Hoffman JP, et al: Is percutaneous nephrostomy for hydronephrosis appropriate in patients with advanced cancer? *Am J Surg* 1988; 156:206–208.

3. Lang EK: Percutaneous management of ureteral strictures. *Semin Intervent Radiol* 1987; 4:79–89.

4. Maillet PJ, Pelle-Francoz D, Leriche A, et al: Fistulas of the upper urinary tract: percutaneous management. *J Urol* 1987; 138:1382–1385.

5. Badlani G, Eshghi M, Smith AD: Percutaneous surgery for ureteropelvic junction obstruction (endopyelotomy): Technique and early results. *J Urol* 1986; 135:26–28.

6. Orihuela E, Smith AD: Percutaneous treatment of transitional cell carcinoma of the upper urinary tract. *Urol Clin North Am* 1988; 15:425–431.

7. Bennett LN, Voegli DR, Crummy AB, et al: Urologic complications following renal transplantation: Role of interventional radiologic procedures. *Radiology* 1986; 160:531–536.

8. Doemeny JM, Banner MP, Shapiro MJ, et al: Percutaneous extraction of renal fungus ball. *AJR* 1988; 150:1331–1332.

9. Gordon RL, Banner MP, Pollack HM: Selected endourologic techniques. *Radiol Clin North Am* 1986; 24:633–649.

10. Dretler SP, Pfister RC: Primary dissolution therapy of struvite calculi. *J Urol* 1984; 131:861–863.

11. Hawkins IF Jr, Hunter P, Leal G, et al: Retrograde nephrostomy for stone removal: Combined cystoscopic/percutaneous technique. *AJR* 1984; 143:299–304.

12. Amendola MA, Banner MP, Pollack HM, et al: Fluoroscopically guided pyeloureteral interventions by using a perurethral transvesical approach. *AJR* 1989; 152:97–102.

13. Coleman CC, Castañeda-Zuñiga W, Miller R, et al: A logical approach to renal stone removal. *AJR* 1984; 143:609–615.

14. Janetschek G, Kunzel KH: Percutaneous nephrolithotomy in horseshoe kidneys: Applied anatomy and clinical experience. *Br J Urol* 1988; 62:117–122.

15. Spies JB, Rosen RJ, Lebowitz AS: Antibiotic prophylaxis in vascular and interventional radiology: A rational approach. *Radiology* 1988; 166:381–387.

16. LeRoy AJ, May GR, Bender CE, et al: Percutaneous nephrostomy for stone removal. *Radiology* 1984; 151:607–612.

17. Lind LJ, Mushlin PS: Sedation, analgesia, and anesthesia for radiologic procedures. *Cardiovasc Intervent Radiol* 1987; 10:247–253.

18. Coleman CC, Castañeda-Zuñiga WR, Amplatz K: Renal anatomy for uroradiologic interventions. *Semin Intervent Radiol* 1987; 4:1–9.

19. Picus D, Weyman PJ, Clayman RV, et al: Intercostal-space nephrostomy for percutaneous stone removal. *AJR* 1986; 147:393–397.

20. Lang EK: Percutaneous nephrostolithotomy and lithotripsy: A multi-institutional survey of complications. *Radiology* 1987; 162:25–30.

21. LeRoy AJ, Williams HJ Jr, Bender CE, et al: Colon perforation following percutaneous nephrostomy and renal calculus removal. *Radiology* 1985; 155:83–88.

22. Hopper KD, Chantelois AE: The retrorenal spleen: Implications for percutaneous left renal invasive procedures. *Invest Radiol* 1989; 42:592–595.

23. Cope C, Zeit RM: Pseudoaneurysms after nephrostomy. *AJR* 1982; 139:255–261.

24. Voegeli DR, Crummy AB, McDermott JC, et al: Percutaneous dilation of ureteral strictures in renal transplant patients. *Radiology* 1988; 169:185–188.

25. Streem SB, Novick AC, Steinmuller DR, et al: Long-term efficacy of ureteral dilatation for transplant ureteral stenosis. *J Urol* 1988; 140:32–35.

26. Shapiro MJ, Banner MP, Amendola MA, et al: Balloon catheter dilatation of ureteroenteric strictures: Long-term results. *Radiology* 1988; 168:385–387.

27. Lee WJ, Badlani GH, Karlin GS, et al: Treatment of ureteropelvic strictures with percutaneous pyelotomy: Experience in 62 patients. *AJR* 1988; 151:515–518.

28. Castañeda F, Castañeda-Zuñiga WR, Hunter DW, et al: New developments in endourology. *Semin Intervent Radiol* 1987; 4:22–25.

29. Uflacker R, Wholey MH: A new low-speed, rotational atherolytic device for ureteral recanalization. *AJR* 1988; 151:1157–1158.

30. Mitty HA, Train JS, Dan SJ: Placement of ureteral stents by antegrade and retrograde techniques. *Radiol Clin North Am* 1986; 24:587–600.

31. Babel SG, McDermott JC, Goldrath DE, et al: Uretero-arterial fistula after balloon dilatation and stent placement: Case report and review of the literature. *J Intervent Radiol* 1988; 3:135–138.

32. Brazzini A, Castañeda F, Castañeda-Zuñiga WR, et al: Urostent designs. *Semin Intervent Radiol* 1987; 4:26–35.

33. Whitaker RH: An evaluation of 170 diagnostic pressure flow studies of the upper urinary tract. *J Urol* 1979; 121:602–604.

34. Jaffe RB, Middleton AW Jr: Whitaker test: Differentiation of obstructive from nonobstructive uropathy. *AJR* 1980; 134:9–15.

35. Spital A, Valvo JR, Segal AJ: Nondilated obstructive uropathy. *Urology* 1988; 31:478–482.

36. Dretler SP: Laser lithotripsy: A review of 20 years of research and clinical applications. *Lasers Surg Med* 1988; 8:341–356.

37. Segura JW: The role of percutaneous surgery in renal and ureteral stone removal. *J Urol* 1989; 141:780–781.

38. Segura JW, Patterson DE, LeRoy AJ, et al: Percutaneous lithotripsy. *J Urol* 1983; 130:1051–1054.

39. Lee WJ, Snyder JA, Smith AD: Staghorn calculi: Endourologic management in 120 patients. *Radiology* 1987; 165:85–88.

40. Graff J, Diedrichs W, Schulze H: Long-term followup in 1,003 extracorporeal shock wave lithotripsy patients. *J Urol* 1988; 140:479–483.

41. Castañeda-Zuñiga WR, Clayman R, Smith A, et al: Nephrostolithotomy: Percutaneous technique for urinary calculus removal. *AJR* 1982; 139:721–726.

42. Clayman RV, Surya V, Hunter D, et al: Renal vascular complications associated with the percutaneous removal of renal calculi. *J Urol* 1984; 132:228–230.

43. Bush WH, Brannen GE, Lewis GP, et al: Upper ureteral calculi: Extraction via percutaneous nephrostomy. *AJR* 1985; 144:795–799.

44. Hunter DW, Castañeda-Zuñiga WR, Young AT, et al: Percutaneous removal of ureteral calculi: Clinical and experimental results. *Radiology* 1985; 156:341–348.

45. Streem SB, Hall P, Zelch MG, et al: Endourologic management of upper and mid ureteral calculi: Percutaneous antegrade extraction vs. transurethral ureteroscopy. *Urology* 1988; 31:34–37.

46. Saltzman B, Khasidy LR, Smith AD: Measurement of renal pelvis pressures during endourologic procedures. *Urology* 1987; 30:472–474.

47. Liedl B, Jocham D, Lunz C, et al: Prävalenz und inzidenz der arteriellen hypertonie bei ESWL-behandelten nierensteinpatienten. *Urologe [A]* 1989; 28:130–133.

48. Pemberton J: Extra-corporeal shock wave lithotripsy. *Postgrad Med J* 1987; 63:1025–1031.

49. Dretler SP, Keating MA, Riley J: An algorithm for the management of ureteral calculi. *J Urol* 1986; 136:1190–1193.

50. Kaude JV, Williams CM, Millner MR, et al: Renal morphology and function immediately after extracorporeal shock-wave lithotripsy. *AJR* 1985; 145:305–313.

51. Williams CM, Kaude JV, Newman RC, et al: Extracorporeal shock-wave lithotripsy: Long-term complications. *AJR* 1988; 150:311–315.

52. Ackaert KSJW, Schröder, FH: Effects of extracorporeal shock wave lithotripsy (ESWL) on renal tissue: A review. *Urol Res* 1989; 17:3–7.

53. Hulbert JC: Percutaneous intrarenal endoscopic surgery. *Semin Intervent Radiol* 1987; 4:109–114.

54. Lund G, Rysavy JA, Castañeda-Zuñiga WR, et al: Techniques for percutaneous mechanical occlusion of the ureter: Experimental evaluation. *Semin Intervent Radiol* 1987; 4:73–78.

55. Banner MP, Amendola MA, Pollack HM: Anastomosed ureters: Fluoroscopically guided transconduit retrograde catheterization. *Radiology* 1989; 170:45–49.

56. Emmett JL, Levine SR, Woolner LB: Coexistence of renal cyst and tumour: Incidence in 1,007 cases. *Br J Urol* 1963; 35:403–410.

57. Sandler CM, Houston GK, Hall JT, et al: Guided cyst puncture and aspiration. *Radiol Clin North Am* 1986; 24:527–537.

58. Castañeda F, Letourneau JG, Reddy P, et al: Alternative treatment of prostatic urethral obstruction secondary to benign prostatic hypertrophy. *ROFO* 1987; 147:426–429.

59. Bookstein JJ, Valji K, Parsons L, et al: Pharmacoarteriography in the evaluation of impotence. *J Urol* 1987; 137:333–337.

60. Delcour C, Wespes E, Vandenbosch G, et al: Impotence: Evaluation with cavernosography. *Radiology* 1986; 161:803–806.

61. Puyau FA, Lewis RW, Balkin P, et al: Dynamic corpus cavernosography: Effect of papaverine injection. *Radiology* 1987; 164:179–182.

62. Pochaczevsky R, Lee WJ, Mallett E: Management of male infertility: Roles of contact thermography, spermatic venography, and embolization. *AJR* 1986; 147:97–102.

63. Zeitler E, Jecht E, Richter EI, et al: Selective sclerotherapy of the internal spermatic vein in patients with variococeles. *Cardiovasc Intervent Radiol* 1980; 3:166–169.

Index

Y